Lecture Notes in Computer Science   10179

Commenced Publication in 1973
Founding and Former Series Editors:
Gerhard Goos, Juris Hartmanis, and Jan van Leeuwen

More information about this series at http://www.springer.com/series/7409

Zhifeng Bao · Goce Trajcevski
Lijun Chang · Wen Hua (Eds.)

Database Systems
for Advanced Applications

DASFAA 2017 International Workshops:
BDMS, BDQM, SeCoP, and DMMOOC
Suzhou, China, March 27–30, 2017
Proceedings

 Springer

Editors
Zhifeng Bao
Royal Melbourne Institute of Technology
Melbourne
Australia

Goce Trajcevski
Northwestern University
Evanston, IL
USA

Lijun Chang
University of New South Wales
Sydney, NSW
Australia

Wen Hua
The University of Queensland
Brisbane, QLD
Australia

ISSN 0302-9743   ISSN 1611-3349 (electronic)
Lecture Notes in Computer Science
ISBN 978-3-319-55704-5   ISBN 978-3-319-55705-2 (eBook)
DOI 10.1007/978-3-319-55705-2

Library of Congress Control Number: 2017934640

LNCS Sublibrary: SL3 – Information Systems and Applications, incl. Internet/Web, and HCI

Preface

Along with the main conference, the DASFAA 2017 workshops provided an international forum for researchers and practitioners to gather and discuss research results and open problems, aiming at more focused problem domains and settings. This year there were four workshops held in conjunction with DASFAA 2017:

- The 4th International Workshop on Big Data Management and Service (BDMS 2017)
- The Second Workshop on Big Data Quality Management (BDQM 2017)
- The 4th International Workshop on Semantic Computing and Personalization (SeCoP 2017)
- The First International Workshop on Data Management and Mining on MOOCs (DMMOOC 2017)

All the workshops were selected after a public call-for-proposals process, and each of them focused on a specific area that contributes to, and complements, the main themes of DASFAA 2017. Each workshop proposal, in addition to the main topics of interest, provided a list of the Organizing Committee members and (a tentative) Program Committee. Once the selected proposals were accepted, each of the workshops proceeded with their own call for papers and reviews of the submissions.

In total, 37 papers were accepted, including 13 papers for BDMS 2017, five papers for BDQM 2017, five papers for SeCoP 2017, and 14 papers (nine full and five short) for DMMOOC 2017.

We would like to thank all of the members of the Organizing Committees of the respective workshops, along with their Program Committee members, for their tremendous effort in making the DASFAA 2017 workshops a success. In addition, we are grateful to the main conference organizers for their generous support as well as the efforts in including the papers from the workshops in the proceedings series.

March 2017

Zhifeng Bao
Goce Trajcevski

Organization

The 4th International Workshop on Big Data Management and Service (BDMS 2017)

Workshop Chairs

Xiaoling Wang   East China Normal University, China
Kai Zheng   Soochow University, China
An Liu   Soochow University, China

Program Committee

Jialong Han   Nanyang Technological University, Singapore
Tieke He   Nanjing University, China
Yuwei Peng   Wuhan University, China
Han Su   University of Electric Science and Technology
                       of China, China
Da Yan   University of Alabama at Birmingham, USA
Kun Yue   Yunnan University, China
Wei Zhang   East China Normal University, China
Xin Zhao   Renmin University of China, China
Bolong Zheng   University of Queensland, Australia
Yaqian Zhou   Fudan University, China
Zhixu Li   Soochow University, China
Ke Sun   King Abdullah University of Science and Technology,
                       Saudi Arabia
Haoran Xie   Education University of Hong Kong, SAR China

The Second Workshop on Big Data Quality Management (BDQM 2017)

Honorable Chair

Jianzhong Li   Harbin Institute of Technology, China

Workshop Chairs

Hongzhi Wang   Harbin Institute of Technology, China
Jing Gao   University at Buffalo, USA/State University
                       of New York, USA

Program Committee

Cheqing Jin   East China Normal University, China
Jiannan Wang   Simon Fraser University, Canada

Lingli Li	Heilongjiang University, China
Rihan Hai	Lehrstuhl Informatik 5, RWTH Aachen University, Germany
Wenjie Zhang	University of New South Wales, Australia
Yingyi Bu	Couchbase, USA
Yueguo Chen	Renmin University of China, China
Zhaonian Zou	Harbin Institute of Technology, China
Zhijing Qin	Pinterest, USA

The 4th International Workshop on Semantic Computing and Personalization (SeCoP 2017)

General Chairs

Haoran Xie	The Education University of Hong Kong, SAR Hong Kong
Fu Lee Wang	Caritas Institute of Higher Education, SAR Hong Kong
Tak-Lam Wong	The Education University of Hong Kong, SAR Hong Kong
Yi Cai	South China University of Technology, China

Organizing Chairs

Wei Chen	Agricultural Information Institute of CAAS, China
Tianyong Hao	Guangdong University of Foreign Studies, China
Zhaoqing Pan	Nanjing University of Information Science and Technology, China

Publicity Chairs

Xiaohui Tao	Southern Queensland University, Australia
Di Zou	The Hong Kong Polytechnic University, SAR China
Xudong Mao	City University of Hong Kong, SAR China
Yanghui Rao	Sun-Yet San University, China
Yunhui Zhuang	City University of Hong Kong, SAR China
Zhenguo Yang	Guangdong University of Technology, China

Program Committee

Zhiwen Yu	South China University of Technology, China
Jian Chen	South Chia University of Technology, China
Raymong Y.K. Lau	City University of Hong Kong, SAR China
Rong Pan	Sun Yat-Sen University, China
Yunjun Gao	Zhejiang University, China
Shaojie Qiao	Southwest Jiaotong University, China
Jianke Zhu	Zhejiang University, China
Neil Y. Yen	University of Aizu, Japan
Derong Shen	Northeastern University, China

Jing Yang	Research Center on Fictitious Economy and Data Science CAS, China
Wen Wu	Hong Kong Baptist University, SAR China
Raymong Wong	Hong Kong University of Science and Technology, SAR China
Wenjuan Cui	China Academy of Sciences, China
Xiaodong Li	Hohai University, China
Xiangping Zhai	Nanjing University of Aeronautics and Astronautics, China
Xu Wang	Shenzhen University, China
Ran Wang	Shenzhen University, China
Debby Dan Wang	Caritas Institute of Higher Education, SAR China
Jianming Lv	South China University of Technology, China
Tao Wang	The University of Southampton, UK
Guangliang Chen	TU Delft, The Netherlands
Wenji Ma	Columbia University, USA
Kai Yang	South China University of Technology, China
Yun Ma	City University of Hong Kong, SAR China

The First International Workshop on Data Management and Mining on MOOCs (DMMOOC 2017)

Workshop Chairs

Wenjun Wu	Beihang University, China
Yan Zhang	Peking University, China
Yongxin Tong	Beihang University, China

Program Committee

Yurong Cheng	Northeastern University, China
Dawei Gao	Beihang University, China
Xiaonan Guo	Stevens Institute of Technology, USA
Di Jiang	Baidu Inc., China
Jun Liu	Xi'an Jiaotong University, China
Xinjun Mao	National University of Defense Technology, China
Rui Meng	Hong Kong University of Science and Technology, SAR China
Longfei Shangguan	Princeton University, USA
Jieying She	Hong Kong University of Science and Technology, SAR China
Tianshu Song	Beihang University, China
Zhiyang Su	Microsoft, China
Jie Tang	Tsinghua University, China
Qian Tao	Beihang University, China
Libin Wang	Beihang University, China
Qiong Wang	Peking University, China

Ting Wang   National University of Defense Technology, China
Wei Xu   Tsinghua University, China
Gang Yin   National University of Defense Technology, China
Xiaolong Zheng   Tsinghua University, China
Zimu Zhou   ETH Zurich, Switzerland

Contents

BDQM

SeCoP

DMMOOC

BDMS

Automatically Classify Chinese Judgment Documents Utilizing Machine Learning Algorithms

Miaomiao Lei, Jidong Ge[✉], Zhongjin Li, Chuanyi Li, Yemao Zhou,
Xiaoyu Zhou, and Bin Luo

State Key Laboratory for Novel Software Technology, Software Institute,
Nanjing University, Nanjing 210093, Jiangsu, China
gjdnju@163.com

Abstract. In law, a judgment is a decision by a court that resolves a controversy and determines the rights and liabilities of parties in a legal action or proceeding. In 2013, China Judgments Online system was launched officially for record keeping and notification, up to now, over 23 million electronic judgment documents are recorded. The huge amount of judgment documents has witnessed the improvement of judicial justice and openness. Document categorization becomes increasingly important for judgments indexing and further analysis. However, it is almost impossible to categorize them manually due to their large volume and rapid growth. In this paper, we propose a machine learning approach to automatically classify Chinese judgment documents using machine learning algorithms including Naive Bayes (NB), Decision Tree (DT), Random Forest (RF) and Support Vector Machine (SVM). A judgment document is represented as vector space model (VSM) using TF-IDF after words segmentation. To improve performance, we construct a set of judicial stop words. Besides, as TF-IDF generates a high dimensional feature vector, which leads to an extremely high time complexity, we utilize three dimensional reduction methods. Based on 6735 pieces of judgment documents, extensive experiments demonstrate the effectiveness and high classification performance of our proposed method.

Keywords: Chinese judgment documents · Text classification · TF-IDF · Support Vector Machine · Naive Bayes · Decision Tree · Random Forest · Judicial stop-words construction · Dimensional reduction

1 Introduction

In law, a judgment is a decision made by a court that resolves a controversy and determines the rights and liabilities of parties in a legal action or proceeding. Most courts now store their judgments electronically. In 2013, China Judgment Online System, the largest judgment documents sharing website around the world, was launched officially. Up to now, over 23 million electronic judgment

© Springer International Publishing AG 2017
Z. Bao et al. (Eds.): DASFAA 2017 Workshops, LNCS 10179, pp. 3–17, 2017.
DOI: 10.1007/978-3-319-55705-2_1

documents are recorded and more than 70K new judgments are indexed every-day. This huge amount of judicial documents is of great importance not only for improving judicial justice and openness, but also for court administrators for record keeping and future reference in decision making and judgment writing. Furthermore, data sharing and deeper analysis on these judgments is the key approach in the process of information construction and legislation system improvement.

In China Judgment Online System, judicial cases are indexed with five major types depending on the cause of action, which are administrative case, criminal case, civil case, compensation case and execution case respectively. Beneath each type usually lies other organization hierarchies. Grouping by keywords is one of the most common used methods, for example, keywords in criminal cases can be illegal possession, surrender, joint offence, penalty, etc. Keywords matching does make a contribution to a better organized judgment documenting system. But, there are also some limits existing, on one hand, a list of keywords must be manually created and maintained. However, enumerating all keywords completely is difficult, thus leading to extra human labor cost. On the other hand, grouping by keywords does not meet all demands in real applications, when new classification requirements are generated, it is almost impossible to categorize all of them manually.

Text classification, as an important task in natural language processing, involves assigning a text document to one of the predefined classes or topics. Text classification has been widely studied in the text mining and information retrieval communities and commonly used in a number of diverse domains such as news automatic categorization, spam detecting and filtering, opinion mining and document indexing, etc. In this paper, we propose a machine learning approach to automatically classify Chinese judgment documents into predefined categories. Our work includes: (1) propose an automated method to construct a list of judicial specific stop-words, (2) propose an effective strategy for Chinese judgment documents representation as well as feature dimensional reduction while keeping as much important information as possible, (3) achieve high performance utilizing machine learning based algorithms to classify Chinese judgment documents.

To evaluate the performance of our approach, we also manually label 6735 pieces of Chinese judgment documents that are related to liabilities of product quality into 13 categories based on the statutory standard of industry division. With the experiment results on this dataset, we will prove the contributions of this paper by answering the research questions as follows:

(1) How is the performance of the classifier improved by domain stop words list construction and text preprocessing in Chinese judgment documents classification?
(2) What kind of features should be selected and how can we benefit from dimensional reduction?
(3) Which machine learning algorithm achieves better performance for Chinese judgment documents classification?

The remainder of this paper is laid out as follows. Section 2 introduces related work of this paper. Section 3 introduces our approach in detail. Section 4 describes our experiments and evaluation results and Sect. 5 concludes with a discussion of future work.

2 Related Work

In recent years, the problem of text classification has gained increasing attention due to the large amounts of text data that are created in many information-centric applications, which is concomitant with tremendous researches on methods and algorithms in text classification. In this section, we provide an overview of key techniques for text classification.

Technically, text data is distinguished from other forms of data such as relational or quantitative data in many aspects. The most important characteristic of text data is that it is sparse and high dimensional [1]. Text data can be analyzed at different levels of representation. Bag-of-words (BOW) simply represent text data as a string of words. TF-IDF takes both word frequency and document frequency into consideration to determine the importance of each word, which is commonly used in the representation of documents. Strzalkowski demonstrated that a proper term weighting is important and different types of terms and terms that are derived from different means should be differentiated [2]. Jiang integrated rich document representations to derive high quality information from unstructured data to improve text classification [3]. Liu studied document representation based on semantic smoothed topic model [4]. Besides, document representations in many different applications have been studied. Yang proposed a novel approach for business document representation in e-commerce [5], and Arguello introduced two document representation models for blog recommendation [6].

The bag of words (BOW) representation can help retain a great deal of useful information, but it is also troublesome because BOW vectors are very high dimensional. To provide an ideal lower-dimensional representation, dimension reduction in many forms are studied to find the semantic space and its relationship to the BOW representation. Two techniques for dimensional reduction stand out. The first one is latent semantic indexing (LSI) which is based on singular vector decomposition to find a latent semantic space and construct a low rank approximation of the original matrix while preserving the similarity between the documents [7]. The other one is topic models that provide a probabilistic framework for the dimension reduction task [8]. Hofmann proposed PLSI that provides a crucial step in topic modeling by extending LSI in a probabilistic context which is a good basis for text analysis but contains a large number of parameters that grows linearly with the number of documents [9]. Latent Dirichlet Allocation (LDA) includes a process that generates topics in each document, therefore, it greatly reduces the number of parameters to be learned and is an improvement of PLSI [10].

A wide variety of machine learning algorithms have been designed for text classification and applied in many applications. Apte proposed an automated learning method of decision rules for text categorization [11]. Baker introduced a text classification method based on distributional clustering of words [12]. Drucker studied support vector machines for spam categorization [13]. Ng and Jordan made comparisons of discriminative classifiers and generative classifiers [14]. Sun presented supervised Latent Semantic Indexing for document categorization [15].

3   Approach

In this section, we present our approach for Chinese judgment documents classification in detail as follows. Section 3.1 presents an overview of the workflow for Chinese judgment documents classification. Section 3.2 introduces text preprocessing. Section 3.3 introduces a method of document representation and approaches to provide a lower-dimensional feature vector after generated by document representation method. Section 3.4 introduces classifiers used in our work.

3.1   Overview

Figure 1 presents the overview of the workflow we use for Chinese judgment documents classification. To analyze Chinese judgment documents in a deep going way, the classification approach starts from setting a clear-out classification goal and then, based on it, accessing Chinese Judgment Online System to obtain amount of to-be-classified judgment documents with a certain kind of cause of action. To build a classification model and evaluate the performance of it, we put a huge amount of human efforts in labeling a proportion of those judgment documents according to the classes we have defined in the goal setting process. After that, labeled judgment documents will be separated into 2 parts, respectively called training dataset and test dataset, one for model training and the other one for performance evaluation. As for how large a portion should be used as training dataset, this problem has been well discussed in [16] and 70% is the number we choose in our work.

Different from documents written in English, Chinese documents require different text preprocessing methods because of the huge distinction in morphology, grammar, syntax, etc. Beyond that, judgment documents also have their own characteristic. Generally, a certain format is used when a court administrator writing a judgment. Therefore, not all of content in a judgment is useful for achieving our classification goal. Due to this reason, we figure out that extracting the content we need for classification only can make a contribution to reducing noise thus improving performance. In Sect. 3.2.1 we will introduce the method we use to extract contents. As mentioned, Chinese documents should be segmented into words before they are available for representation. For this purpose, we employ techniques in natural language processing, also called Chinese Word Segmentation, whose detail will be clarified in Sect. 3.2.2. While it is fairly easy to use a published set of stop words list, in many cases, using such stop words is

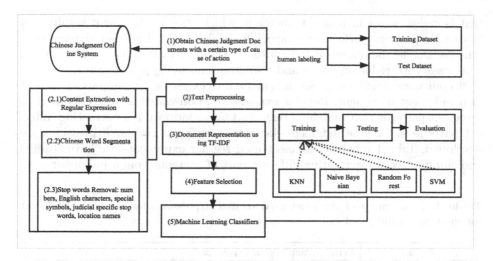

Fig. 1. Overview of the workflow for Chinese judgment documents classification

completely insufficient for certain applications. To improve the performance of classifying Chinese judgment documents, we propose an algorithm to construct a judgment domain specific stop words list in Sect. 3.2.3, and then we remove all stop words in Sect. 3.2.4.

After the step of text preprocessing, we employ TF-IDF to represent documents, or so called feature extraction, which is widely used in document classification research. To further reduce feature dimensions and improve performance, we also apply feature selection methods. Section 3.3 will introduce the methods in detail. Supervised machine learning algorithms, Naive Bayes (NB), Decision Tree, Random Forest and Support Vector Machine (SVM) are used in building classification model.

3.2 Text Preprocessing

In this section, we mainly focus on introducing the methods used to solve the problem of how to preprocess Chinese judgment documents. As Chinese documents are different from English-written ones in terms of morphology, grammar, syntax, etc. extensive preprocessing methods are required.

3.2.1 Content Extraction

After accessing Chinese Judgment Online System and obtaining an amount of judgment documents, we investigated a lot of documents and found that no matter what category a judgment belongs, a certain format exists in all of them as court administrators usually use a similar pattern when writing judgments. In addition, not the complete content in a judgment is related to a certain object of classification, we are concerned with a certain part of the contents for different purposes. Moreover, irrelevant information existing in a to-be-analyzed text is called noises in text classification, which effects experiment results and leads to unpleasant performance especially in short texts such as judgment documents. Due to the reasons listed above, we think extracting particular contents is necessary for reducing noises and therefore, improving performance.

In our work, regular expression is utilized to extract a certain part of content in a judgment document. Regular expression, abbreviated as regex, is used to represent patterns that matching text need to conform to. For extracting different parts from Chinese judgment documents, different regular expressions are developed. Table 1 presents the most used regular expressions we have studied and summarized for extracting different part of contents from a judgment.

Table 1. A summary of the most used regular expressions for extracting different contents from a judgment

Request Content	Corresponding Regular Expression
Case Detail	原告.*?诉称[\S\s]*(?=经审理查明)
Plaintiff Libel	原告.*?诉称[\S\s]*(?=被告.*?辩称)
Defendant Argument	被告.*? 辩称[\S\s]*(?=经审理查明)
Court Decision	本院认为[\S\s]*(?=^审判员)

3.2.2 Word Segmentation

Tokenization of raw text is a standard preprocessing step for many natural language processing (NLP) tasks, tokenization usually involves punctuation splitting and separation of some affixes like possessives. While unlike English, Chinese language requires more extensive preprocessing, known as word segmentation. As the most fundamental task in Chinese NLP tasks, word segmentation has been studied for several years. Chinese word segmentation involves splitting a paragraph of text into a sequence of words, sometimes includes part-of-speech tagging, semantic dependency relationship mining and name entity recognition.

Currently, there exist a number of Chinese word segmentation systems, including Jieba, ICTCLAS, SCWS, LTP, NLPIR, etc. Most of them can achieve a satisfying performance. LTP [17], known as Language Technology Platform, has the highest accuracy among all and offers multithread processing service, but it has some disadvantages as follows: (1) results are given in xml format which need another processing, thus leading to extra time cost especially for large datasets, (2) sometimes tasks are possible to fail. Jieba is one of the most efficient systems and can be easily integrated in python systems, but Jieba has lower accuracy. Taking all factors into consideration, we combine Jieba and LTP and utilize them in our work, LTP is regarded as a main tool while Jieba works as a complement in case some of the tasks fail. To improve performance, all segmented documents will be stored in our database.

3.2.3 Judicial Specific Stop-Words

While it is easy to use a published set of stop words, in the task of preprocessing judgment documents, using such stop words is completely insufficient. For example, in judgments, terms like plaintiff, defendant, argue, libel, court, etc. occur almost in every document. So, these terms should be regarded as potential stop words in judgment documents retrieval and classification. However, the common used stop words list does not contain such specific terms. To construct a judicial stop words list, we first, use 6735 pieces of documents as our corpus, and then we use 2 methods as follows to do the job:

1. Use the terms that occur frequently across judgment documents (low IDF terms) as stop words. Inverse Document Frequency (IDF) refers to the inverse fraction of documents in whole collection that contain a specific term, IDF is often used to represent the importance of a term. In other words, if a term occurs almost in every document of a collection, that means it is not important and can be regard as potential stop word. After word segmentation, we sum the term frequency of each unique word by scanning all documents. And then sort the terms in descending order. Top N terms have been chosen as stop words. The number N is chosen manually after human scanning. To make sure the set of stop words be judicial related, we filter all common used Chinese words in advance. The benefit of this approach is that it is very intuitive and easy to implement.
2. Use the least frequent terms as stop words. Terms are extremely infrequent may not be useful for text mining and retrieval. For example, these terms in judgments may be location names, human names, specific feeling expressions, etc. which may not be relevant for a judgment classification object. Removing these terms can significantly reduce overall feature space.

3.2.4 Stop-Words Removal

Removing as many stop words as possible can significantly reduce noises and make great contributions to a better classification performance. Given different text mining goals, different text preprocessing steps are in need. After a raw document is segmented into a list of terms, each term is regarded as data stream, which would go through every procedure defined by text preprocessing workflow module. In specific, each procedure returns null if a term is recognized as a stop word with correspondent type, otherwise returns itself. In this way, each word can be filtered, therefore, none stop words are left for further analysis. In specific, 7 kinds of stop words removal procedures are considered for Chinese judgment documents classification as follows:

(1) Numbers, Chinese numbers, English letters and special symbols. As none of word segmentation systems can provide accuracy of 100%, some errors can occur in this step. As a result, words with numbers or Chinese numbers, English letters and special symbols can be found in word segmentation results, which is not useful for judgment classification.

(2) Judicial stop words. After a set of judicial stop words has been constructed as presented in last section, we remove all words in every document that is matched as a judicial stop word.

(3) Location names. By investigating an amount of judgment documents, location has occurred in almost every judgment, such as province, city, district, village, street, etc. By accessing a location name database, we can remove most of them. For the rest, we match the last word of a term, see if it represents a location, if is, remove the term.

(4) Human names. Human name recognition is one of the Name Entity Recognition (NER) tasks in natural language processing. Some NLP systems provide such services to accurately label human names occur in a given text. However, in judgments, human names are also considered as least frequent terms, as a step of stop words removal, using such systems for this purpose is a waste of resource. Therefore, by removing the least frequent terms, we can remove human names together.

3.3   Feature Extraction

The input of text classification is the content of judgments, however, a sequence of words can not be fed directly to the machine learning algorithms. In order to address this, text should be transformed into other formats which can be mathematically computed, this process is called document representation. In this section, we select TF-IDF for document representation and feature reduction methods will be introduced at the end of the section.

3.3.1   Document Representation

Although stop words removal help remove an amount of noises, there still exists a large amount of words carrying very little meaningful information. Therefore, TF-IDF is utilized in our approach to calculate the amount of meaning each term carries.

TF-IDF, stands for term frequency-inversed document frequency, is one of the most used algorithms to transform a text into a feature vector in text mining and retrieval. Typically, the TF-IDF weight is composed of two terms: Term Frequency (TF) computes the number of times a word appears in a document, Inversed Document Frequency (IDF) measures how important a term is. Since every document is different in length, it is possible that a term would appear much more times in a long document than shorter ones. Thus, Term Frequency is often divided by the total number of words in a document for normalization. While computing Term Frequency, all terms are considered important equally. However, Inversed Document Frequency is aimed at weight down the frequent terms (stop words included) while scale up the rare ones. The TF-IDF of a word w in document d is calculated as: TF-IDF(w,d) = TF(w,d) * IDF(w), where TF(w,d) = (frequency of w in d)/(total number of words in d), and IDF(w) = log_e(total number of documents in the corpus/number of documents with w in it). By calculating the TF-IDF weight, each term is regarded as a feature of the document, and the corresponding value is its TF-IDF weight. Thus all documents are transformed into a feature vector.

3.3.2   Feature Reduction

Similar with document presentation for other kinds of documents, TF-IDF always generates a large feature vector as it computes every term occurs in every document. As a result, feeding the feature vector generated by TF-IDF directly to machine learning classifiers is cost inefficient. Hence, we study feature reduction for Chinese judgment classification in order to achieve better performance while keeping as much important information as possible. Three feature reduction methods have been studied and utilized in our work as follows:

1. Minimum document frequency. Document Frequency (DF) represents the number of documents with a word in it. If the DF of a term is extremely low, it might not be a meaningful word for text classification. By adjusting minimum document frequency requirement in document representation, we can easily filter out a large amount of features, thus, provide a lower dimensional feature vector.
2. Principle component analysis (PCA). PCA is a statistical procedure that uses an orthogonal transformation to convert a set of possibly correlated variables into a set of values of linearly uncorrelated variables called principle components. This transformation is defined in such a way that the first principle component has the largest possible variance. The first K principle components are chosen as the new vector basis. In this way, feature dimensions can be reduced greatly while maintaining as much information as possible.
3. Truncated Singular Value Decomposition (SVD). The dimensionality of documents is reduced by projecting the bag-of-words vectors into a semantic space. In specific, SVD construct a low rank approximation of the original matrix while preserving the similarity between the documents.

3.4   Classifiers

At present, the most used machine learning algorithms are Nave Bayes, Decision Tree, Random Forest and Support Vector Machine (SVM). In order to study the performance that different classifiers can achieve in classifying Chinese judgment documents, after documentation representation and dimensional reduction, we apply off-the-shelf machine learning algorithms to train a classification model that can be used to classify an unseen record as belonging to one of the predefined categories. In our work, we will evaluate all these most used machine learning algorithms by training corresponding learned classifiers and compare their results together.

4   Evaluation

To evaluate the performance of Chinese judgment documents classification, we experiment on further separating judgments related to liabilities for product quality into their specific industries. In this section, we introduce the experiment dataset and evaluation metrics. At the end of this section, experiments and results are presented.

4.1 Dataset

In order to train a classification model and test the accuracy and performance of it, a golden standard dataset for Chinese judgment documents is required. As there exist no such datasets, we make a lot of efforts to manually label 6735 pieces of judgment documents which are related to liabilities for product quality into 13 categories based on the statutory standard of industry division.

Hold-out method is utilized in dataset separation. To avoid introducing extra errors in splitting datasets and to maintain the consistency of data distribution, we use stratified sampling method other than random sampling to separate training dataset and test dataset. With stratified sampling method, the ratio of each category in training dataset and test dataset maintain the same. However, experiment results using singular hold-out may not be stable and reliable. Therefore, multiple stratified sampling is used and the average of all results are given to evaluate performance. As for the setting of ratio value, there exists a trade-off, since the more samples training dataset contains, the closer the classification model is to the one trained by the overall dataset, but the less samples test dataset will contain, which may lead to inaccurate and unstable results. While the larger test dataset is, the smaller training set will be, thus leading to a more distorted classification model. No perfect solution exists for this problem, common way to tackle it is using 2/3–4/5 samples as training dataset while the rest as test dataset. In our work, we set the ratio as 70%. Figure 2 illustrates the distribution of our datasets for Chinese judgment documents classification. Among all labeled judgment documents, 70% is used for building classification model, 30% for performance evaluation.

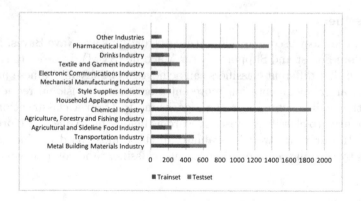

Fig. 2. Datasets for Chinese judgment documents classification

4.2 Evaluation Metrics

To evaluate the performance of Chinese judgment documents classification, three evaluation metrics are employed as follows:

1. Overall accuracy. Overall accuracy represents the percentage of records in the test dataset which are classified correctly.
2. Precision and recall for each category. Given a category c, its precision is the percentage of records classified by the algorithm as c that indeed belong to c. And its recall it the percentage of records belongs to c that are correctly classified by the classifier.
3. F-measure. For each category, F-measure is calculated by 2*precision*recall/ (precision + recall). F-measure represents the balance between precision and recall, the higher the F-measure of a category is, the better the performance of the classifier on this category is.

4.3 Experiments and Results

This section presents the experiment results to answer our research problems. NB stands for Nave Bayes, DT stands for Decision Tree, RFC stands for Random Forest Classifier, and SVM stands for Support Vector Machine.

RQ (1): How is the performance of the classifier improved by domain stop words list construction and text preprocessing in Chinese judgment documents classification?

We first analyze the document frequency of each word, Fig. 3 presents the statistics results of words with certain document frequencies. There are 54633 words that only exists in one document, 9804 words exists only in two documents, 315 words exists in 1000–6700 documents, etc. Learning from the data, words with extremely high document frequency only takes a little space. Based on the words sorted by document frequency, we construct a list of judicial stop words. Among them, 2170 judicial stop words with high document frequency have been filtered out.

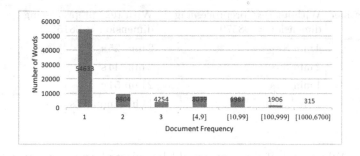

Fig. 3. Number of words with certain document frequency

With stop words removal in text preprocessing, the number of feature dimensions is reduced from 98750 to 68155. Figure 4 presents the overall accuracies of classifiers with text preprocessing and without text preprocessing. As text pre-processing removes a large amount of noises significantly, thus provides a lower dimensional feature vector. As Fig. 4 illustrates, the overall accuracy is improved greatly no matter which classifier is used. In specific, the overall accuracy of NB improves 0.89%, of DT improves 6.22%, of RFC improves 7.2%, and of SVM improves 1.14%.

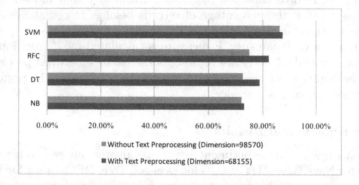

Fig. 4. Overall accuracies of classifiers with text preprocessing and without text pre-processing

When it comes to efficiency, as Table 2 presents, the time cost of classifi-cation model training is reduced on average, especially when using SVM, the performance is greatly improved when feature dimensions get lower. Taken both overall accuracy and time cost into consideration, text preprocessing does great work in improving the performance of Chinese judgment classification.

Table 2. Time cost of classifiers with text preprocessing and without text preprocessing

Classifier	Without text preprocessing dimension = 98570	With text preprocessing dimension = 68155
NB	4 min 8 s	2 min 25 s
DT	2 min 35 s	2 min 44 s
RFC	1 min 46 s	2 min 51 s
SVM	15 h 32 min 32 s	11 h 44 min 17 s

RQ (2): What kind of features should be selected and how can we benefit from dimensional reduction?

By utilizing 3 kinds of dimensional reduction methods can we reduce feature dimensions. Figure 3 illustrates the words with low document frequency, which may not be meaningful for classification, occupy most of the places. We have done experiments on adjusting the parameter of minimum document frequency requirement in feature extraction step, and tried PCA and Truncated SVD for further dimensional reduction. D presents the abbreviation of feature dimensions after dimensional reduction. Figure 5(a) presents the overall accuracies of each classifier with different approaches. Figure 5(b) presents the corresponding running time cost.

As Fig. 5(a) illustrates, for each classifier, the overall accuracies do not change a lot accordingly with the change of minimum document frequency. However, SVM shows a decreasing trend when feature dimensions get lower in general. The overall accuracy of RFC reaches peak when D is 9654, Min_df is 5 and D is 500 with PCA or SVD. With the increasing trend RFC shows, PCA and SVD can help improve the accuracy of RFC. But for DT, PCA and SVD do not show positive effects on improving overall classification accuracy. However, with Min_df, DT improves in overall accuracy when feature dimensions are reduced. For NB, SVD works better than PCA in general, but both of them do not show improvement in overall accuracy.

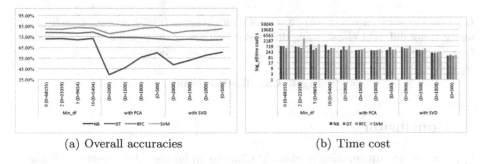

(a) Overall accuracies   (b) Time cost

Fig. 5. Classification results using different dimensional reduction approaches

Meanwhile, when it comes to the time costs of utilizing each classifier for classification model training, the time cost of SVM reduces accordingly with the reduction of feature dimensions. For other classifiers, the performance improves at a certain degree with PCA and SVD. Taken both overall accuracy and time cost into consideration, SVM achieves the best overall accuracy comparing to the others, but it costs more time in model training, with minimum document frequency and PCA or SVD for further dimensional reduction, better performance can be achieved with a little cost of accuracy.

RQ (3): Which machine learning algorithm achieves better performance for Chinese judgment documents classification?

Figure 6(a), (b), (c) illustrates the precision, recall, F1-score of each category respectively when using different classifier. In each figure, the last cluster represents the average result. When sorting the performance of classifiers by average precision, SVM takes the first place, RFC is better than DT, and DT is better than NB. Meanwhile, when sorting by average recall, we have the same results.

Since F1-score measures the balance of precision and recall, the higher F1-score is, the better performance. Therefore, as Fig. 6(c) presents, for Chinese judgment documents classification, we have the overall performance order as: SVM > RFC > DT > NB. Drilling down to each category, the distinction of the performance of classifiers can be tracked back to the different training dataset size of each category, also known as value of support. With categories with larger amount of dataset, i.e. Pharmaceutical Industry and Chemical Industry, SVM can achieve better results than others. However, with smaller datasets, RFC can have more stable performance. Based on the experiment results, SVM achieves better performance with F1-score at 87%.

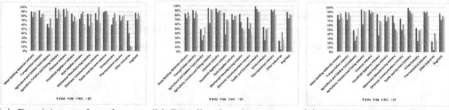

(a) Precisions of each category using different classifiers.

(b) Recalls of each category using different classifiers.

(c) F1-scores of each category using different classifiers.

Fig. 6. Classification results

5   Conclusion

Approaches to automatically classify Chinese judgment documents utilizing a wide variety of machine learning algorithms are explored in this paper. Different from other documents, Chinese judgments have a certain kind of format, as a result, for different classification goal, related content extraction from original judgments is necessary for removing unrelated information. To improve the performance of classification, first, we construct a list of judicial stop words by statistically analyzing words that occur frequently across all documents as well as words with least frequencies. Second, we utilize three different dimensional reduction methods to reduce feature dimensions while keeping as much information as possible, which include minimum document frequency, PCA and Truncated SVD. Results of experiments demonstrate the effectiveness of those dimensional reduction approaches for improving the performance in Chinese judgments classification. Third, four machine learning algorithms are applied for document classification, SVM achieves better performance compared to others, with average F1-score at 87%.

These methods can be easily applied in other judgments classification. To realize a more openness, justice and functional judicial system, more classification tasks in other kinds of judgments and deeper analysis on judgments should be carried out. As a judgment could belong to multiple classes, for achieving better performance, multi-labeled text classification methods need to be explored. Besides, a more functional search engine based on automatic classification for Chinese judgment documents should be studied and developed.

Acknowledgement. This work was supported by the Key Program of Research and Development of China (2016YFC0800803), the National Natural Science Foundation, China (No. 61572162, 61572251), the Fundamental Research Funds for the Central Universities.

References

1. Aggarwal, C.C., Zhai, C.X.: An introduction to text mining. In: Mining Text Data, pp. 1–10 (2012)
2. Strzalkowski, T.: Document representation in natural language text retrieval. In: Proceedings of the Workshop on Human Language Technology, pp. 364–369 (1994)
3. Jiang, S., Lewris, J., Voltmer, M.: Integrating rich document representations for text classification. In: Systems and Information Engineering Design Symposium (SIEDS) (2016)
4. Liu, Y., Song, W., Liu, L.: Document representation based on semantic smoothed topic model. In: International Conference on Software Engineering, Artificial Intelligence, Networking and Parallel/Distributed Computing (SNPD) (2016)
5. Yang, S., Guo, J.: A novel approach for business document representation and processing without semantic ambiguity in e-commerce. In: 6th IEEE Conference on Software Engineering and Service Science (ICSESS) (2015)
6. Arguello, J., Elsas, J.L., Callan, J., Carbonell, J.G.: Document representation and query expansion models for blog recommendation. In: Proceedings of the 2nd International Conference on Weblogs and Social Media (ICWSM) (2008)
7. Berry, M.: Large-scale sparse singular value computations. Int. J. Supercomput. Appl. **6**(1), 13–49 (1992)
8. Blei, D., Lafferty, J.: Dynamic topic models. In: ICML, pp. 113–120 (2006)
9. Hofmann, T.: Probabilistic latent semantic analysis. In: UAI, p. 21 (1999)
10. Blei, D., Ng, A., Jordan, M.: Latent Dirichlet allocation. J. Mach. Learn. **3**, 993–1022 (2003)
11. Apte, C., Damerau, F., Weiss, S.: Automated learning of decision rules for text categorization. ACM Trans. Inf. Syst. **12**(3), 233–251 (1994)
12. Baker, L., McCallum, A.: Distributional clustering of words for text classification. In: ACM SIGIR Conference (1998)
13. Drucker, H., Wu, D., Vapnik, V.: Support vector machines for spam categorization. IEEE Trans. Neural Netw. **10**(5), 1048–1054 (1999)
14. Ng, A.Y., Jordan, M.I., On discriminative vs. generative classifiers: a comparison of logistic regression and Naive Bayes. In: NIPS, pp. 841–848 (2001)
15. Sun, J.-T., Chen, Z., Zeng, H.-J., Lu, Y., Shi, C.-Y., Ma, W.-Y.: Supervised latent semantic indexing for document categorization. In: ICDM Conference (2004)
16. Zhou, Z.: Machine Learning (2015)
17. Che, W., Li, Z., Liu, T.: LTP: a Chinese language technology platform. In: Proceedings of the COLING: Demonstrations, Beijing, China, pp. 13–16, August 2010

A Partitioning Scheme for Big Dynamic Trees

Atsushi Sudoh, Tatsuo Tsuji[✉], and Ken Higuchi

Information Science Department, Faculty of Engineering, University of Fukui,
Bunkyo 3-9-1, Fukui City, 910-8507, Japan
{sudou,tsuji,higuchi}@pear.fuis.u-fukui.ac.jp

Abstract. In this paper, we propose a scheme for partitioning dynamically growing big trees and its implementation. The scheme is based on the history-pattern encoding scheme for dynamic multidimensional datasets. Our scheme of handling big dynamic trees is relying on the history-pattern encoding, by which large scale datasets can be treated efficiently. In order to partition these dynamic trees efficiently, the encoding scheme will be improved and adapted to the partitioning. In our partitioning scheme of a tree T, the path from the T's root node to the root node of a partitioned subtree, is treated as an index for selecting the subtree. The path is split into a shared path and the local path in the subtree and each path is encoded by using the history-pattern encoding. The partitioning scheme contributes to the reduction of the storage cost and the improvement of the retrieval cost. In this paper, after our tree encoding scheme designed for the partitioning is described, some problems caused in the encoding are addressed and their countermeasure is presented. Finally, an implemented prototype system is described and evaluated.

1 Introduction

Tree graph is an extremely useful graph structure and is employed as a basic data structure in wide application domains such as sorting/searching, information retrieval, data mining, XML document processing, and so on. In these application domains, tree graphs are used for data/knowledge representation or control structure. It has been widely recognized that efficient data analysis and processing are enabled due to their simple and powerful operational capabilities.

Along with the increase of the application domains such as XMLDB or Web mining, in which dynamic growth of graph size or change of graph structures are caused frequently, highly efficient processing capability for such kind of dynamicity is strongly desired. Many researches have been conducted concerning on the schemes for clustering and partitioning a tree graph. The aim of these researches often includes improvement of query processing (e.g., [4]) or performance improvement obtained by parallel processing (e.g., [5]).

In this paper, we will propose an encoding scheme for large scale dynamic tree graphs with which re-encoding is not necessary even if a tree structure dynamically grows. This encoding scheme is based on the history-pattern encoding scheme [6, 7] designed for dynamically increasing multidimensional datasets.

© Springer International Publishing AG 2017
Z. Bao et al. (Eds.): DASFAA 2017 Workshops, LNCS 10179, pp. 18–34, 2017.
DOI: 10.1007/978-3-319-55705-2_2

Implementation of dynamic trees based on the history-pattern encoding suffers from its own problem that the storage cost might be degraded when the insertion of new nodes occur at the level lower than the current highest level. In order to avoid this deficiency, a tree is partitioned into a set of subtrees. In our partitioning scheme, the path from the root node of a tree to a node n works as an index for selecting the partitioned subtree whose root node is n. This enables the reduction of the storage cost and the improvement of the retrieval cost. We present an history-pattern encoding scheme adapted to this partitioned tree. Note that by placing each partitioned subtree to separate machines on the network, our partitioning scheme can be well adapted to the distributed environment.

In the rest of this paper, after the related work is explained, the history-pattern encoding is outlined. Next the implementation of tree graphs is described. Then the countermeasure against the above problem will be presented. Subsequently, we propose a scheme for partitioning dynamically growing trees along with the improvement and adaptation of the history-pattern encoding to this partitioning scheme. Finally, an implemented prototype system is described and evaluated.

2 Related Work

As an important work related to the history-pattern encoding scheme, notion of *array database* [10, 11] for scientific applications can be listed. It provides a data model handling arrays as its main data structure. Scientific phenomena captured by arrays can be treated easily in this data model. Array databases can be well utilized in the various array oriented application domains other than scientific application. In addition to these array oriented applications, the history-pattern encoding is also based on array data structure, it can efficiently encode and handle a dynamic tree by embedding it to an dynamically extendible array. Therefore our approach can also cover tree-oriented application domains.

In order to handle dynamically growing trees, it is important to provide an efficient encoding scheme for each tree node n that can well capture and reflect the position in the tree using the path from the root node to node n. For example, [1, 2, 8] provide a encoding scheme for dynamic XML tree preserving the *document order* even if a new node is inserted in the tree.

However, these schemes have a problem in handling big XML trees. In [1, 2], the label length become so large and the storage cost is high [8] is based on the encoding scheme [9] similar to the history-pattern encoding, but the encoded result should be within single machine word; e.g., 64 bits. This makes it impossible to encode a big tree. In contrast the history-pattern encoding scheme can handle very big data with only small metadata for each tuple. In fact only with one byte metadata, a tuple whose encoded size is up to 256 bits can be represented and treated. At the same time, a dynamically increasing tuple dataset up to 2^{256} tuples can be handled. Therefore very big trees in their height and breadth can be implemented and handled.

3 History-Pattern Encoding

In this section, we describe the *history-pattern* encoding scheme for dynamic multidimensional datasets.

3.1 Data Structures for History-Pattern Encoding

Figure 1 illustrates the required data structures for the history-pattern encoding scheme. When an n-dimensional extendible array A extends, a fixed size sub-array equal to the size of the current A in every dimension is attached to the extended dimension. The number enclosed in Fig. 1 represents the insertion order of the element whose coordinate is corresponding to the tuple.

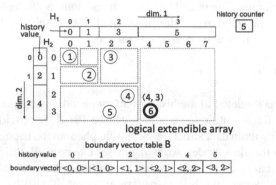

Fig. 1. Data structures for history-pattern encoding

The data structures for A consist of *history tables* and a *boundary vector table* that maintain extension history of A.

History Table

For each dimension i ($i = 1, \ldots, n$), the history table H_i is maintained. Each history value h in H_i represents the extension order of A along the i-th dimension and identifies the past *shape* of A when the history counter value was h; the counter value is initialized to 0 and incremented by one each time A is extended. H_i is a one-dimensional array, and each subscript k ($k > 0$) of H_i corresponds to the subscript range from 2^{k-1} to $2^k - 1$ of the i-th dimension of A covered by the sub-array S_h; S_h was attached to A at the extension along the i-th dimension when the history counter value was h.

For example, consider the coordinate (7, 10). Because $(7, 10) = (111_{(2)}, 1010_{(2)})$, the bit pattern sizes are 3 and 4, respectively. Therefore, $H_1[3]$ and $H_2[4]$ keep the history counter values when A was extended along the first and second dimension, respectively. It can also be determined that the subscript ranges of A covered by the attached subarrays are 4–7 for the first dimension and 8–15 for the second dimension.

Boundary Vector Table

The *boundary vector table* B is a single one-dimensional array whose subscript is a history value. Each element of B maintains the past *shape* of A represented by the corresponding boundary vector when the history counter was a given value. Together with the boundary vector, B also maintains the dimension of A extended at the given history counter value. At initialization A includes only the element $(0, 0, ..., 0)$, and the history counter is initialized to 0. B[0] includes $<0, 0, ...,0>$ as its boundary vector.

Assume that the current history counter value is h, and B[h] includes $<b_1, b_2, ..., b_i, ..., b_n>$ as its boundary vector. When the current A extends along the i-th dimension, B[$h + 1$] includes $<b_1, b_2, ..., b_i + 1, ..., b_n>$ as its boundary vector.

In the history-pattern encoding, an extendible array A has two size types: real and logical. Assume that the tuples in an n-dimensional dataset M have been converted to the set of coordinates. Let s be the largest subscript of dimension k, and $b(s)$ be the bit size of s. Then, the real size of dimension k is $s + 1$, and the logical size is $2^{b(s)}$. The real size is the cardinality of the k-th attribute. In Fig. 1, the real size and the logical size are [4, 5] and [4, 8] respectively.

3.2 Array Extension

Suppose that a tuple, whose k-th attribute value is new, is inserted in M. This insertion increases the real size of A in dimension k by one. If the increased *real size + 1* of dimension k does not exceed the current logical size $2^{b(s)}$, then A is not physically extended, and neither history table H_k nor the boundary vector table B is updated. However, if the *real size* of dimension k exceeds the current logical size, then A is logically extended. That is, the current history counter value h is incremented to $h + 1$, and this value is set to $H_k[b(s + 1)]$. Moreover, the boundary vector in B[h] is copied to B[$h + 1$], and the dimension k of the boundary vector is incremented (Fig. 1).

Note that h has one-to-one correspondence with its boundary vector in B[h] and uniquely identifies the past (logical) shape of A when the history counter value was h. To be more precise, for the history value $h > 0$, if the boundary vector in B[h] is $<b_1, b_2, ..., b_n>$, the shape of A at h was $[2^{b1}, 2^{b2}, ..., 2^{bn}]$. For example, as shown in Fig. 1, because the boundary vector for the history value 3 is $<2, 1>$, the shape of A when the history counter value was 3 is $[2^2, 2^1] = [4, 2]$. Note that h also uniquely identifies the sub-array that is attached to A at extension when the history counter value was $h - 1$, and vice versa.

3.3 Encoding/Decoding

Using the data structures described in Sect. 3.1, an n-dimensional coordinate $I = (i_1, i_2, ..., i_n)$ can be encoded to the pair $<h, p>$ of *history value* h and *bit pattern* p. The history tables H_i ($i = 1, ..., n$) and the boundary vector table B are used for the encoding. The history value h is determined as the maximum value in $\{H_k[b(i_k)] \mid 1 \leq k \leq n\}$, where $b(i_k)$ is the bit size of the subscript i_k in I. For each history value h, the boundary vector in B[h] gives the bit pattern size of each subscript in I. According to this boundary vector, the coordinate bit pattern p can be obtained by concatenating the subscript bit pattern of each dimension in descending order (from the lower to the higher bits of p). The storage unit for p can be one word length, i.e., 64 bits.

In Fig. 1, the history-pattern encoding $<h, p>$ of array element (4,3) is shown as an example. $H_1[b(4)] = H_1[b(100_{(2)})] = H_1[3] = 5$ and $H_2[b(3)] = H_2[b(11_{(2)})] = H_2[2] = 4$. Since $H_1[b(4)] > H_2[b(3)]$, h is $H_1[3] = 5$. So element (4,3) is known to be included in the *sub-array* on dimension 1 at history value 5. Therefore, the boundary vector to be used is $<3, 2>$ in $B[5]$. In (4, 3) to be encoded, the subscript 4 of the first dimension and the subscript 3 of the second dimension form the upper 3 bits and lower 2 bits of p, respectively. Therefore p becomes $10011_{(2)} = 19$. Eventually, the element (4, 3) is encoded to $<5, 19>$. Generally, the bit size of history value h is rather small compared to that of pattern p; if the storage size for the pair is assumed to be 16 bits, typically the upper 4 bits are for h, and the lower 12 bits are for p.

Conversely, to decode the encoded pair $<h, p>$ to the original n-dimensional coordinate $I = (i_1, i_2,..., i_n)$, the boundary vector in $B[h]$ is known. Then, the subscript value of each dimension is sliced out from p according to the boundary vector. Note that the procedure for extending an extendible array described in Sect. 3.2 assures the following important property on $<h, p>$.

[**Property 1**] History value h denotes the bit size of its coordinate bit pattern p.

It should be noted that due to this property, h can be used as the header of the en coded tuple $<h, p>$. This enables the output file of the encoded results to be a sequential file of packed variable length records.

3.4 Implementation of a Multidimensional Dataset

As well as the core data structures (the history tables and the boundary vector table) presented in Sect. 3.1, three additional types of data structures are required to implement a multidimensional dataset M using history-pattern encoding.

CVT$_i$ (attribute subscript conversion tree) converts the attribute values of dimension $i (i = 1 ,..., n)$ to their subscript values of the corresponding extendible array. CVT$_i$ is implemented using a B$^+$ tree. CVT$_i$ is used for tuple encoding.

AVT$_i$ (attribute value table) is a one-dimensional array for dimension i $(i = 1 ,..., n)$. If attribute value v is mapped by CVT$_i$ to subscript k, the k-th element of AVT$_i$ keeps v. AVT$_i$ is used for tuple decoding.

ETF (encoded tuple file) is an output file of the encoded results for M.

We call the implementation scheme of M using the core data structures in Fig. 1 together with the above data structures as HPMD (History Pattern implementation of Multidimensional Datasets).

Figure 2 shows encoding of the tuple set M at the left side using HPMD, and Fig. 3 is the produced CVTs and AVTs. In fact, Fig. 1 is the core data structures constructed from this example. Tuples in M are encoded sequentially in the input order. Each attribute value is converted to its coordinate subscript using the CVT of the corresponding dimension. The subscript values are assigned from 0 in the order that the new attribute value appears.

	Company	Item		coordinate	boundary vec	\<history, pattern\>
1	D	Pencil	→	(0, 0)	\<0,0\>	\<0, .\> = \<0,0\>
2	B	Scissors	→	(1, 1)	\<1,1\> ○	\<2, 1.1\> = \<2,3\>
3	C	Pencil	→	(2,0)	\<2,1\> ○	\<3,10.0\> = \<3,4\>
4	A	Ruler	→	(3,2)	\<2,2\> ○	\<4,11.10\> = \<4,14\>
5	C	Eraser	→	(2,3)	\<2,2\>	\<4,10.11\> = \<4,11\>
6	E	Scissors	→	(4,1)	\<3,2\> ○	\<5,100.01\> = \<5,17\>

Remark : ○ denotes that the insertion of the tuple causes the extension of the logical size of the le array

Fig. 2. An encoding example of a multidimensional dataset

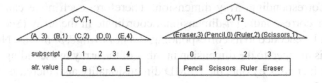

Fig. 3. Produced CVTs and AVTs for Fig. 2

3.5   Adding a New Attribute

In HPMD, the cost of adding a new attribute dynamically after the schema definition is very small. For example consider the case where a new attribute value "Price" is added to the tuple set shown in Fig. 2. Dimensional extension of HPMD is done from two dimensions to three dimensions; for the third dimension "Price", a history table H_3 is created and added to the core data structures shown in Fig. 1, and attribute subscript conversion tree CVT_3 and attribute value table AVT_3 are created and added for the new dimension attribute.

Moreover, boundary vectors in the boundary vector table become three dimensional as is shown in Fig. 4. Existing \<history, pattern\>s already encoded according to the two dimensional boundary vectors are not necessary to be re-encoded, and also the boundary vector table is unnecessary to be reorganized. This is because that the boundary vectors before extension can be treated as three dimensional after the dimensional extension; the value of the dimension three of the vector can be treated as 0 and this implies that the bit string for the subscript of the dimension three is a null string.

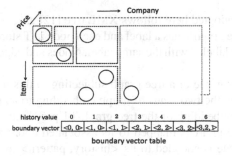

Fig. 4 Dimensional extension of Fig. 1

4   Storage Scheme for Dynamic Tree

In this section, we describe the mapping scheme of a dynamic tree to HPMD.

4.1   Mapping Dynamic Tree

Each depth level of a tree is mapped to a dimension of an extendible array. The number that represents each node's relative position among its siblings is mapped to the subscript value of the corresponding array dimension. Therefore, each node can be uniquely specified by its corresponding n-dimensional coordinate in the array (See Fig. 5). The coordinate can be encoded to its corresponding *<history value, pattern>*. Note that each level of a tree is mapped to the dimension number of the array in ascending order. When the tree height increases by one, the HPMD dimensionality also increases by one, as is shown in Fig. 5.

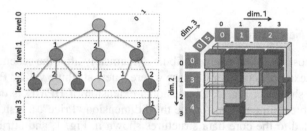

Fig. 5. Mapping dynamic tree to an extendible array

We can see from Fig. 5 that horizontal nodes insertion along some level in a tree graph results in the extension along the corresponding dimension. Also we can observe that the tree increases its height by one, a new dimension is added to the corresponding extendible array and this dimensional extension can be performed efficiently as was described in Sect. 3.5.

4.2   Handling Labeled Tree Graphs

The tree graph model adopted here follows XML tree [12] as :

"Each edge of a tree graph T has a label and each node in T stores its own data. More than one edge might be labeled with the same label. Nodes and edges can be dynamically added or deleted."

Figure 6(a) is an example of a tree graph subjecting to this model. A tree graph T will be implemented by the two kinds of graph embedding to an extendible array. See Fig. 6(b). One is the embedding of the tree graph T1 produced by dropping the edge labels of T. Each node in T1 is mapped to its associated coordinate of an extendible array and this coordinate is encoded to its <history, pattern> as in Fig. 5. Due to the mapping strategy, we can see that this encoded result hp keeps the position information of all its ancestor nodes and hp also its position among its brother nodes in T1. According to these position information we can perform the structural traversal in T1.

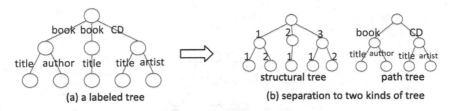

Fig. 6. Separation of a tree graph to two kinds of tree

The other kind of graph is for path retrievals in T. A path from the root node r of T to node n is the concatenation of the edge labels on the path. In our tree graph model, more than one edge might be labeled with the same label. Therefore a path in T might access more than one node. A tree graph T2 is a functional graph produced from T. T2 For each node n in T2 a path from the root node to n cannot access any other nodes. Note that T2 is *dataguides* in [3]. T2 is embedded in a HPMD; edge labels on the level k of T2 are mapped to their corresponding subscripts by CVT_k and each node in T2 is encoded by history-pattern encoding as in T1. T1 and T2 will be called structural tree and path tree respectively.

Both the structural information and path information of each node in T can be compactly aggregated in <history, pattern> in structural tree and path tree, respectively. This would provide the advantage of our implementation scheme of T both in storage cost and retrieval costs. In the following, the encoded result <history, pattern> for each node in the structural tree and the path tree will be called as nID (node ID) and pID (path ID), respectively.

5 Problems on Our Storage Scheme

There are following two problems on our storage scheme for tree graphs presented in Sect. 4.

5.1 Duplication of Ancestor Path

The first problem is the duplication of the ancestor path among the child nodes of the same parent node. This is an inherent problem in tree structures. Consider the example shown in Fig. 7. There exist three nodes on level 5. The coordinates of these nodes in the corresponding extendible array are $(2, 3, 2, 1, 1)$, $(2, 3, 2, 1, 2)$, $(2, 3, 2, 1, 3)$, and the ancestor patterns corresponding to $(2, 3, 2, 1)$ are stored in duplicate. If the tree grows vertically, the storage overhead caused by such duplication of ancestor patterns would be increased.

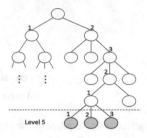

Fig. 7. Duplication of ancestors

5.2 Problem Caused by Expansion Order of Dynamic Tree

The second problem concerns the one caused by employing HPMD for implementing dynamic trees. Namely, depending on the order of vertical and horizontal growing of the trees, the higher dimensions of the coordinate pattern might be occupied by the redundant 0's and the storage cost would be degraded. Such redundancy might arise when the extendible array used for tree mapping extends its size along the dimension other than the maximum one; i.e., when a node insertion at a level lower than the maximum level of the tree causes an extension of the extendible array.

Figure 8(a) shows an example of growing tree and the boundary vector table of the related extendible array. The number of a tree node represents its insertion order. Figure 8(b) and (c) show the history tables and the boundary vector table of the extendible array for Fig. 8(a) respectively. Figure 9 shows the encoding sequence of the tree nodes according to the input order.

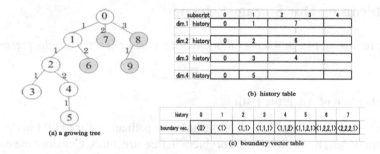

Fig. 8. A growing tree

	coordinate	boundary vec.	<history, pattern>		
node 0 →	(0)	<0>	<0, .>	=	<0,0>
node 1 →	(1)	<1> ○	<1,1>	=	<1,1>
node 2 →	(1, 1)	<1,1> ○	<2,1.1>	=	<2,3>
node 3 →	(1, 1, 1)	<1,1,1> ○	<3,1.1.1>	=	<3,7>
node 4 →	(1, 1, 2)	<1,1,2>○	<4,1.1.10>	=	<4,14>
node 5 →	(1, 1, 2, 1)	<1,1,2,1> ○	<5,1.1.10.1>	=	<5,29>
node 6 →	(1, 2, 0, 0)	<1,2,2,1>○	<6,1.10.00.0>	=	<6,48>
node 7 →	(2, 0, 0, 0)	<2,2,2,1>○	<7,10.00.00.0>	=	<7,64>
node 8 →	(3, 0, 0, 0)	<2,2,2,1>	<7,11.00.00.0>	=	<7,96>
node 9 →	(3, 1, 0, 0)	<2,2,2,1>	<7,11.01.00.0>	=	<7,104>

Fig. 9. The encoding of the tree nodes in Fig. 8

Since the insertions of node 1, 2, 3 increase the height of the tree, dimensional extensions occur and the related extendible array becomes 3 dimensional. At this point, the history counter value is 3. The insertion of node 4 extends the array along the dimension 3, since the subscript bit size of this dimension increases by 1. The insertion of node 5 causes dimensional extension and the maximum tree level becomes 4.

The insertions of node 6 and node 7 cause array extensions along dimension 2, and dimension 1 respectively, since both of the subscript bit sizes of these dimensions increase from 1 bit to 2 bits. At this point the dimension bit sizes of the array becomes <2, 2, 2, 1> and the history counter value becomes 7. The insertions of node 8 and node 9 do not exceed these dimension bit sizes, so no array extensions occur by these insertions. The levels (i.e., dimensions) of node 6, 7, 8, 9 are all under this maximum level.

As we can see from Fig. 9, the coordinates for node 6, 7, 8, 9 are encoded to the <history, pattern> s as <6,1.10.00.0>, <7,10.00.00.0>, <7,11.00.00.0>, <7,11.01.00.0> respectively using their boundary vectors. Note that the history value of the coordinate $(i_1, i_2, \ldots, i_k, 0, 0, \ldots, 0)$ where $i_j \neq 0$ $(1 \leq j \leq k)$, can be determined by $(i_1, i_2, \ldots, i_k)$ by comparing history values in $H_j[i_j]$ $(1 \leq j \leq k)$ as explained in Sect. 3.3.

Note also that higher dimensions of the pattern parts of these encoded results might be occupied by redundant 0's. Such redundancies arise frequently when the tree gains its size in the vertical direction at its early stage of growth as in Fig. 8(a). Such situation is typical in the case when an XML document is transformed to its XML tree in *preorder* and loaded into main memory. According to our prior experimentation for constructing HPMD from a tree graph, greater part of the output ETF is proved to be occupied by these redundant 0's.

6   Encoding Methods for Eliminating Redundancy

In this section, we will present two kinds of an encoding method to eliminate the redundancy arising in the tree encoding described in Sect. 5. Since the path trees are generally very small compared with the structural trees, we will apply our improved schemes to only structural trees.

6.1   Horizontal Partitioning

This method separates the output file ETF of the encoded results by tree level. Let the current maximum level of a tree be n. If the level of a tree node is k $(1 \leq k \leq n)$, the encoded result of the node is stored in ETF_k. When the current maximum level of the tree increases, ETF_{n+1} is created.

For the nodes 7, 8 in Fig. 8(a), their encoded results become <7, 10.> = <7, 2> and <7, 11.> = <7, 3> being the redundant 0's on dimensions 3, 4, 5 dropped out and stored in ETF_1 sequentially. For the nodes 6, 9, their encoded results become <6, 1.10> = <6, 6> and <7, 11.01> = <7, 13> being the redundant 0's on dimensions 4, 5 dropped out and are stored in ETF_2.

Decoding of each $<h, p>$ in ETF_k can be done by knowing the boundary vector in $B[h] = <b_1, b_2,..., b_k,......>$; pattern p can be decoded to its k dimensional coordinate $(i_1, i_2,..., i_k)$ by using $b_1, b_2,..., b_k$. For example, in Fig. 9, for the encode $<6, 6>$ in ETF_2, first $B[6] = <1,2,2,1>$ is inspected. Since $<6, 6> = <6, 110_{(2)}>$ is stored in ETF_2, the tree level of the node is 2. Therefore the first and the second elements of the boundary vector is used for decoding, so the pattern part $110_{(2)}$ is split into $1.10_{(2)}$, namely 1 bit and 2 bits. Hence the $<6, 6>$ is decoded to the coordinate $(1,2)$ which specifies node 6. In Fig. 9(b), we can know that the encode of node 6 is actually encoded to $<6, 48> = <6,1.10.00.0>$ according to the usual encoding scheme shown in 2.3; 3 redundant bits 0's.

This method of separating ETF has the advantages:

(1) Storage overhead by the redundant 0's does not arise.
(2) When the levels of the retrieval target can be narrowed in advance, only their corresponding separate ETF files are sufficient to be searched.

Conversely, the disadvantage of this method includes:

(3) Retrieval that requires traversing tree structure might be slow, since the horizontal partitioning of ETF breaks down the paths from the root node.

Organization of ETF in the Horizontal Partitioning
Each ETF_k above is organized to reflect the cross reference e between a node of structural tree and its node of path tree in Fig. 6(b). For the cross reference, ETF_k is simply organized as a sequential file keeping the pairs of nID (node ID) and its corresponding pID (path ID). See Sect. 4.2 on nID and pID.

6.2 Attaching Level of a Node as Meta Data

Let k be the current maximum level of a tree and $(i_1, i_2,..., i_k)$ be the coordinate of a tree node, we will add the tree level l of the node to the coordinate as meta data like $(l, i_1, i_2,..., i_l, i_{l+1},..., i_k)$ (for $l \leq j \leq k, i_j = 0$). This coordinate is encoded to its $<h, p>$ using the corresponding HPMD. If the boundary vector $B[h] = <b_l, b_1, b_2,..., b_k>$, the coordinate $(l, i_1, i_2,..., i_l)$ is encoded according to the boundary vector $<b_l, b_1, b_2,..., b_l>$. Figure 10 shows the encoding of the tree nodes in Fig. 8(a) attaching the node level. For example, the encoded results of node 6, 7, 8, 9 in Fig. 9 are $<9, 21>$, $<10, 6>$, $<10, 7>$, $<10, 45>$ respectively with the redundant 0's being dropped out.

Conversely, the decoding of $<h, p>$ can be done as follows. The boundary vector $B[h] = <b_l, b_1, b_2,..., b_k>$ is inspected and the upper b_l bits are sliced out from p. If the value of this b_l bits is l, the rest of p is decoded according to the boundary vector $<b_1,b_2,...,b_l>$ to obtain its coordinate $(i_1, i_2,..., i_l)$. For example, consider the node whose encode is $<10, 6>$. Since $B[10] = <3, 2, 2, 2, 1>$ and 6 is divided to 001.10, the upper 3 bits of 6 is $001_{(2)}$, so the level of the node is known to be 1 and decoded as $(2, 0, 0, 0)$. Note that $2 + 2 + 1 = 5$ bits of 0 can be saved in the encode.

This method of attaching level of a node has the advantages:

	coordinate	boundary vec.	<history, pattern>		
node 0 →	(0, 0)	<0,0>	<0, . >	=	<0,0>
node 1 →	(1, 1)	<1,1> O	<2, 1.1>	=	<2,3>
node 2 →	(2, 1, 1)	<2,1,1> O	<4, 10.1.1>	=	<4,11>
node 3 →	(3, 1, 1, 1)	<2,1,1,1> O	<5, 11.1.1.1>	=	<5,31>
node 4 →	(3, 1, 1, 2)	<2,1,1,2>O	<6, 11.1.1.10>	=	<6,62>
node 5 →	(4, 1, 1, 2, 1)	<3,1,1,2,1> O	<8, 100.1.1.10.1>	=	<8,157>
node 6 →	(2, 1, 2, 0, 0)	<3,1,2,2,1>O	<9, 010.1.10.00.0>	=	<9,21>
node 7 →	(1, 2, 0, 0, 0)	<3,2,2,2,1>O	10, 001.10.00.00.0:		<10,6>
node 8 →	(1, 3, 0, 0, 0)	<3,2,2,2,1>	10, 001.11.00.00.0:	=	<10,7>
node 9 →	(2, 3, 1, 0, 0)	<3,2,2,2,1>	10, 010.11.01.00.0:	=	<10,45>

Fig. 10. The encoding of the tree nodes in Fig. 8(a) with meta data

(1) Storage overhead by the redundant 0's does not arise.
(2) Retrieval along a tree path might be efficient since the method preserves the tree structure. Conversely, the disadvantage of this method includes:
(3) Storage cost for keeping level of a node would arise.

7 Vertical Partitioning

The disadvantage of horizontal partitioning method stated in Sect. 6.1 is serious for structural oriented retrieval along tree paths as will be confirmed in Sect. 8.3. Here we present another method for partitioning large scale dynamic tree into a set of subtrees using the method stated in Sect. 6.2. This avoids both of the two problems stated in Sect. 5.

7.1 Vertical Partitioning of a Structural Tree

Let T be a structural tree. T is partitioned into a set of subtrees $\{T_0, T_1, ..., T_m\}$. The scheme might split a path from the root node of T at the root node of some subtree T_i ($1 \leq_i \leq n$). Figure 11 shows an example of vertical partitioning of T. Two kinds of HPMD would be generated according to the organization of ETF, namely the output file of the encoded tuple file. One is for T_0 and the other is for $\{T_1, ..., T_m\}$. Here T_0 functions as an index for accessing the vertically partitioned subtrees and implemented by using a B +tree. T_i ($1 \leq i \leq m$) is implemented by using a sequential file.

7.2 Splitting a Path

Let r be a root node of T and l be a level of T. Let $a_1\_a_2 \cdots\_ a_{l-1}\_a_l$ be a subscript path reaching to the node n_l, which is the root node of the subtree T_i ($1 \leq i \leq m$); namely $(a_1, a_2, \cdots, a_l)$ is the coordinate of the node n_l. Note that T_i has more than one node including its root node n_l. For a node n_p in T_i, the subscript path $p = ra_1\_a_2 \cdots\_a_l\_a_{l+1}\_\cdots\_a_p$ that reaches n_p can be split into two sub-paths; $sp = ra_1\_a_2 \cdots\_a_l$ and $lp = n_l a_{l+1}\_\cdots\_a_{p.}$. Note that the sp can be shared among the path from r to an arbitrary node in T_i. Here the node

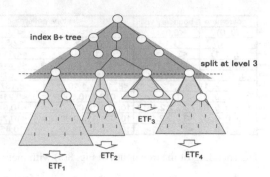

Fig. 11. Vertical partitioning of a dynamic tree

n_l will be called as the split point of the subscript path p. sp and lp are called as shared path and local path for T_i respectively. Each root node in T_i $(1 \le i \le m)$ is a split point.

Figure 12 shows an example showing this splitting scheme at level 3. In the example, the path $r2\_3\_2\_1\_9$ is split into two sub-paths $r2\_3\_2$ and n_l1_9, where $n_l = r2\_3\_2$ is the split point.

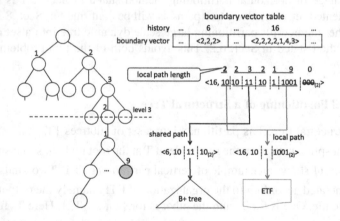

Fig. 12 Encoding shared path and local path

7.3   Encoding Shared Path and Local Path

For a node n_p in the subtree T_i, let the subscript path from the root node r of T be $ra_1\_a_2\_\cdots\_a_l\_a_{l+1}\_\cdots\_a_p$, where $ra_1\_a_2\_\cdots\_a_l$ be the shared path. The encoding of the shared path is straightforward. The coordinate $(a_1, a_2, \cdots, a_l)$ of the shared path is encoded to its <history, pattern>, which is stored in the index B+tree in the key part; the data part is the, reference to the corresponding subtree. On the other hand, the local path n_la_l $_{+1}\_\cdots\_a_p$ is encoded according to the scheme stated in Sect. 6.2 in order to avoid the redundant storage occupied by the trailing 0's in the pattern part. Namely, the local path length p-l is attached to the coordinate $(a_{l+1}, a_{l+2}, \cdots, a_p)$ as its meta information, so actually $(p$-$l, a_{l+1}, a_{l+2}, \cdots, a_p)$ is encoded.

Figure 12 shows the actual implementation and encodings for a shared path and a local path. Note that the local path length is attached on the top of the full path coordinate and this length is used in encoding the local path. Note also that the redundant trailing 0's are not necessary to be stored due to this local path length.

In Fig. 12, the coordinate of the full path $r2\_3\_2\_1\_9$ is $(2, 2, 3, 2, 1, 9, 0)$; the first 2 is the local path length and the last 0 is the redundant trailing 0. Let this coordinate be encoded to the history value 16 and the boundary vector B [16] be $<2, 2, 2, 2, 1, 4, 3>$. The coordinate of the shared path is encoded to $<6, 10.11.10_{(2)}> = <6, 46>$ and this is stored in the index B+tree with its reference to the corresponding subtree. The coordinate of the local path should be $(1, 9, \mathbf{0})$ due to the second problem in Sect. 5.2. But, in order to suppress the redundant 0 on the third subscript, the first 2 of the full path coordinate is attached to the local path coordinate. Hence the local path coordinate becomes $(2, 1, 9)$. This coordinate is encoded to $<16.10.1.1001>$ by referring the boundary vector B [16], and the result is entered into the corresponding ETF file.

8 Experimental Evaluations

Using the constructed prototype systems, experimental evaluations were conducted on our partitioning schemes for dynamic trees.

8.1 Evaluation Environment

(1) Used machine:
   CPU: Intel Core i7 920, 2.67 GHz
   Memory: 48 GB
   OS: Cent OS 5.7 (Linux 2.6.18)
(2) Evaluated implementations:
   For the horizontal partitioning scheme explained in Sect. 6.1, the following implementation is used. We denote this implementation as HP in the following.
   (a) Let n be the maximum level of the current dynamic tree T. For each horizontal level k of $(1 \leq k \leq n,)$ of T, the encoded <history, pattern>s of all the nodes on the level k are stored in the single ETF_k.
       For the vertical partitioning scheme presented in 7, the following two kinds of implementation were used for evaluations, depending on the organization of ETF (Encoded Tuple File). See Fig. 11.
   (b) The encoded <history, pattern>s for nodes in each subtree T_i $(1 \leq i \leq m)$ are stored in the single ETF_i.
   (c) The encoded <history, pattern>s for all nodes in all subtrees T_i $(1 \leq i \leq m)$ are stored in a single ETF file. In this ETF file, the encoded <history, pattern>s of the nodes in each subtree T_i are kept in the same single page list.
       We denote the implementations (b) and (c) above as VP_1, and VP_2 respectively in the following.

(3) Input tree data file:

The input tree data file was produced from an XML file, which was generated by using *xmlgen* program provided by XMark [13]. The produced data file consists of only the pairs of *<element_name>* and *</element_name>* in the generated XML file by removing all attributes and texts. The produced tree data file is:

file size: 332,807,472 Bytes

number of nodes: 16,701,210

maximum tree level: 12

8.2 Storage Cost and Construction Cost

For HP in Sect. 8.1, the total size of ETF_k $(1 \leq k \leq n)$, for VP-1, the total size of index B+tree and ETF_k $(1 \leq k \leq n)$, and for VP_2, the total size of index B+tree and ETF were measured. See Fig. 13. Since HP is the simplest organization, the storage size is the smaller than VP_1 and VP_2, both of which keeps the index B+tree. Vertical partitioning of the structural tree enables sharing ancestor paths contributes to suppress the storage size, however index B+tree occupies larger storage size. VP_2 is larger than VP_1 due to the internal fragments in the page list.

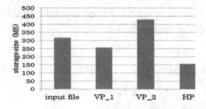

Fig. 13. Storage cost

Figure 14 shows the constructing HPMD data structures from the input tree data file. Since constructing index B+tree is time consuming, VP_1 and VP_2 are much slower than HP.

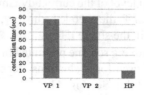

Fig. 14. Construction time

8.3 Retrieval Time

One percent of the total number of the structural tree nodes were randomly selected for retrieval.

Structural Retrieval

Each selected node is set as a current node, and various structural retrievals are performed. It can be observed in Fig. 15 that both VP_1 and VP_2 are much faster compared with HP. While HP must read every nID in ETF_k files of the related level sequentially, both VP_1 and VP_2 are suitable for structural retrieval; they only read the nIDs of the related subtrees The disadvantage of HP is distinct in the retrieval of *descendent* axis. See Fig. 15.

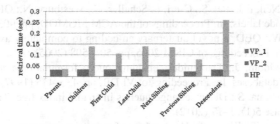

Fig. 15. Structural retrieval time

Path Retrieval

For the label path expressions to the selected structural tree nodes, their retrieval times are measured and averaged. See Fig. 16. VP_1 and VP_2 are both negligibly small compared with HP. This is because that in HP, if a label path expression is $l_1\_l_2\_..._\_l_k$ ($1 \leq k \leq n$), all the ETF_i ($1 \leq i \leq k$) files should be searched sequentially. On the contrary, in VP_1 and VP_2, only search the related subtrees as in the structural retrieval. See Fig. 16.

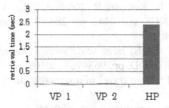

Fig. 16. Path retrieval time

9 Conclusion

We have presented and described partitioning schemes for big dynamic trees based on the history-pattern encoding. We have also evaluated them and confirmed that the horizontal partitioning can be compactly stored, but the traversal along the tree structure is unsuitable, and both the structural and the path retrievals are time consuming. On the other hand, the scheme of the vertical partitioning by path splitting is suitable for the tree traversal and the retrieval is very fast. We are now designing the load balancing scheme among the partitioned ETFs in order to adapt our partitioning scheme to the distributed environments.

Acknowledgments. This work was supported by JSPS KAKENHI Grant Number JP26330132.

References

1. O'Neil, P.E., O'Neil, E.J., Pal. S., Cseri, I., Schaller, G., Westbury, N.: ORDPATHs: insert-friendly XML node labels. In: Proceedings of the ACM SIGMOD, pp. 903–908 (2004)
2. Li, C., Ling, T.W.: QED: a novel quaternary en-coding to completely avoid re-labeling in XML updates. In: Proceedings of CIKM 2005, pp. 501–508 (2005)
3. Goldman, R., Widom, J.: Dataguides: enabling query formulation and optimization in semistrucutured databases. In: Proceedings of VLDB, pp. 436–445 (1997)
4. Kanbara, T., Ueshima, S.: Distance range queries in spatial index tree. Trans. Inf. Process. Soc. Jpn.: Database **5**(1), 1–17 (2012)
5. Kido, K., Amagasa, T., Kitagawa, H.: A scheme for parallel processing of XML data using PC clusters. In: Proceedings of Database Engineering Workshop, 6B-oi3 (2006)
6. Makino, M., Tsuji, T., Higuchi, K.: History-pattern implementation for large-scale dynamic multidimensional datasets and its evaluations. In: Renz, M., Shahabi, C., Zhou, X., Cheema, M.A. (eds.) DASFAA 2015. LNCS, vol. 9050, pp. 275–291. Springer, Heidelberg (2015). doi:10.1007/978-3-319-18123-3_17
7. Makino, M., Tsuji, T., Higuchi, K.: History-pattern encoding for large-scale dynamic multidimensional datasets and its evaluations. IEICE Trans. **E99-D**(4), 989–999 (2016). (extended version of [6])
8. Li, B., Kawaguchi, K., Tsuji, T., Higuchi, K.: A labeling scheme for dynamic XML trees based on history-offset encoding. Trans. Inf. Process. Soc. Jpn.: Database **3**(1), 1–17 (2010)
9. Hasan, K.M.A., Tsuji, T., Higuchi, K.: An efficient implementation for MOLAP basic data structure and its evaluation. In: Kotagiri, R., Krishna, P.R., Mohania, M., Nantajeewarawat, E. (eds.) DASFAA 2007. LNCS, vol. 4443, pp. 288–299. Springer, Heidelberg (2007). doi: 10.1007/978-3-540-71703-4_26
10. Brown, P.G.: Overview of sciDB: large scale array storage, processing and analysis. In: Proceedings of the ACM SIGMOD, pp. 963–968 (2010)
11. Soroush, E., Balazinska, M.: Time travel in a scientific array database. In: Proceedings of ICDE, pp. 98–109 (2013)
12. Extensible Markup Language. http://www.w3.org/XML/
13. XMark – An XML Benchmark Project. http://www.xml-benchmark.org/

Optimization Factor Analysis of Large-Scale Join Queries on Different Platforms

Chao Yang[1], Qian Wang[1], Qing Yang[3], Huibing Zhang[1], Jingwei Zhang[1,2(✉)], and Ya Zhou[1]

[1] Guangxi Key Laboratory of Trusted Software, Guilin University of Electronic Technology, Guilin 541004, China
437843196@qq.com, 1369815448@qq.com, 352131667@qq.com, gtzjw@hotmail.com, ccyzhou@guet.edu.cn
[2] Guangxi Cooperative Innovation Center of Cloud Computing and Big Data, Guilin University of Electronic Technology, Guilin 541004, China
[3] Guangxi Key Laboratory of Automatic Measurement Technology and Instrument, Guilin University of Electronic Technology, Guilin 541004, China
gtyqing@hotmail.com

Abstract. Popular big data computing platforms, such as Spark, provide new computing paradigm for traditional database operations, such as queries. Except for the management ability of large-scale data, big data platforms earn the reputation for their simple programming interface and good performance of scaling out. But traditional databases have intrinsic optimization mechanisms for fundamental operators, which supports efficient and flexible data processing. It is very valuable to give a comprehensive view of these two kinds of platforms on data processing performance. In this paper, we focus on join operation, a primary and frequently used operator for both databases and big data analysis, design and conduct extensive experiments to test the performance of the two classic platforms under unified datasets and hardware, which will disclose the performance influence on computing schema, storage media, etc. Based on the experimental analysis, we also put forwards our advice on computing platform onsideration for different application scenarios.

Keywords: Join query · Large-scale data · Performance analysis

1 Introduction

A variety of data management platforms, from traditional DataBase Management Systems to popular big data platforms, provide us many choices for managing and analyzing data. From a performance perspective, users always hope that their data computation problems can achieve performance improvement as much as possible under the help of the above platforms. This requires us to have a holistic understanding of these different data management platforms, especially the performance influence of their data organization schema, storage medium, computation paradigms, etc. Aiming at those problems, we should conduct a

© Springer International Publishing AG 2017
Z. Bao et al. (Eds.): DASFAA 2017 Workshops, LNCS 10179, pp. 35–46, 2017.
DOI: 10.1007/978-3-319-55705-2_3

performance evaluation on some primary operators to find how these operators are computed, the performance difference under various storage medium, and so on.

Join is a fundamental and frequent operator for both databases and big data analysis. For example, we can execute a join query on two relations to find those tuples satisfying the designated conditions. Obviously, the traditional relational databases, such as MySQL [8], Oracle, etc., and big data platforms, such as Hadoop [10], Spark [6,7], etc., provide different execution strategy for join. Since frequent join operations have a great impact on the performance of data analysis, it will be very significant to test its performance on different platforms so that we can build an intuitive view for various application scenarios, especially for those large-scale data applications.

Usually, big data platforms aim at scaling out and simple data processing schema, traditional relational databases pay more attention on complex optimization strategies. Considering the concrete platforms for more specific discussions, we choose two classic platforms, MySQL and Spark, as the representatives of traditional relational databases and big data platforms separately. MySQL is widely applied and has a good performance on structured data management. According to the DB-Engines Ranking [5], MySQL is still very popular in the practical applications. Spark is a distributed memory processing platform, which has four major components like Spark SQL [11], Mlib [12], Spark Streaming, GraphX and makes full use of memory and MapReduce programming paradigm. They can help bring up a detailed performance analysis at platform level, including data organization schema, storage media, implementation strategies [17], etc.

This paper focuses on the experimental comparison and analysis of large-scale data processing performance on traditional relational databases and big data platforms, which takes join as a specific example. Section 2 reviews related work on join optimization and platform features. Sections 3 and 4 present the join query and its implementation strategies on different platforms. Extensive experiments are designed and executed for performance evaluation in Sect. 5, including comparative analysis and application proposals. Section 6 concludes this study.

2   Related Work

Traditional relational databases, such as MySQL, use centralized storage and index-based data access optimization mechanism. Large-scale data processing platforms utilize large memory data computing architecture [1], and make distributed file system, such as HDFS [10], KFS and Amazon S3, as their storage media. For join query on large-scale platforms, there are four kinds of implementation strategies, which are Map-side join implementation, Reduce-side join implementation, improved join strategy on MapReduce and join implementation on indexes. [2] introduces an implementation of Map-side join query. [4,13] design Reduce-side join algorithms for equijoin on two relations. [3] exploits

MapReduce to contribute a join method of multiple relations. [14] puts forwards a novel join framework, Map-Reduce-Merge, which attaches an additional phase, namely Merge, to support more efficient join operations. [7] introduces Join process between Map and Reduce tasks, which allows to complete join query on multiple relations in a single MapReduce process. [15] integrates index with MapReduce, which brings performance improvement by removing redundant data through data pruning and preprocessing. Hadoop++ [16] allows users to customize indexes, which provides the efficiency and flexibility of join queries for different applications. Oriented for different applications and platforms, an more elaborate design for join queries on large-scale data sets can bring more performance gains.

3   Join Query

A join query is composed of three parts, query input, query conditions and query output. Query input are one or more relations, query conditions can be represented as relational or logical expressions, query output is only one final relation that are filtered out by query conditions and computed according to the specific functions. A simple join of two relations is illustrated in relational algebra as Formula 1,

$$\pi_{C.brand, C.price}(C \rhd \lhd_{C.dealerID = D.ID \wedge D.name = 'Penfold'} D) \qquad (1)$$

where we use C and D to represent two relations, namely *Cars(CID, brand, model, price, dealerID)* and *Dealers(ID, name, address, contacts)*, which are query input. $C.dealerID = D.ID \wedge D.name = 'Penfold'$ corresponds to query conditions. The query output is a relation whose schema is *(C.brand, C.price)*, which holds the related information of all cars sold by the dealer named 'Penfold'.

Based on the query condition expressions and the filtration of query results, there are three basic types of join, namely equijoin, non-equijoin and natural join. For a clear clarification and detailed discussion, the equijoin query is taken as a further demonstration. For equijoin query, there are three kinds of implementation strategies, which are nested loop join, sort merge join and hash join. The first strategy is to try all possible pairs for matching, which is not suitable to practice in distributed platforms because of its huge time overhead for big datasets. The second can stop matching check immediately once one mismatching happens, but a prior sorting is its extra cost. The third may be the optimal one, but a passable hash method may cause some data loading imbalance. The latter two can be applied in distributed environments, but a global sorting is still a time-consuming task in a distributed platform.

4   Join Query Strategy and Implementation on Different Platforms

Traditional relational databases, such as MySQL, and big data platforms, such as Spark, have different data processing paradigms, which contribute to different implementation and performance on data computation and analysis. For example, though big data platforms have better scalability for large-scale data, they will also lead to a large amount of communication overhead between nodes since all data need to store in separate nodes. Therefore, we need to conduct the performance evaluation on the two typical kinds of platforms by mechanism analysis and experimental analysis so that we can find some specific performance factors to provide both the proof of platform selection and optimization suggestions for different applications. In this section, we take equijoin query on two relations as an example to elaborate their primary implementation mechanism on those two platforms.

4.1   Equijoin Query Strategy on MySQL

MySQL, a popular open source relational database management system, covers all the above three equijoin query strategies described in Sect. 3. MySQL has natural advantages on managing relational data and query optimization for its built-in semantic schema and index functions. For nested loop join, MySQL provides two implementations, namely Simple Nested Loop Join (*SNLJ*) presented in Fig. 1 and Block-based Nested Loop Join (*BNLJ*) presented in Fig. 2. The operating mode for *SNLJ* is to input one tuple of relation C each time, and then let each tuple in relation D to match with this tuple, which requires relation D to be scanned n times. *BNLJ* read $k(k > 1)$ tuples of relation C each time, and let each tuple of relation D to match with k tuples simultaneously, which only requires relation D to be scanned $\lceil \frac{n}{k} \rceil$ times. Sort-merge join is a combination of pre-sorting and nested loop join, which can guide nested loop join to end as early as possible. Index-based join can also be applied when the indexed fields are just the fields for join, including hash index.

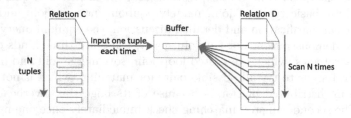

Fig. 1. Simple nested loop join illustration.

4.2   Equijoin Query Strategy in Spark

Spark is a memory-based distributed computing framework and uses RDD [7] as its primary data structure, which also supports the data transformation from HDFS to RDD. For equijoin query on Spark, Map-side join and Reduce-side join are two common strategies since its programming paradigm is similar to MapReduce. The Map-side join query is more suitable for those join operations

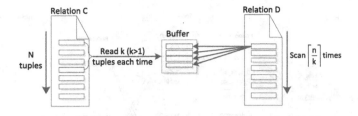

Fig. 2. Block-based nested loop join illustration.

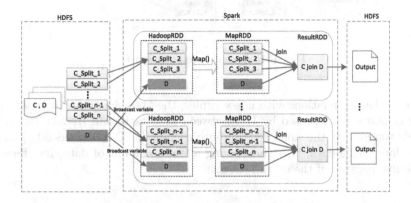

Fig. 3. Map-side join on Spark.

between a relation with a large number of tuples, such as Cars (abbr. as C), and a relation with less tuples, such as Dealers (abbr. as D). If the relation D is small enough to be resided in the local memory of each node, we can utilize the Map process to partition the relation C into different nodes and complete the join query, which saves network overhead from Map task to Reduce task since it avoids the Shuffle and Reduce stage. Simultaneously, Spark broadcast variables can be distributed through the network variable or file to each node, which will effectively enhance the efficiency of the join query. The whole join process is presented in Fig. 3. The Reduce-side join query is prepared for such join between two relations both of which have a large number of tuples. Since any relation cannot be stored in memory wholly, we need to utilize both Map and Reduce process to complete the join query. First, the Map process is responsible for partitioning C and D into different nodes based on the preset mapping function, which should ensure that those tuples with the same key in both C and D will be dispatched to a same Reduce task. Second, the Reduce processes in each node are called to combine those tuples with a same key value and to complete the join query. An illustration of Reduce-side join query is shown in Fig. 4.

The Map-side join and the Reduce-side join have their own data preferences in order to achieve best performance though they can be qualified for many

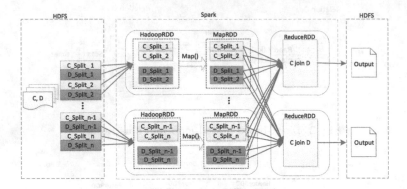

Fig. 4. Reduce-side join on Spark.

situations. Facing relations with a few tuples, the Map-side join can avoid much of the time overhead caused by I/O between Map tasks and Reduce tasks, but it needs to replicate the small relation into each node. The Reduce-side join can handle large-scale data, but when facing a small amount of data, its efficiency will be low because of the extra I/O cost.

5   Experiments

In this section, we will design and execute several groups of experiments on a unified dataset and hardware set to test the join query performance of both the traditional relational databases represented by MySQL and distributed memory computing platforms represented by Spark, and further to summary their applicability for different requirements, which mainly focus on the processing framework, including centralized data processing and distributed data processing, external data storage management and memory data management, as well as indexing.

5.1   Experiment Setup and Datasets

In order to analyze the validity of join query of different platforms and strategies, a unified hardware platform is set up to deploy MySQL and Spark. Each node has a 2.53 GHz dual-core CPU, 500G local disk and 4 GB memory. The MySQL version is MySQL-5.1.66. The Spark-1.5.1 is deployed on a 3-node cluster by the standalone mode, and is not accompanied with any resource management system. There is one master node and two slave nodes in the cluster. The physical storage layer is HDFS. We take the TPC-H [9] to create our experimental dataset, which is composed of two relations, *ORDERS* and *LINEITEM*. For our queries, we use the *O_ORDERKEY* attribute of *ORDERS* and the *L_ORDERKEY* attribute of *LINEITEM* to establish the query condition expression. The details of the dataset is shown in Table 1, whose size will change under different test requirements. In addition, the number of tuples in *LINEITEM* is approximately four times than the number of tuples in *ORDERS*, and the join rate is 100%. Namely, for each record in *ORDERS*, there is at least one record in *LINEITEM* that satisfies the query condition, on the other side, for each record in *LINEITEM*, there is only one record in *ORDERS* that satisfies the query condition.

Table 1. Details of generated data set

Relations	Number of tuples	Size
ORDERS	10,000–22,500,000	1 MB–3 GB
LINEITEM	40,000–90,000,000	4 MB–12 GB

5.2 Evaluation Metrics

In order to measure the performance of join query on MySQL and Spark, we take the average time of query completion as an evaluation criterion. Formulas 2 and 4 also provide another criterion, the number of processed tuples in unit time, where N_{MySQL} and N_{Spark} denote their respective average time of query completion, Num_Result is the number of tuples covered by query output, T_{MySQL} and T_{Spark} denote the total time of query completion. Since Spark cluster has three nodes, T_{Ave_spark} is used to hold the performance of one node, expressed as Formula 3. The performance ratio between Spark and MySQL is also defined in Formula 5, which can present a more intuitive view on the performance of the two platforms.

$$N_{MySQL} = \frac{Num\_Result}{T_{MySQL}} \qquad (2)$$

$$T_{Ave\_Spark} = 3 * T_{Spark} \qquad (3)$$

$$N_{Spark} = \frac{Num\_Result}{T_{Ave\_Spark}} \qquad (4)$$

$$R = \frac{N_{MySQL}}{N_{Spark}} \qquad (5)$$

5.3 Experiment Design and Results

In this section, we design several groups of experiments based on the above data set to evaluate platform performance as defined in Sect. 5.2 and provide the comparison analysis.

Experiment 1: join query performance comparison between Map-side and Reduce-side strategies. We gradually increase the data size and ensure both the fixed proportion relationship and the join rate between *ORDERS* and *LINEITEM* to execute the experiments, whose dataset details are shown in Table 2. The experimental result is presented in Fig. 5. Facing a small size of data set, the Map-side join and the Reduce-side join have the similar performance, but when increasing the size of data set, the time overhead of Map-side join presents a fast growth, even Map-side join cannot complete the query task because of the memory overflow. For different size of data sets, the query time of Reduce-side join grows slowly and its performance is stable. Considering the performance of Map-side join, the specific reason is that there are three stages for join on a distributed platform, which are partitioning the first relation, partitioning the second relation, and executing join, these three stages can only be executed serially since it is only allowed to run on Map-side. Reduce-side join support their concurrent execution and present a good performance than Map-side join for large-scale datasets. The following experiments run on Spark will adopt Reduce-side join strategy as a comparison.

Table 2. The dataset details of experiment 1

Group	*ORDERS* (10^3)	*LINEITEM* (10^3)
1	10	40
2	25	100
...	...	...
13	6000	24000

Experiment 2: Join query performance test for different size of execution memory. Spark is known for the advantages of memory-based computing, memory size is an important factor that affects join performance. In default configuration, the computing node uses 60% of the memory to cache RDD data, and 40% of the memory for task computation, which can avoid the problem of low efficiency caused by multiple I/O in traditional MapReduce. We conduct join query on Spark with different memory size and observe its impact on query performance. The memory size varies from 500 MB to 4 GB, the data details are shown in Table 3, and the experimental results are presented in Fig. 6.

Table 3. Data set details for testing memory influence on query performance

Group	*ORDERS* (10^6)	*LINEITEM* (10^6)
A	1.5	6
B	4.5	18
C	7.5	30

When inputting more data, the size of Spark running memory has a significant impact on the query performance because the memory of Spark is not only for the computing tasks but also for the RDD caching, especially for big datasets. Big memory is helpful to improve the processing ability and RDD caching. The group of experiments also give us a new hint, excessive memory does not bring sustained performance improvement, which has the following two factors, one is that the parallelism of Spark is determined by the number of CPU cores in the cluster, a big exclusive memory for a single Spark process will reduce the parallelism, the other is that Spark will wait for a status of RDD caching to start data processing, a big exclusive memory for a single Spark process will cause a long waiting time. The above two factors should be considered comprehensively to ensure a good parallelism. Taking into account query performance maximization of our experiments, the running memory of each Spark node is set at 2GB in the follow-up comparative experiments.

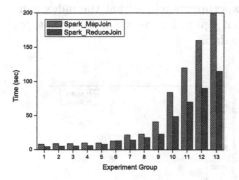

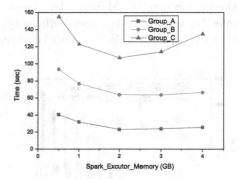

Fig. 5. Time performance comparison between Map-side join and Reduce-side join on Spark.

Fig. 6. Query performance on different memory size.

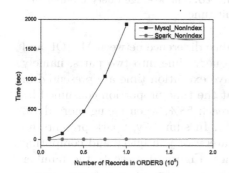

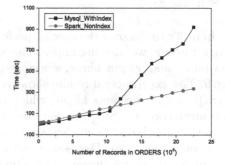

Fig. 7. Performance comparison between MySQL (no index) and Spark (no index).

Fig. 8. Performance comparison between MySQL (with index) and Spark (no index).

Experiment 3: The join performance comparison between MySQL and Spark. In this group of experiments, we compare the query execution between MySQL and Spark, the former will work in two statuses, namely with index and without index. Figures 7 and 8 present the related experimental results. In Fig. 7, the query performance on MySQL without index rises sharply, which coincides with its time complexity $O(|ORDERS| * |LINEITEM|)$. Figure 8 shows that the join query on MySQL with index present a better performance than Spark when the number of tuples in $ORDERS$ is less than 10 million. There are two factors for MySQL to win, one is the index structure which helps to locate tuples quickly, the other is that Spark needs to shift data between nodes to produce the final results, whose overhead is a big proportion for small-scale data join. With adding more tuples, the query time on MySQL grows sharply, but Spark has only an approximately linear growth. When the number of tuples in $ORDERS$ reaches 15 million, there are $T_{MySQL} > T_{Spark}$, $T_{MySQL} > T_{Ave\_Spark}$, which means that the join query performance of any node in the Spark cluster has exceeded the performance of MySQL. The above experiments tell us that the index can effectively improve the query efficiency, but Spark has an absolute advantage over MySQL when dealing with large-scale data.

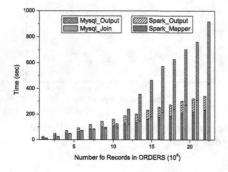

Fig. 9. Query time decomposition of MySQL and Spark.

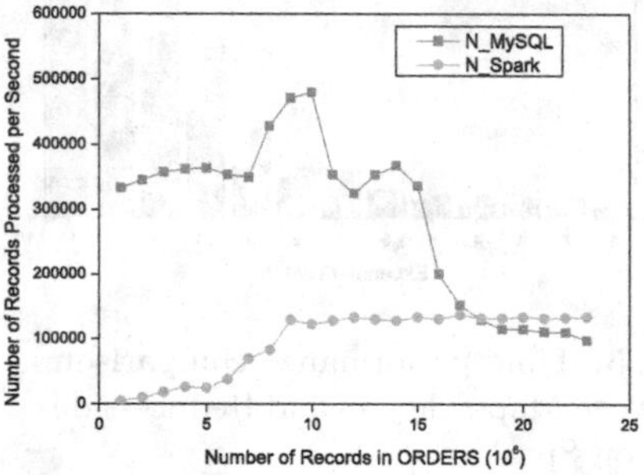

Fig. 10. The average query number in unit time.

In order to discover the specific performance difference between MySQL and Spark further, we also decompose the join query time into two parts, namely join phase and output phase, whose respective execution time are presented in Fig. 9. The experimental results show us that the time proportion consumed by MySQL output becomes larger, which can reach 58%, when facing more data. The proportion for Spark is only about 32%. In summary, Spark presents its advantages on both the implementation strategy of join query and computing based on memory for large-scale join queries. Figure 10 presents the number of completed queries in unit time. Obviously, as the number of tuples grows, Spark has a stable performance, which corresponds to the good scalability on large-scale data. In addition, the performance ratio between MySQL and Spark, R, decreases from initial 76 times to 0.7, which also proves that single-node efficiency of Spark covers MySQL.

6   Conclusions

The study provides a comprehensive view of join operations on large-scale datasets by its performance evaluation on heterogeneous platforms, including the traditional relational database, MySQL, and the distributed memory computing system, Spark. Focusing on discovering the concrete factors for the performance improvement of join, this paper designs and conducts extensive experiments under different join strategies, based on which the advantages and disadvantages of both Spark and MySQL on large-scale join query operation and a further analysis for the specific factors affecting query performance are summarized, which provides the basis for both join query optimization on different platforms and the platform selection of the real application scenarios.

Through the experimental analysis, MySQL has some advantages in join query, especially works with index when queries only require a handful of results. Spark contributes two different join strategies, namely join by Map task and join by both Map and Reduce task, the former requires that one relation for join is small enough to reside in memory and the latter is a general join strategy. For large-scale datasets, Spark presents a stable performance because of its interior processing logic. In addition, the size of memory has also an influence on the join performance, but an oversize memory cannot always bring performance gains for Spark. When the dataset is large enough, both the join time and the output time on Spark will dominate the performance on MySQL. For satisfying the computing requirements of real-time applications, such as the matching between users and items in location-based services, Spark will be an eligible platform. There are two ways for a further optimization consideration on Spark, one is to design well-focused index structures for different computing requirements, the other is to provide an optimal memory allocation based on tasks when multiple slaves are deployed on a single physical node.

Acknowledgment. This study is supported by the National Natural Science Foundation of China (Nos. 61363005, 61462017, U1501252, 61662013), Guangxi Natural Science Foundation of China (Nos. 2014GXNSFAA118353, 2014GXNSFAA118390, 2014GXNSFDA118036), Guangxi Key Laboratory of Automatic Detection Technology and Instrument Foundation (YQ15110), Guangxi Cooperative Innovation Center of Cloud Computing and Big Data, and the High Level Innovation Team of Colleges and Universities in Guangxi and Outstanding Scholars Program Funding.

References

1. Jiang, D., Tung, A.K.H., Gang, C.: Map-Join-Reduce: toward scalable and efficient data analysis on large clusters. IEEE Trans. Knowl. Data Eng. **23**(9), 1299–1311 (2011)
2. Zhou, M., Zhang, R., Zeng, D., et al.: Join optimization in the MapReduce environment for column wise data store. In: Proceedings of 6th International Conference on Semantics Knowledge, Girds (SKG 2010), Los Alamitos, CA, 2011 Observation of Strains, pp. 97–104. IEEE Computer Society (2010)

3. Afrati, F.N., Ullman, J.D.: Optimizing multiway joins in a MapReduce environment. IEEE Trans. Knowl. Data Eng. **23**(9), 1282–1298 (2011)
4. Zhao, Y.-R., Wang, W.-P.: Efficient join query processing algorithm CHMJ based on hadoop. J. Softw. **23**(8), 2032–2041 (2012)
5. DB-Engines Ranking. http://db-engines.com/en/ranking/relational+dbms
6. Zaharia, M., Chowdhury, M., Franklin, M.J., Shenker, S., Stoica, I.: Spark: cluster computing with working sets. In: HotCloud 2010, June 2010
7. Zaharia, M., Chowdhury, M., Das, T., Dave, A., Ma, J., McCauley, M., Franklin, M.J., Shenker, S., Stoica, I.: Resilient distributed datasets: a fault-tolerant abstraction for in-memory cluster computing. In: NSDI 2012, April 2012
8. MySQL. http://www.mysql.com
9. TPC-H. http://www.tpc.org/tpch/
10. Dean, J., Ghemawat, S.: MapReduce: simplified data processing on large clusters. CACM **51**(1), 107–113 (2008)
11. Armbrust, M., Xin, R.S., Lian, C., Huai, Y., Liu, D., Bradley, J.K., Meng, X., Kaftan, T., Franklin, M.J., Ghodsi, A., Zaharia, M.: Spark SQL: relational data processing in spark. In: SIGMOD Conference, pp. 1383–1394 (2015)
12. Meng, X., Bradley, J.K., Yavuz, B., Sparks, E.R., Venkataraman, S., Liu, D., Jeremy Freeman, D.B., Tsai, M.A., Owen, S., Xin, D., Xin, R., Franklin, M.J., Zadeh, R., Zaharia, M., Talwalkar, A.: MLlib: machine learning in apache spark (2015). CoRR arXiv:1505.06807
13. Blanas, S., Patel, J.M., Ercegovac, V., et al.: A comparison of join algorithms for log processing in MapReduce. In: Proceedings of 2010 ACM SIGMOD International Conference on Management of data, pp. 975–986. ACM (2010)
14. Yang, H.C., Dasdan, A., Hsiao, R.L., et al.: Map-Reduce-Merge: simplified relational data processing on large clusters. In: ACM SIGMOD International Conference on Management of Data, pp. 1029–1040. ACM (2007)
15. Yang, H., Parker, D.S.: Traverse: simplified indexing on large Map-Reduce-Merge clusters. In: Zhou, X., Yokota, H., Deng, K., Liu, Q. (eds.) DASFAA 2009. LNCS, vol. 5463, pp. 308–322. Springer, Heidelberg (2009). doi:10.1007/978-3-642-00887-0_27
16. Dittrich, J., Quian Ruiz, J.A., et al.: Hadoop++ making a yellow elephant run like a cheetah (without it even noticing). Proc. VLDB Endow. **3**(12), 518–529 (2010)
17. Agrawal, D., et al.: SparkBench – a spark performance testing suite. In: Nambiar, R., Poess, M. (eds.) TPCTC 2015. LNCS, vol. 9508, pp. 26–44. Springer, Heidelberg (2016). doi:10.1007/978-3-319-31409-9_3

Which Mapping Service Should We Select in China?

Detian Zhang[✉], Jia-ao Wang, and Fei Chen

School of Digital Media, Jiangnan University, Wuxi, China
detian.cs@gmail.com, sarawangja@163.com, chenf@jiangnan.edu.cn

Abstract. The mapping services have been widely used in our daily lives. People can use the services to find his/her nearest POI (Point of Interests), the shortest travel route from a source location to a destination location, and even life services like booking hotels, calling taxis and so on. Consequently, more and more mapping service providers have emerged in the past years in China, like Baidu Maps, Amap and Sogou Maps. However, there is no existing study on how to select the suitable one for users/developers when they facing so many different mapping services, which is the problem that we focus on in this paper. We first design a questionnaire and analyze the results to show the current mapping service situation in China; then, we introduce and summarize the most three popular native mapping APIs in China, e.g., Baidu Maps API, Amap API and Sogou Maps API, to give readers a brief guider for selecting their suitable mapping services.

Keywords: Mapping services · Mapping API · Baidu Maps · Amap · Sogou Maps

1 Introduction

With the development and ubiquity of GPS-enabled equipments and wireless Internet access, the mapping service becomes more and more popular and important in our daily lives. This is because a user can easily find his/her nearby POIs (Point of Interests), the shortest route from a source location to a destination location, the public transportation stations and routes, and other life services (e.g., booking hotels, calling taxis, searching Groupons) with the help of mapping services.

Based on the "Report on Mobile Map Marketing in China" in the first half of 2014 published by iiMedia Research on July 18, 2014 [1], the number of mobile users in China was 448 million in 2014Q1 and it reached to 472 million in 2014Q2. According to the data released by Analysys [2], it is expected that the number of mobile map users in China will be 457 million, 28.7% increased in 2016 compared with 2015, and mobile mapping services will cover 642 million users by 2018 (see Fig. 1). This is because the mapping service not only provides basic functions, like map browsing, POI search, navigation, and so on, but also provides many

© Springer International Publishing AG 2017
Z. Bao et al. (Eds.): DASFAA 2017 Workshops, LNCS 10179, pp. 47–59, 2017.
DOI: 10.1007/978-3-319-55705-2_4

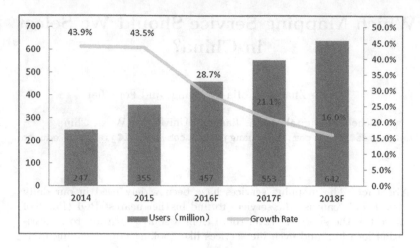

Fig. 1. The number of mobile map users and growth rate in China (from Analysys)

other useful life services, such as hotel booking and taxi services, which prompts the rapid growth of the mobile map users (Fig. 2).

Currently, the most popular mapping service providers in China are Baidu Maps, Amap (also known as Gaode Maps), Sogou Maps, Google Maps, Bing Maps and Tencent Maps. Analysys recently released another report on mobile map market of the fourth quarter of 2015 in China [3]. It shows that Baidu Maps ranks the first with 70.8% market share in active users, Amap ranks the second with 23.7% market share, and Sogou Maps ranks the third with 5.8% market share. Besides the coverage rate of active users, the number of service start-up and the using time are another two crucial metrics. According to the report, among those mapping service providers, Baidu Maps still ranks the first with 66.2% market share in service start-up, followed by Amap and Sogou Maps, which own 22.4% and 3.2% respectively. Furthermore, Baidu Maps is also in the leading position in terms of the service using time as high as 64.4%. Amap and Sogou Map possess 23.3% and 3.3%, respectively. Even though Google Maps and Bing Maps are very popular in English-speaking countries (especially in U.S.), they are not that famous in China (as shown in Sect. 3.2). This is because they do not have sufficient map data as local mapping service providers, and provide very limited life services. Even for native Baidu Maps, Amap, Sogou Maps, and Tencent Maps in China, they may have different types of users. For example, Amap is famous for its accuracy and good route suggestions, so it is widely used in navigation area; while Tencent Maps is famous for its street views, so it is very attractive for travelers. Therefore, it is necessary to do a study on how to select suitable mapping services.

In this paper, we study how to select right mapping services for users, especially for developers who want to build applications or web sites based on mapping APIs. We first design a questionnaire and analyze the results to show the current mapping service situation in China, including map popularity, mapping

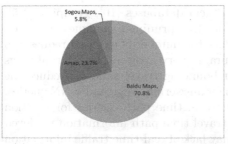

(a) The market share in active users

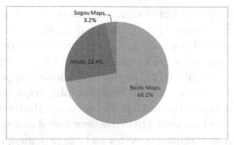

(b) The market share in service start-up

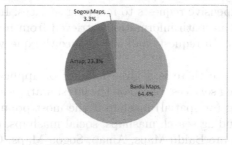

(c) The market share in using time

Fig. 2. The market shares of Baidu Maps, Amap, Sogou Maps in China (from Analysys)

services, usage frequency, factors that affect users selecting mapping service providers, and life services provided by mapping services. Then, we introduce and summarize the most three popular mapping APIs in China, e.g., Baidu Maps API, Amap API and Sogou Maps API, to give readers (especially developers) a brief guider for selecting their suitable mapping services.

The remainder of this paper is organized as follows. Section 2 highlights related work. Section 3 describes our designed questionnaire and analyses the results. In Sect. 4, we introduce three popular native mapping APIs in China, i.e., Baidu Maps API, Amap API, and Sogou Maps API. Finally, Sect. 5 concludes this paper.

2   Related Work

Mapping services are not only very useful and popular in our daily lives, but also very important in industrial and academic. Typical example using mapping services is spatial (or GIS/mapping) mashup. A spatial (or GIS/mapping) mashup [4–9] provides a cost-effective way for a Web or mobile application that combines data, representation, and/or functionality from at least one mapping service and other local/external services to create a new application.

In [4–6], the authors proposed k-NN query processing algorithms based on spatial mashups, where the distance metric (i.e., travel time) is retrieved from mapping services. Since accessing travel time from mapping services is much more expensive than accessing data from local databases [10], grouping [4,5], direction sharing [5], shared query execution [4,5], pruning techniques [4–6], and parallel requesting [6] are employed to reduce the number of external requests to mapping services and provide highly accurate query answers. In [9], route logs are employed to derive tight lower/upper bounding travel times to reduce the number of external Web mapping requests for answering range and k-NN queries. In [7,8], the authors focused on shortest travel time path queries for location based services (LBS), where the shortest travel time path information retrieved from mapping services as LBS providers may lack of real-time traffic information to compute them. They utilize waypoints in an external route request to reduce the number of such expensive requests to mapping services. In [11], the authors proposed a cache to store path information retrieved from mapping services to reduce those external path requests and also reduce the query response time to users.

Therefore, we can see there is a large number of applications or research works based on mapping services. Based on the latest statistics of Programmable Web [12], the mapping (or spatial) mashup is the most popular one among all types of mashups including search mashups, social mashups, etc. Typical mapping services in China are Baidu Maps, Amap, Sogou Maps, Google Maps, and Microsoft Bing Maps. Even though mapping mashup is the most popular type and more and more applications and Web sites have emerged out, there is no existing work on giving a comparison or summarization of mapping services in China. On the contrary, there exists several works on comparing popular mapping services in U.S., such as Google Maps, Apple Maps, Bing Maps, MapQuest Maps, etc. For example, the authors in [13] gave a thorough comparison in the aspect of the user experience of current popular mapping services, including maps in landscape, portrait, different zoom levels, navigation/directions mode and so on. Particularly, they discussed static presentation consistency issues [14] and dynamic presentation consistency issues [15] in smartphone mapping applications, respectively, which serve as a mobile map application design guider for developers and researchers.

Since all of those research work [13–15] are focus on non-Chinese mapping services, e.g., Google Maps, Bing Maps, and MapQuest Maps, which are not widely used in China. Therefore, we briefly compare and summarize Chinese native mapping services in this paper, e.g., Baidu Maps, Amap, and Sogou Maps, to give users/developers a simple guider.

3 Questionnaire and Result Analysis

3.1 Questionnaire Design

This questionnaire focuses on college students and the young work people who born after 1970. These people are the major part of the Internet and map users. Specifically, we choose 60 college students, where 50 of them are from Jiangnan University and 10 of them are from other universities. We also choose 40 workers,

The Questionnaire of Map Usage and Preferences

Q1: Which of the following maps have you heard of or used?

□Baidu Maps   □Amap   □Sogou Maps   □Bing Maps   □Google Maps

□Bar Maps   □Tencent Maps □Soso Maps

Q2: Which map do you use most?   _____

Q3: The reasons you choose this one among various map products:

□Friendly UI   □Accuracy of positioning and navigation

□Nice routing suggestion   □Performance

□Sufficient and complete map data   □Easy operation

□Accurate search result   □Others

Q4: The purposes you use the mobile map:

□POI search   □Navigation   □Positioning

□Check-in and location sharing □Transportation routing   □Real-time traffic routing

□Traffic station inquiry   □Searching nearby hot spots □Map browsing

□Real-time traffic condition   □Others

Q5: What is the frequency for you to use the mobile map?

□Every day   □At least once a week   □At least once two weeks

□At least once a month   □Haven't been used more than one month

Q6: What functions do you hope to be added in the map app? _____

Q7: What the biggest problem have you met when you use the mobile map?_____

Q8: What kind of life information do you search most on the mobile map?

□Food & restaurant   □Bank & ATM   □Entertainment   □Hotel   □Supermarket

□Groupon   □Discount   □Taxi   □Others

Fig. 3. The designed questionnaire.

where 30 of them work in state-owned or foreign enterprises and 10 of them are individuals or freelancers. All of them are from 20 to 50 years old.

Figure 3 shows the detailed information of the questionnaire. From the questionnaire result (see Sect. 3.2), we can preliminarily learn which mapping service providers and mapping services are popular, map usage frequency, factors that affect users selecting mapping service providers, popular life services provided by mapping services, and so on.

3.2 Result Analysis

Map Popularity. According to the statistical results of the questionnaire, we can have a preliminary knowledge of the map popularity in China. As we can see in Fig. 4, with its powerful search technology and big data platform, Baidu Maps becomes the most popular and well-known mapping service provider in China. The company has also cooperated with all sorts of industries and integrated multiple resources, e.g., cooperating with Uber and acquiring Qunar.com, which further consolidates its advantage in map area. Amap is also familiar to a wide range of users because of its professional and better service. As the most globalized mapping service provider, Google Maps also has a good reputation in China. Besides, Sogou Maps, Tencent Maps and other mapping service providers also have their own users and market shares.

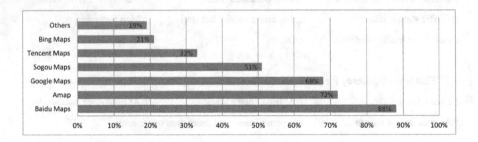

Fig. 4. Map popularity.

Mapping Services. From Fig. 5, we can see what mapping services that users use most. Obviously, users mainly focus on the basic mapping services such as POI search, navigation, and transportation routing. What is more, the current mapping service providers also provide users with map interaction and life services. For example, there are a multitude of services, which have been realized in the mobile map, such as querying bus information, real-time traffic condition, check-in, location sharing and so on.

Usage Frequency. As shown in Fig. 6, the result also shows that most of users use mobile map softwares at least once a week, and some users even once a day, which demonstrates that the mobile map has wholly integrated into our daily lives. With the combination of mobile maps and life services, the usage frequency may continue increasing.

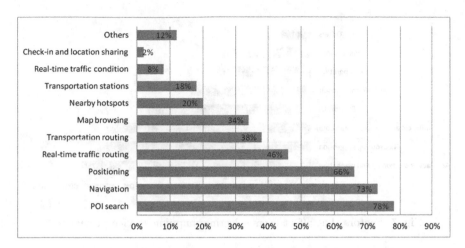

Fig. 5. Mapping services.

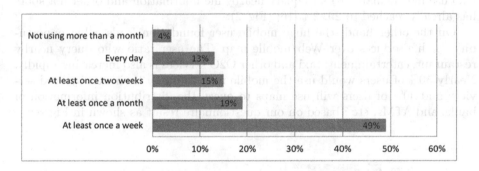

Fig. 6. Usage frequency.

Factors that Affect Users Selecting Mapping Service Providers. Facing to multifarious mobile map applications in the APP market, how to select suitable mobile maps and what kinds of factors should be considered by users are also very important. The survey manifests that the most crucial consideration would be the accuracy of positioning and navigation, which is also the basic mapping service. In addition, during map navigation, accuracy, routing suggestion and map data information are also considered by most of users. As a result, the excellent mapping service provider should not only have accurate positioning and map data, but also should provide convenient operation and rapid response to improve user experience.

Life Services. According to our and others' studies, although half of the users mainly focus on basic functions such as route navigation and POI search when they use mapping services, it is worth noting that with the development of O2O

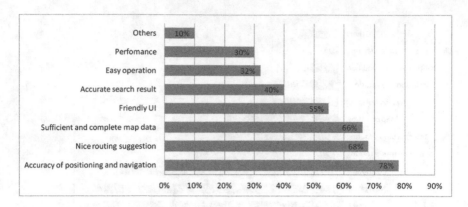

Fig. 7. Factors that affect users selecting mapping service providers.

(Online To Offline) and the easy access of mapping API, the proportion of users who use mobile map APPs to query nearby life information and other hot spots has already reached to 29. 2% [16] (Fig. 7).

On the other hand, the huge mobile user foundation leads a big opportunity for life services over Web/mobile map. The user ratio who query nearby restaurant, entertainment, taxi and other O2O services is also increasing rapidly. Nearly 50% of users would use the mobile map to query their nearby food services and 40% of users will use maps to query the distribution information of banks and ATMs, etc., based on our questionnaire result as shown in Fig. 8.

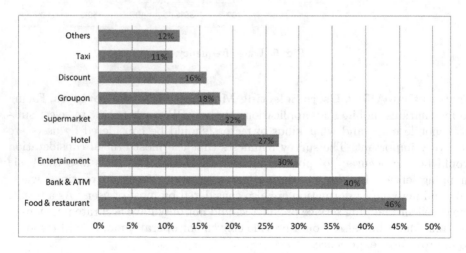

Fig. 8. Popular life services provided by current maps.

4 Popular Mapping APIs

4.1 Baidu Maps API

Baidu Maps is a desktop and mobile web mapping service application and technology provided by Baidu, offering satellite imagery, street maps, street view and indoor view perspectives, as well as functions such as a route planner for traveling by foot, car, or with public transportation. Android and iOS applications are available.

Table 1. Types of Baidu Maps API.

Web development	Common JavaScript API
	JavaScript API high-speed edition
	Web component API
Android development	Android map SDK
	Android location SDK
	Android navigation SDK
	Android panorama SDK
iOS development	iOS Map SDK
	iOS navigation SDK
	iOS panorama SDK
Service API	LBS Cloud
	Web Service API
	Static imagery API
	Panoramic static API
	Internet of vehicles API
	URI API
	Bird's Eye
Tool support	API console
	LBS cloud visualization controller
	Coordinate collection
	Map generator
	Map card
	Groupon plug-in
	Zero-cost switching tool
	Map editing tools
	Development sources

On April 23th. 2010, Baidu Maps officially announced to open mapping API, which is free for majority of developers. Baidu mapping API is a set of application interface based on Baidu Maps services freely for developers, including JavaScript

API, Web Service API, Android SDK, iOS SDK, Positioning SDK, Internet of
Vehicles API, LBS Cloud and many other development tools and services. It
provides basic functions of the map, such as display, search, positioning, reverse-
geocoding, geocoding, routing, LBS cloud storage and retrieval. It is suitable
for PC client, mobile client, servers and other equipment and it is also the map
application development under multiple operating systems. Table 1 shows all of
the types of Baidu Maps API, developers can choose the right APIs based on
their applications. What is more, Baidu Maps APIs also have the maximum
access times and access frequency to avoid abuse as shown in Table 2.

Table 2. Applicability of Baidu Maps API.

Applicability	Coverage
Browser	IE 6. 0+, Firefox 3. 0+, Opera 9. 0+, Safari 3. 0+, Chrome
Operating system	Windows, Mac, Linux
Facility type	PC, mobile phones and other mobile clients
Calling mode	Online calling
Access times and access frequency	Place API and Place suggestion API 100,000 times/day/KEY
	Direction API 100,000 times/day/KEY
	Geocoding API 100,000 times/day
	Coordinate conversion API no access limit
	The number of concurrent users: 1000–1500 times/sec to calculate according to the initial JS per second
	Search services: 800 times/sec
	Bus, car service: 600 times/sec
	Geocoding performance: 200 times/sec
	Bandwidth limit: full support for services (due to the small amount of API data)
	Coordinate conversion interface: single IP 50 times/sec (if more than 100 times, returning back to 403 error)
	Static imagery: independent IP 10/sec

4.2   Amap API

Amap is the China's leading provider that devotes itself to digital map contents,
navigation and location-based solutions. The company entered the global excellent market of NASDAQ in America. Amap possesses the three class-A qualifications whose high-quality electronic map database has become its core competitiveness. The three class-A qualifications are surveying and mapping qualification
of Navigating Electronic Map qualification class-A qualification, mapping aer-
ial photography qualification class-A qualification and Internet mapping service
class-A qualification.

Table 3. Types of Amap API.

Web and server development	JavaScript API
	Nephogram API
	Map component
	Static mapping API
	URI API
Android development	Android SDK
	Android nephogram SDK
	Android navigation SDK
	Android street view SDK
	Android location SDK
iOS development	iOS SDK
	iOS navigation SDK
	iOS nephogram SDK
Map tools	Coordinate collection
	Quick map generator
	Map card
	Easy map website building
In the laboratory	Vector web maps
	Indoor maps
	iOS street view
Windows platform	Windows phone SDK

Table 4. Access limits of Amap API.

Types	Access limits per day	Access limits per 10 min
Geocoding/Reverse-geocoding	100000/Key	10000/Key
Place search	100000/Key	50000/Key
Inputting hints	100000/Key	50000/Key
Navigation	100000/Key	5000/Key
Road search	25000/Key	2500/Key
Static map	25000/Key	2500/Key
Positioning	100000/Key	5000/Key

The LBS open platform of Amap is subordinated to the Alibaba, which is a leading e-commerce company in China (even in the world). In October 2013, Amap cooperated with Ali Cloud to issue the LBS cloud strategy together. Amap LBS cloud platform has already connected with the Ali cloud platform. Since Amap has been the first batch of Internet Service Provider to provide mapping API, it nearly has 10 years of LBS technology experience. In addition, Amap

LBS open platform owns more than 300,000 developers and partners and its services are called by over 10,000 applications every day.

More specifically, the types and applicability of Amap API are summarized in Tables 3 and 4, respectively.

4.3 Sogou Maps API

Compared with the powerful API of Baidu Maps and Amap, Sogou Maps only provide JavaScript API. It makes users to construct map applications with simple operation and abundant functions on their own websites. There are various interfaces that have basic functions for map-construction, including location search, nearby hotspot search, navigation and so on. The provided JavaScript API is compatible with the following browsers:

- IE 6.0+ (Windows) and other browsers with IE kernel, such as Sogou Highspeed Browser, Maxthon Browser, 360 Browser and The World Browser, etc.
- Firefox 2.0+ (Window—Mac—Linux).
- Safari 3.1+ (Mac—Windows).
- Chrome (Windows).
- Opera 10+ (Windows).

The data of Sogou Maps are as follows:

- Coverage: the data of Sogou Maps nearly covers 400 medium and large cities, hot tourism cities and 3,000 districts in China, which support for national urban life information, public transportation transfer, and inquiry of driving route. What is more, it provides services of multi-city real-time transportation, satellite images and 3D urban display.
- Updating frequency: mass data acquisition is updated every half year. All kinds of information update on a daily iteration.
- Support for the city real-time traffic status information: it supports four cities currently, including Beijing, Shanghai, Guangzhou and Shenzhen.

5   Conclusion

In this paper, we first briefly discussed current mapping service situation in China by a questionnaire and its result analyses, including which mapping service providers and mapping services are popular, map usage frequency, factors that affect users selecting mapping service providers, popular life services provided by mapping services, and so on. Then, we introduce and summarize three popular native mapping APIs in China, e.g., Baidu Maps API, Amap API and Sogou Maps API. We hope that this paper can give users/developers a simple and useful guider when they plan to choose a suitable mapping service.

Acknowledgments. Research reported in this publication was partially supported by the Fundamental Research Funds for the Central Universities in China (Project No. JUSRP11557), and the Natural Science Foundation of Jiangsu Province (Project No. BK20160191).

References

1. iiMedia Research (2014). www.199it.com/archives/256990.html
2. Analysys (2016). http://www.askci.com/news/chanye/2016/03/02/154313qve2. shtml
3. Analysys (2016). http://www.cctime.com/html/2016-2-24/1140899.htm
4. Zhang, D., Chow, C.-Y., Li, Q., Zhang, X., Xu, Y.: Efficient evaluation of k-NN queries using spatial mashups. In: Pfoser, D., Tao, Y., Mouratidis, K., Nascimento, M.A., Mokbel, M., Shekhar, S., Huang, Y. (eds.) SSTD 2011. LNCS, vol. 6849, pp. 348–366. Springer, Heidelberg (2011). doi:10.1007/978-3-642-22922-0_21
5. Zhang, D., Chow, C.Y., Li, Q., Zhang, X., Xu, Y.: SMashQ: Spatial mashup framework for k-NN queries in time-dependent road networks. Distrib. Parallel Databases, DAPD **31**(2), 259–287 (2013)
6. Zhang, D., Chow, C.Y., Li, Q., Zhang, X., Xu, Y.: A spatial mashup service for efficient evaluation of concurrent k-NN queries. IEEE Trans. Comput. **65**(8), 2428–2442 (2016)
7. Zhang, D., Chow, C.-Y., Li, Q., Liu, A.: Efficient evaluation of shortest travel-time path queries in road networks by optimizing waypoints in route requests through spatial mashups. In: Li, F., Shim, K., Zheng, K., Liu, G. (eds.) APWeb 2016. LNCS, vol. 9931, pp. 104–115. Springer, Heidelberg (2016). doi:10.1007/978-3-319-45814-4_9
8. Zhang, D., Chow, C.Y., Liu, A., Zhang, X., Ding, Q., Li, Q.: Efficient evaluation of shortest travel-time path queries through spatial mashups. GeoInformatica (2017). doi:10.1007/s10707-016-0288-4
9. Li, Y., Yiu, M.L.: Route-saver: leveraging route apis for accurate and efficient query processing at location-based services. IEEE TKDE **27**(1), 235–249 (2015)
10. Levandoski, J.J., Mokbel, M.F., Khalefa, M.E.: Preference query evaluation over expensive attributes. In: CIKM (2010)
11. Thomsen, J.R., Yiu, M.L., Jensen, C.S.: Effective caching of shortest paths for location-based services. In: ACM SIGMOD (2012)
12. ProgrammableWeb. http://www.programmableweb.com/category-api
13. Samet, H., Fruin, B.C., Nutanong, S.: Duking it out at the smartphone mobile app mapping API corral: Apple, Google, and the competition. In: ACM MobiGIS, pp. 41–48 (2012)
14. Samet, H., Nutanong, S., Fruin, B.C.: Static presentation consistency issues in smartphone mapping apps. Commun. ACM **59**(5), 88–98 (2016)
15. Samet, H., Nutanong, S., Fruin, B.C.: Dynamic presentation consistency issues in smartphone mapping apps. Commun. ACM **59**(9), 58–67 (2016)
16. CNNIC (2013). http://www.199it.com/archives/126997.html

An Online Prediction Framework for Dynamic Service-Generated QoS Big Data

Jianlong Xu[1(✉)], Changsheng Zhu[1], and Qi Xie[2]

[1] Shantou University, Shantou, China
{xujianlong,cszhu}@stu.edu.cn
[2] Southwest University of Nationalities, Chengdu, China
qi.xie.swun@gmail.com

Abstract. With the prevalence of service computing, cloud computing, and Internet of Things (IoT), various service compositions are emerging on the Internet based on Service-Oriented Architecture (SOA). To evaluate the performance attribute of these service compositions, dynamic Quality of Service (QoS) data are generated abundantly, named service-generated QoS Big Data. Selecting optimal services to build high quality SOA systems can be based on these data. However, a mass of service-generated QoS data are unknown and it is become a challenge to predict these data. In this paper, we present a framework for service-generated QoS big data prediction, named DSPMF. Under this framework, we present an optimization objective function and employ online stochastic gradient descent algorithm to solve this function. Extensive experiments are conducted to verify the effectiveness and efficiency of our proposed approach.

Keywords: Service computing · Online learning · QoS prediction · Matrix factorization · Service-generated big data

1 Introduction

In the modern IT era, big data has attracted worldwide attentions as a widely deployed technique with the support of large-scale distributed systems, such as cloud applications, Internet-of-Things, and many other domains. In these domains, very-large data are generated from videos, audios, images, emails, logs, search queries, social networking interactions, science data, sensors and mobile phones and their applications [1]. These data sets are characterized by high volume, high velocity, high variety, high veracity, and high variability. With the prevalence of service computing and cloud computing, more and more modern services are emerging and deployed in cloud infrastructures based on service-oriented architecture (SOA) to provide various kinds of functionalities. These services and service users are widely distributed around the world. As the quantity of the services and users increases rapidly, enormous data are explosively generated by these service systems over time, such as trace logs, Quality of service (QoS) information, service relationship, etc. [2]. These data are valuable for building and optimizing high quality SOA system (e.g. service selection, service composition, service adaptation, etc.). How to exploit these service-generated big data to obtain more useful information

Z. Bao et al. (Eds.): DASFAA 2017 Workshops, LNCS 10179, pp. 60–74, 2017.
DOI: 10.1007/978-3-319-55705-2_5

has become a promising research area. In this paper, we focus on QoS information prediction for service-generated big data.

Typically, in large-scale service-oriented systems, many services are composed in a loosely-coupled way to fulfill complex application systems. As the number of services is booming exponentially, service users (cloud platforms or applications that invoke the services) are faced with a massive set of candidate services providing equivalent or similar functionalities. In order to select proper services, each user have to consider the QoS attributes of these service, such as response time, throughput, failure probability, availability, etc. [3]. These QoS attributes are generated from the users who had invoked the services. These users observe the QoS data when they invoke the services. Then a user can decide whether to choose a service according to the QoS data which are observed by other users, which can obtain personalized QoS attributes for the user. Typically, the characteristic of QoS data is as follows. (1) Due to the fast increase of system scale, the size of service-generated QoS data becomes more and more, QoS datasets become high volume. (2) Due to the impact of varying server workload and dynamic network conditions, QoS delivered to users may vary widely during different time periods. (3) Ensuring the veracity of QoS is fundamental for service composition, since inaccurate QoS data may lead to improper execution.

In this environment of service computing, to acquire personalized QoS attributes for each user, each user has to obtain all QoS data for a specific service, and other users have to invoke all services and observe the corresponding QoS data. However, it is infeasible actually. For one thing, it is impractical and high cost that a user invokes all the services at a time slice. Each user only invoked a few services according to the need. Therefore many user-observed QoS data are unknown. For another, the number of users and services may vary, and the user-observed QoS data are dynamic overtime. More-over, the working services are invoked frequently, while candidate services are invoked in lower probability. Actually, it is unnecessary for one user to invoke all services and generate all observed QoS data, since the unknown observed QoS data can be obtain by predicting. There are already some popular QoS prediction approaches [4–6], which are based on collaborative filtering. However, most methods are just suitable for offline condition and low effective for dynamic service-generated QoS data in online condition. Therefore, in order to provide each user with online personalized QoS information, it becomes an urgent task to explore effective approaches to accurately and efficiently predict the unknown QoS data of candidate services without requiring direct invocations.

To address the problems above, in this paper, we propose an online personalized QoS prediction approach for dynamic service-generated QoS big data. Our approach utilizes the QoS information of real-world services from geographically distributed service users to estimate the QoS data of candidate services at each time slice. Motivating from the collaborative filtering (CF) model, we use matrix factorization (MF) to predict the unknown QoS data for service-generated QoS big data, namely dynamic service-generated QoS big data prediction approach based on matrix factorization (DSPMF). We built an online MF objective function and use online stochastic gradient descent algorithm to solve the function. Noteworthily, to evaluate our approach, we conduct extensive experiments in real world public datasets in comparison with other methods. The experimental results show the effectiveness and efficiency of our approach.

Briefly summarized, the main contributions of this paper can be concluded as the following aspects: (1) We identify the problem of QoS personalized prediction for service-generated QoS big data. (2) We propose a dynamic service-generated QoS big data prediction method which employs online learning technique and matrix factorization. (3) We conduct sufficient experiments on real-world datasets, which verify the effectiveness of our models under various experimental cases.

The rest of this paper is organized as follows: Sect. 2 reviews the related work. Section 3 presents the framework of QoS prediction. Then we describe our DSPMF approach for QoS prediction in Sect. 4, and demonstrate the experimental results in Sect. 5. Finally, we conclude the whole paper and discuss the future work.

2 Related Work

Service-generated QoS data have played an important role in service discovery, selection and composition, which have lead QoS-centered approaches being studied widely in service computing area [6, 7]. In these approaches, QoS information is assumed to be known and accurate. However, as we discussed above, due to service-generated QoS data are sparse, there are many unknown QoS data in the real situations. Therefore, in order to obtain the unknown QoS data, a fundamental process before QoS-based service selection is to predict these data.

At present, a number of QoS prediction approaches have been proposed for Web services, cloud services and applications. Collaborative filtering (CF) is a popular approach to predict QoS values in most of the existing reports [4, 5]. CF can be divided into memory-based CF, model-based CF, and other hybrid CF. The memory-based CF employs similarity computation based on past usage experiences to find similar users and services for making the QoS value, including user-based approaches [8], item-based approaches [9], and their fusion [10, 11]. Shao et al. introduced CF to make Web service QoS prediction [4]. Zheng in [11, 12] proposed a hybrid QoS prediction method by combining user-based CF approach and service-based CF approach. However, when the data are very sparse, the prediction accuracy is not good. Moreover, due to the similarity calculation, the time complexity of these approaches is all quadratic to the data size. Hence, memory-based CF approaches are not suitable to be used on very large datasets, especially in real-time status where the data are growing exponentially. Thus more efficient models need to be explored to overcome the shortcomings.

Compared with memory-based CF approaches, model-based approaches use the known QoS data to learn a predictive model. A typical model-based CF approach is matrix factorization (MF), which trains a model according to the available QoS data in the user-item matrix. Relative to memory-based CF, MF is accurate and scalable in many applications. Yin et al. [13] proposed three service-neighbourhood enhanced prediction models based on probabilistic matrix factorization model. Lo et al. [14] proposed an extended matrix factorization (EMF) framework with relational regularization to make unknown QoS values prediction. In our previous work [15], we present a reputation-based Matrix Factorization (RMF) for predicting the unknown Web service QoS values. Memory-based methods are easy to implement and understand. Relative to neighborhood-based CF, MF can achieve better performance [16]. Hence, we focus on MF in this paper.

In recent years, many researchers focus on integrating online learning with MF. Based on sequential order data, this method can quickly adjust the model to reflect the change of the data timely and improve the online prediction accuracy [17, 18]. Different from batch learning techniques, online learning generate the best predictor by learning on streaming data instead of the entire training data set Online learning has been gained comprehensive attention, but there are only limited investigations in matrix factorization. Mairal et al. [19] proposed online learning for matrix factorization, which is applied in computer vision area. Zhu et al. [20] proposed a framework for online nonlinear Nonnegative matrix factorization (NMF), exploring the stochastic gradient descent and the mini-batch strategies. Zhao et al. [21] proposed a framework for performing online nonnegative matrix factorization in the presence of outliers. The framework is suited to large-scale data. In this paper, we study online algorithms to solve the issues facing batch-trained MF algorithms and propose online learning MF algorithms for service-generated QoS data.

3   The Prediction Framework

We propose a prediction framework for service-generated QoS big data as illustrated in Fig. 1. In this framework, Users are defined as applications which invoke the services. The details steps are as follows: Firstly, many users invoke Web services or cloud services. Due to the differences of geographical distribution with users, different user invokes the same service will generate different QoS data. In other words, QoS data observed by different users may be different. It is noted that each user may invoke some services instead of all the services. Therefore, though the amount of users and services is large and the volume of QoS data is big, the QoS data for each user are still sparse. Secondly, users submit their observed QoS data to the QoS prediction server. These data are important to make prediction. Other unknown QoS data will be predicted based on these

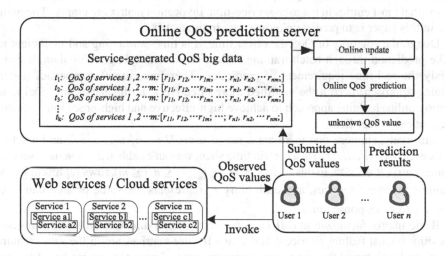

Fig. 1. Prediction framework for service-generated QoS big data

data. Thirdly, the prediction server collects the user-contributed QoS data and saves them to database. If new data comes, the database will perform update. Fourthly, service-generated QoS prediction is conducted by prediction server. In this step, the prediction server online updates unknown QoS data according to the new coming QoS data. Finally, based on the known QoS data and prediction results with certain service, target users can judge the optimal QoS data and invoke the corresponding service.

Noteworthily, this paper mainly focuses on user-side QoS properties (e.g., response time, throughput, failure probability). These user-side QoS data amount is big and vary over time. Moreover, with respect to server-side QoS data, user-side QoS data are more realistic and personalized which can provide different property data for different users [22, 23].

4   Service-Generated QoS Big Data Prediction

In this section, we first introduce the QoS data prediction problem in Sect. 4.1. Then we present our dynamic service-generated QoS big data prediction approach based on matrix factorization (DSPMF) models in detail in Sect. 4.2.

4.1   Problem Description

As shown in Fig. 1, users invoke services and obtain the service-generated QoS data (e.g., response-time). These data compose a user-service-time matrix, where each entry in this matrix represents the QoS value of a certain service (e.g., s_1 to s_m) observed by a service user (e.g., u_1 to u_n) at each time slice (e.g., t_1 to t_k). Although the service-generated QoS data amount is big, actually, a user may not invoke all services and a service may not be invoked by all users. There also exist some users who do not invoke any services and some candidate services which do not be invoked by any user. Therefore, the intractable issue in QoS prediction is data sparsity at each time slice, which means that most entries in the user-service-time invocation matrix are empty. The main task in this paper is to predict unknown data in the matrix.

Due to the volatility of QoS data over time, it is time-consuming and inefficient to make prediction through batch training. When faced with a large scale data, it can't satisfy the real-time requirement of prediction. Therefore current approaches (batch training approach) need to be modified more effectively. In this paper, our goal is to employ online learning approach to achieve high-effective and high-precision performances at each time slice.

Intuitively, suppose that there are a set of users $U = \{u_1, u_2, \ldots, u_n\}$, and a set of services $S = \{s_1, s_2, \ldots, s_m\}$ at a certain time slice, we can establish a $n \times m$ user-service sparse matrix $R \in \mathbb{R}^{n \times m}$. In this matrix, $R = \{r_{ij}\}_{n \times m} (i \leq n, j \leq m)$, rows represent users, columns represent services, and each entry r_{ij} represents the value of a certain QoS property (e.g., response time).

Basic matrix factorization (MF) can factorize a high dimensional matrix into two low dimensional feature matrices. These two feature matrices are in the same feature space, each column of the feature matrices represents the user or service latent feature vector. Suppose $U \in \mathbb{R}^{l \times n}$ and $S \in \mathbb{R}^{l \times m}$ represent user and service latent feature matrices, respectively. l is the number of latent factors (called dimensionality [23]), which is far less than m and n. Each column of U or S performs as a "factor vector". R can be factorized as the product of U and S approximately and expressed as follows:

$$R \approx \tilde{R} = U^T S \tag{1}$$

where $\tilde{R} = \left\{ U_i^T S_j \right\}_{n \times m} = \left\{ \tilde{r}_{ij} \right\}_{n \times m} (1 \leq i \leq n, \; 1 \leq j \leq m)$ is the approximate matrix of R, in which U_i and S_j are the ith and j^{th} column of U and S, $\tilde{r}_{ij}$ is the predicted value of r_{ij}. In the user-service matrix, unknown entry (QoS data) is calculated as the inner product of the corresponding user feature vector and service feature vector. This process of calculation is essentially a process of optimization, which aims to reduce the total the difference between each pair of r_{ij} and $\tilde{r}_{ij}$. The problem of prediction approach for dynamic service-generated QoS big data can be stated as: how to online predict the unknown QoS data by approximately reconstructing from the user-observed data.

4.2 Dynamic Service-Generated QoS Big Data Prediction Approach

In basic matrix factorization, feature spaces are obtained by batch training. In real system conditions, service-generated QoS data are dynamic and varying, which required to keep continuous and incremental updating using the sequentially QoS data. Moreover, because only part of user may invoke the service at a time slice and not every use-observed QoS data changes greatly, so batch training is not timely and inefficient. For this purpose, we present dynamic service-generated QoS big data prediction based on online matrix factorization (DSPMF). Unlike the basic matrix factorization, DSPMF adjusts the prediction model stochastically by taking into account the user' observed QoS data which are coming sequentially at each time slice. Suppose the new coming QoS data is $(U_i, S_j, r_{ij}, t_{ij}) \in R_t$, for each QoS data r_{ij} observed by user U_i invoking service S_j, the objective function can be described as following:

$$\mathcal{L}(U_i, S_j) = \frac{1}{2} I_{ij} (r_{ij} - U_i^T S_j)^2 + \frac{\lambda_u}{2} \|U_i\|_2^2 + \frac{\lambda_s}{2} \|S_j\|_2^2, \tag{2}$$

where I_{ij} plays as an indicator. $I_{ij} = 1$ means R_{ij} is known and $I_{ij} = 0$ means otherwise. $\|\cdot\|_F^2$ is the Frobenius norm. λ_u and λ_s are positive to control the importance of over-fitting issue during the learning process. The first term in Eq. (2) is the squared error between the observation and predicted data, and the remaining two terms are the corresponding regularizations which are employed to avoid over-fitting problem. To acquire a local minimum within iterative loops, we employ stochastic gradient descent (SGD) [24], a classic online learning algorithm, to train our model. The update feature space U_i and S_j, as following:

$$U_i \leftarrow U_i - \alpha(I_{ij}(U_i^T S_j - r_{ij})(U_i^T S_j)' S_j + \lambda_u U_i), \tag{3}$$

$$S_j \leftarrow S_j - \alpha(I_{ij}(U_i^T S_j - r_{ij})(U_i^T S_j)' U_i + \lambda_s S_j), \tag{4}$$

where α is the learning rate which is used to control how much change to make at each step. This online algorithm is stochastic every time, it can be described that every time when a new data sample $(U_i, S_j, r_{ij}, t_{ij})$ comes, online updating can be performed on its corresponding factors U_i and S_j using Eqs. 3 and 4.

The pseudo code of dynamic service-generated QoS big data prediction algorithm based on DSPMF is provided in Algorithm 1. In Algorithm 1, firstly, U_i and S_j are initialized with random numbers. Then, at each iteration, we check the coming QoS data. If a new datum $(U_i, S_j, r_{ij}, t_{ij})$ comes, we update the model, else if the QoS datum is not new, we reinitialize U_i and S_j. This algorithm will wait until new QoS data come under the convergent conditions. Finally, the prediction result can be obtained by U_i and S_j.

Algorithm1 DSPMF-based dynamic service-generated QoS big data prediction

Input: R, $(U_i, S_j, r_{ij}, t_{ij})$, all the model parameters

Output: $\tilde{r}_{ij} \leftarrow (U_i, S_j)$

1:     **repeat**
2   Collect service-generated QoS big data;
3:          **if** a new datum $(U_i, S_j, r_{ij}, t_{ij})$ comes **then**
4:               $I_{ij}=1$;
5:               initialize U_i and S_j;
6:               update r_{ij}, t_{ij};
7:               online update U_i, S_j by Eq. (3) and Eq. (4);
8:          **else**
9:               $I_{ij}=0$;
10:        **end if**
11:        **if** convergence **then**
12:             wait until new QoS data come;
13:        **end if**
14:    **until forever**

---

5   Experiment

In this section, we conduct experiments based on a real-world Web service QoS dataset to validate the rationality of our approach, including accuracy comparison, impact of parameters, efficiency analysis.

5.1   Experimental Dataset Description

In this paper, service-generated QoS big data are from a real-world Web service QoS dataset [25]. These QoS data are generated based on PlanetLab platform, which is a distributed testbed made up of computers all over the world. On this platform, 142 users (PlanetLab nodes) invoke 4,500 Web services for 64 consecutive time slices, at an interval of 15 min, generating QoS data which includes both response time and throughput data. Each dataset can be translated into a $142 \times 4,500 \times 64$ three-dimensional matrix structure. In our experiments, we use the response time dataset which stands for the time duration between user sending out a request to a service and receiving a response.

5.2   Evaluation Metrics

Due to the large variance of QoS data, we use MRE (Median Relative Error) and 90% NPRE (ninety percentile relative error) as our measurement criterion of prediction accuracy. MRE and NPRE are more appropriate to evaluate the QoS prediction optimization efforts than other metrics (e.g. mean absolute error). MRE and NPRE are defined as:

$$MRE = \underset{I_{ij}=0}{median} \left(\left| \tilde{r}_{ij} - r_{ij} \right| / r_{ij} \right) \tag{5}$$

$$NPRE = 90\% \times \left| \tilde{r}_{ij} - r_{ij} \right| / r_{ij}, \tag{6}$$

where r_{ij} is the response time of Web service j observed by user i, $\tilde{r}_{ij}$ denotes the predicted response time.

5.3   Performance Comparison

In this section, we conduct extensive experiments and compare our method with PMF. PMF is a probabilistic linear model with Gaussian observation noise [16]. In our experiments, PMF employs batch learning, while our approach employs online learning. In this experiment, we use different matrix density which defined as the density of the training dataset. each QoS prediction method is run on 5 different matrices, whose densities are 10% to 50% at a step increase of 10%. Additionally, λ_u and λ_s are set to 30 with PMF, while 0.001 with DSPMF in this experiment. The learning rate is set to 0.01. The maximum iterations are both set to 100, and the dimensionality is both set to 10. There are 64 time slice, at each time slice, each approach is performed 20 times and taken the average. Figure 2 shows the MRE and NPRE results of different methods with different density.

From Fig. 2, we can observe that no matter what the matrix density is, DSPMF approach can achieve smaller MRE and NPRE values than PMF. Concretely, in average sense, DSPMF method can acquire about 43.8% improvement in MRE. Meanwhile, the NPRE value gets about 70% improvement compared with PMF model. It illustrates that our approach has further higher prediction performance.

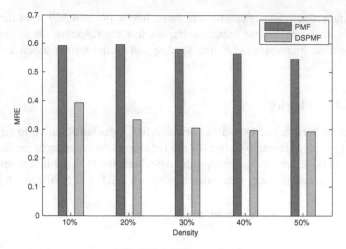

(a) MRE at different density

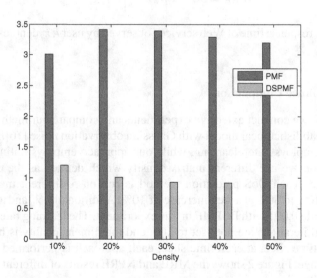

(b) NPRE at different density

Fig. 2. Accuracy comparison on response time

5.4 Impact of Matrix Density

In real world dataset, the matrix density is changing with the time. To study the impact of matrix density on MRE and NPRE, we vary the density of matrix from 5% to 45% with a step value of 10%. Other key parameters are set as: dimensionality = 10, and $\lambda_u = \lambda_s = 0.001$, $\alpha = 0.01$. The average MRE and NPRE under different density matrix for 64 time slices are illustrated in Fig. 3.

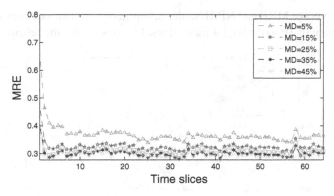

a) MRE at different time slice

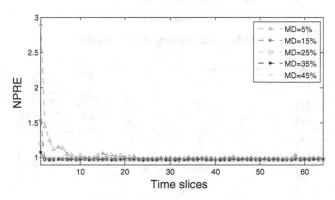

b) NPRE at different time slice

Fig. 3. Impact of matrix density

From Fig. 3, we can see that: (1) When matrix density increases from 5% to 45%, the MRE and NPRE values appear a decreasing trend, which means prediction accuracy is improved. (2) During 64 time slices, it seems that the smaller the matrix density, the greater change of the error over time. When the matrix density is low (e.g., 5%), the error becomes big. With the matrix density become denser, the error seems to be more stable and smaller. Since the prediction model is easier to fall into over-fitting problem as the matrix density become more sparse (less known QoS data contributing to the training phase). At each time slice, this problem weakens when the matrix density increases.

5.5 Impact of Dimensionality

To study the impact of dimensionality on prediction results, we assess how many potential dimensionalities in the model learning is enough to character user and service latent features. We conduct experiments using different number of latent feature in the model by varying the value of dimensionality from 2 to 30. For each dimensionality, we calculate the average MRE and NPRE under different density matrix (from 10% to 50% with a step value of 10%) during 64 time slices. Figure 4 shows the results.

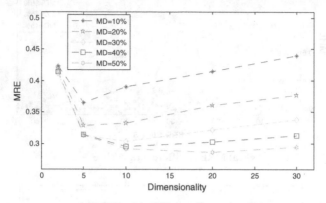

(a) MRE with different dimensionality

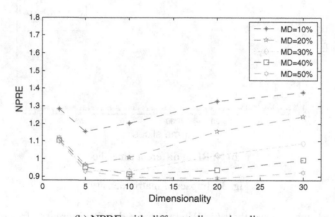

(b) NPRE with different dimensionality

Fig. 4. Impact of dimensionality

Figure 4 shows the impact of dimensionality on MRE and NPRE, respectively. We can observe that MRE and NPRE drop quickly when the dimensionality increases from 2 to 10. When the dimensionality is larger than 10, MRE and RMSE increase slowly. Generally, a higher dimensionality means the more latent features are used to characterize users and services for training the prediction model, which may enhance the prediction performance. Actually, too many latent features might cause the over-fitting problem which will do harm to the performance. Furthermore, the higher value of dimensionality means the more time of learning these features. Therefore, too small or too large dimensionality value will reduce the prediction accuracy and efficiency. In our other experiments, we chose dimensionality = 10.

5.6 Impact of λ_u and λ_s

The regularization parameters λ_u and λ_s are employed to avoid over-fitting through controlling the proportion of the two regularization terms which are used to in Eq. (2). To evaluate the impact of λ_u and λ_s, we perform experiments on the response time data. We set the dimensionality to 10, set λ_u and λ_s from 0.0001 to 0.01, vary the density from 10% to 50% with a step value of 10% for each matrix corresponding each time slices. Figure 5 shows that the experimental result.

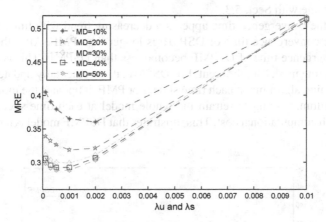

(a) MRE with different λ_u and λ_s

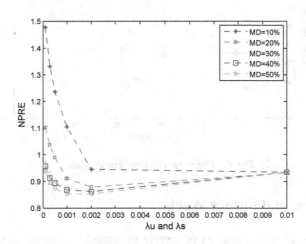

(b) NPRE with different λ_u and λ_s

Fig. 5. Impact of λ_u and λ_s

From Fig. 5, when the matrix density increases, performances of MRE and NPRE both improve. With the increase of λ_u and λ_s, the overall prediction accuracy first increases, reach an optimal value and then drops. When λ_u and λ_s are about 0.001, both

MRE and NPRE reach the optimal value with different matrix density. If λ_u and λ_s are large (e.g., $\lambda_u = \lambda_s = 0.01$) or too small (e.g., $\lambda_u = \lambda_s = 0.0001$), the prediction accuracy will be unsatisfactory. In our other experiments, we chose $\lambda_u = \lambda_s = 0.001$.

5.7 Convergence Time Analysis

In this section, to evaluate the efficiency of our DSPMF method, we conduct experiments on convergence time and compare our DSPMF method with PMF. The main parameters setting is the same with Sect. 5.3.

In Fig. 6, the convergence time appears a decreasing trend with time. For the first time slice, the convergence time of DSPMF is longer than PMF. After the third time slice, the convergence time of DSPMF become less than DSPMF. This is because the modes of updating model are different. For DSPMF, it incrementally updates the model by online learning algorithm at each time slice. For PMF, it updates the model by batch learning algorithm, leading to retrain the whole model at each time slice. Therefore, PMF need high computational cost. This illustrates that DSPMF model is more effective than PMF.

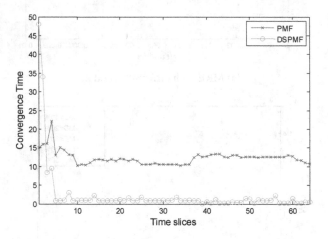

Fig. 6. Convergence time comparison

6   Conclusion and Future Work

Due to the time-variant and sparsity of service-generated user-observed QoS data, it is a challenge to predict the unknown QoS data. In this paper, we present a dynamic service-generated QoS big data prediction approach, we built an objective function and use online stochastic gradient descent algorithm to solve the function. Our proposed method can effectively and efficiently solve the challenge and achieve completive prediction accuracy as demonstrated in the experiments. In our future work, we will take more relational factors (e.g., user context information) into consideration and apply other better online learning method to improve our prediction performance.

Acknowledgment. The work described in this paper was supported by the Guangdong High-Level University Project "Green Technologies for Marine Industries", Guangdong Common Colleges Young Innovative Talents Project, and Shantou University National Fund breeding project (No. NFC16001).

References

1. Assunção, M.D., Calheiros, R.N., Bianchi, S., Netto, M.A., Buyya, R.: Big Data computing and clouds: trends and future directions. J. Parallel Distrib. Comput. **79**, 3–15 (2015)
2. Zheng, Z., Zhu, J., Lyu, M.R.: Service-generated big data and big data-as-a-service: an overview. In: 2013 IEEE International Congress on Big Data (BigData Congress), Santa Clara Marriott, CA, pp. 403–410 (2013)
3. Suchithra, M., Ramakrishnan, M.: Non functional QoS criterion based web service ranking. In: Suresh, L.P., Panigrahi, B.K. (eds.) Proceedings of the International Conference on Soft Computing Systems. AISC, vol. 398, pp. 79–90. Springer, Heidelberg (2016). doi: 10.1007/978-81-322-2674-1_8
4. Shao, L., Zhang, J., Wei, Y., Zhao, J., Xie, B., Mei, H.: Personalized QoS prediction for web services via collaborative filtering. In: IEEE International Conference on Web Services (ICWS 2007), pp. 439–446 (2007)
5. Wang, X., Zhu, J., Shen, Y.: Network-aware QoS prediction for service composition using geolocation. IEEE Trans. Serv. Comput. **8**(4), 630–643 (2015)
6. Zheng, Z., Ma, H., Lyu, M.R., King, I.: Collaborative web service QoS prediction via neighborhood integrated matrix factorization. IEEE Trans. Serv. Comput. **6**(3), 289–299 (2013)
7. Lo, W., Yin, J., Li, Y., et al.: Efficient web service QoS prediction using local neighborhood matrix factorization. Eng. Appl. Artif. Intell. **38**, 14–23 (2015)
8. Jin, R., Chai, J.Y., Si, L.: An automatic weighting scheme for collaborative filtering. In: The 27th Annual International ACM SIGIR Conference on Research and Development in Information Retrieval (SIGIR), pp. 337–344. ACM Press, Sheffield (2004)
9. Deshpande, M., Karypis, G.: Item-based top-n recommendation. ACM Trans. Inf. Syst. **22**(1), 143–177 (2004)
10. Wu, J., Chen, L., Feng, Y., Zheng, Z., Zhou, M., Wu, Z.: Predicting QoS for service selection by neighborhood-based collaborative filtering. IEEE Trans. Syst. Man Cybern. Part A **43**(2), 428–439 (2013)
11. Zheng, Z., Ma, H., Lyu, M.R., King, I.: QoS-aware web service recommendation by collaborative filtering. IEEE Trans. Serv. Comput. **4**(2), 140–152 (2011)
12. Zheng, Z., Ma, H., Lyu, M.R., King, I.: WSRec: a collaborative filtering based web service recommender system. In: 16th International Conference on Web Services, pp. 437–444, Los Angeles, CA (2009)
13. Yin, J., Xu, Y.: Personalised QoS-based web service recommendation with service neighbourhood-enhanced matrix factorization. Int. J. Web Grid Serv. **11**(1), 39–56 (2015). Special Issue
14. Lo, W., Yin, J., Deng, S., Li, Y., Wu, Z.: An extended matrix factorization approach for QoS prediction in service selection. In: International Conference on Services Computing (SCC), pp. 162–169 (2012)
15. Xu, J., Zheng, Z., Lyu, M.R.: Web service personalized QoS prediction via reputation-based matrix factorization. IEEE Trans. Reliab. **65**(1), 28–37 (2016)

16. Salakhutdinov, R., Mnih, A.: Probabilistic matrix factorization. In: Advances in Neural Information Processing Systems, pp. 1257–1264 (2015)
17. Shalev-Shwartz, S.: Online learning and online convex optimization. Found. Trends Mach. Learn. **4**(2), 107–194 (2011)
18. Bertsekas, D.P.: Incremental gradient, subgradient, and proximal methods for convex optimization: a survey. Optimization **2010**(2), 691–717 (2015)
19. Mairal, J., Bach, F., Ponce, J., Sapiro, G.: Online learning for matrix factorization and sparse coding. J. Mach. Learn. Res. **11**, 19–60 (2010)
20. Zhu, F., Honeine, P.: Online kernel nonnegative matrix factorization. Sig. Process. **131**, 141–153 (2016)
21. Zhao, R., Tan, V.Y.F.: Online nonnegative matrix factorization with outliers. IEEE Trans. Sig. Process. **65**(3), 555–570 (2017)
22. Zheng, Z., Lyu, M.R.: WS-DREAM: a distributed reliability assessment mechanism for web services. In: The 38th Annual IEEE/IFIP International Conference on Dependable Systems and Networks (DSN 2008), Anchorage, Alaska, pp. 392–397 (2008)
23. Zheng, Z., Lyu, M.R.: Personalized reliability prediction of web services. ACM Trans. Softw. Eng. Methodol. **22**(2), 1–28 (2013)
24. Bottou, L.: Large-scale machine learning with stochastic gradient descent. In: Lechevallier, Y., Saporta, G. (eds.) Proceedings of COMPSTAT 2010, pp. 177–186. Physica-Verlag HD, Heidelberg (2010). doi:10.1007/978-3-7908-2604-3_16
25. Zhang, Y., Zheng, Z., Lyu, M.R.: WSPred: a time-aware personalized QoS prediction framework for web services. In: The 22nd IEEE Symposium on Software Reliability Engineering (ISSRE), Los Alamitos, California, pp. 210–219 (2011)

Discovering Interesting Co-location Patterns Interactively Using Ontologies

Xuguang Bao and Lizhen Wang[✉]

Department of Computer Science and Engineering,
School of Information Science and Engineering, Yunnan University, Kunming 650091, China
bbaaooxx@163.com, lzhwang@ynu.edu.cn

Abstract. Co-location pattern mining, which discovers feature types that frequently appear in a nearby geographic region, plays an important role in spatial data mining. Common frameworks for mining co-location patterns generate numerous redundant patterns. Thus, several methods were proposed to overcome this drawback. However, most of these methods did not guarantee that the extracted co-location patterns were interesting for being generally based on statistical information. Thus, it is crucial to help the decision-maker choose interesting co-location patterns with an efficient interactive procedure. This paper proposed an interactive approach to discover interesting co-location patterns. First, ontologies were used to improve the integration of user knowledge. Second, an interactive process was designed to collaborate with the user to find interesting co-location patterns efficiently. Finally, a filter was designed to reduce the number of discovered co-location patterns in the result set further. The experimental results on both synthetic and real data sets demonstrated the effectiveness of our approach.

Keywords: Co-location pattern mining · Interactive · Ontology · Filter · Post-mining

1 Introduction

Co-location pattern mining is a new branch studied in the spatial data mining recently. A spatial co-location pattern represents a subset of spatial features whose instances are frequently located in spatial neighborhood. Spatial co-location patterns may yield important insights in many applications. For example, a mobile service provider may be interested in mobile service patterns frequently requested by geographical neighboring users. Botanists may be interested in symbiotic plant species in an area, etc. Other application domains include Earth science, public health, biology, transportation, etc. [1].

Discovering interesting co-location patterns is an important task in spatial data mining. However, a common problem in most co-location mining algorithms is that there are too many co-location patterns in the output while only a few of them is really interesting to a user.

To overcome this drawback, several methods were proposed in the literature such as co-location pattern concise representations [2, 3], redundancy reduction [4], and post

© Springer International Publishing AG 2017
Z. Bao et al. (Eds.): DASFAA 2017 Workshops, LNCS 10179, pp. 75–89, 2017.
DOI: 10.1007/978-3-319-55705-2_6

processing [5, 6]. However, most of these methods are generally based on statistical information in the database. Since pattern interestingness strongly depends on the user's knowledge and goals, these methods do not guarantee that interesting co-location patterns will be extracted. A pattern could be interesting to one user but not to another.

The representation of user knowledge is an important issue. The more the knowledge is represented in a flexible, expressive, and accurate formalism, the more the interesting co-location patterns selection is efficient. In the Semantic Web field, ontology is considered as the most appropriate representation to express the complexity of the user knowledge [6].

This paper proposed a new interactive approach, OICM (Ontology-based Interesting Co-location Miner), to find interesting co-location patterns. We assume a set of candidates (i.e., prevalent co-location patterns) has already been mined. Our goal is to help a particular user interactively discover interesting co-location patterns according to his/her real interest. Instead of requiring the user to explicitly express his/her real interesting co-location patterns, we alleviate the user's burden by only asking him/her to choose a small set of sample co-location patterns according to his/her interest for several rounds.

As shown in Fig. 1, OICM takes a set of candidates as input. In each round, a small collection (e.g., 10) of sample co-location patterns is selected from the candidates (step 1) and then the samples are provided for the user (step 2). The user feedbacks his/her preference for each sample (step 3), and the feedback information will be used by the ontology to update the candidates (step 4.1), meanwhile the interesting co-location patterns will be added to the result set (step 4.2), and then OICM decides which co-location patterns to be selected for the next feedback. The interaction process (step 1–4) continues for several rounds unless the candidates are empty. Finally, a filter based on the anti-monotone [1] of co-location patterns is designed to reduce the number of interesting co-location patterns in the result set (step 5).

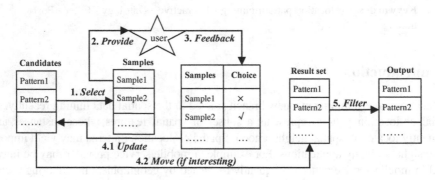

Fig. 1. Framework description

There are two basic research questions in discovering interesting co-location patterns through interactive feedback. First, how do ontologies be used to update the candidates? Second, how does the system select sample co-location patterns to minimize the times for iteration? In this paper, ontologies are used to measure the similarity between co-location patterns. And based on this measurement, an update algorithm is proposed to reduce the number of candidates effectively. In order to reduce the number of interactions, a greedy selection algorithm is proposed to provide sample co-location patterns having the most differences.

This paper is structured as follows: Sect. 2 introduces our problem definition and the related work. Section 3 discusses the method of updating candidates using the ontology. Strategies for selecting samples and reducing the number of co-location patterns from the result set are discussed in Sect. 4. Section 5 gives the complete algorithm. The experimental results are presented in Sect. 6. Section 7 ends the paper with the summary.

2 Problem Statement and Related Work

We present our problem statement with the basic concept of spatial co-location mining and ontology, and then discuss the related work.

2.1 Problem Definition

Suppose F is a set of spatial features, S is a set of their instances and r is a spatial neighbor relationship over S. If the Euclidean metric is used as the neighbor relationship r, two spatial instances are neighbors if the ordinary distance between them is no more than a given threshold d. Figure 2(a) shows an example of spatial data set which has 5 spatial features named with A, B, C, D and E and their instances. A.1 stands for the first instance of feature A. The two spatial instances who are neighbors are connected by a line in Fig. 2(a). A **co-location pattern** $c(c \subseteq F)$ is a subset of spatial features whose instances form cliques under the neighbor relationship frequently. For example, in Fig. 2(a), {A, B, D} is a co-location pattern because A.2, B.2 and D.1 form a clique.

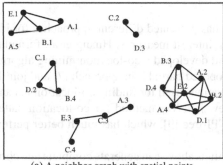
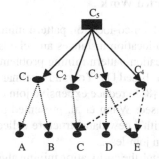

(a) A neighbor graph with spatial points
and their relations

(b) Visualization of the ontology (H of O)
created based on Fig. 2(a)

Fig. 2. A spatial dataset with neighbor relationships and the visualization of the ontology created based on it

In co-location pattern mining, participation index (PI) is often used as a measure of the prevalence of a co-location pattern c. The PI measure indicates wherever a feature in c is observed, with a probability of at least PI(c), all other features in c can be observed in its neighborhood. Given a minimum prevalence threshold *min_prev*, a co-location c is a prevalent co-location if $\mathrm{PI}(c) \geq min\_prev$ holds.

An **ontology** was defined by Gruber as *a formal, explicit specification of a shared conceptualization*. Formally, an ontology is defined as a quintuple $O = \{C, R, I, H, A\}$, $C = \{ C_1, C_2, \ldots, C_n\}$ is a set of *ontology concepts* and $R = \{ R_1, R_2, \ldots, R_m\}$ is a set of relations defined over *ontology concepts*. I is a set of instances of *ontology concepts* and H is a Directed Acyclic Graph (DAG) defined by the inclusion relation (*is-a* relation, $\leq$) between *ontology concepts*. C_2 *is-a* C_1 if the *ontology concept* C_1 includes the *ontology concept* C_2. A is a set of axioms bringing additional constraints on the ontology.

We use 3 types of *ontology concepts* C of $O = \{C, R, I, H, A\}$: *leaf-concepts*, *generalized concepts* from the inclusion relation ($\leq$), and *restriction concepts* proposed only by ontology. *Leaf-concepts* are the leaf nodes in H of O, each feature in a spatial database will be considered as a *leaf-concept*; *generalized concepts* are described as the *ontology concepts* that include other *ontology concepts* in the ontology; and *restriction concepts* are described using logical expressions defined over *leaf concepts* (features).

For example, Fig. 2(b) shows an H of O. Let us consider the ontology presented in Fig. 2(b) as being the ontology constructed over features of a spatial database of Fig. 2(a) and described as follows:

The *ontology concepts* of the ontology are $\{C_1, C_2, C_3, C_4, C_5, A, B, C, D, E\}$, and the three types of *ontology concepts* are: *leaf-concepts*: $\{A, B, C, D, E\}$, *generalized concepts*: $\{C_1, C_2, C_3, C_5\}$, and *restriction concepts*: $\{C_4\}$. Note that in co-location mining, each feature is considered as a *leaf-concept*.

This paper focuses on finding interesting co-location patterns interactively using ontologies.

2.2 Related Work

Most work on co-location pattern mining has presented different approaches for identifying co-location instances and choosing interest measures. Huang et al. formulated the co-location pattern mining problem and developed a co-location mining algorithm called join-based [1]. Yoo and Shekhar proposed partial join approach [7] and join-less approach [8] to reduce expensive join operations used for finding co-location instances in join-based. Wang et al. presented a new join-less method for co-location patterns mining with a new data structure called CPI-tree [9], which has much better performances than join-less.

Although the co-location mining algorithms above can generate complete and correct prevalent co-locations, the enormous number of the output makes it hard to be well analyzed by the decision-maker. Hence mined co-location patterns are required to be compacted with no losses. Yoo proposed an algorithm to mine top-k closed co-location patterns based on FP-growth method [2], and Wang et al. gave an order-clique-based approach [3] for mining maximal co-location patterns.

The conception of ontology was first proposed by Gruber in the early 1990s. After that, several other definitions were proposed in the literature. For instance, ontology is viewed as *a logical theory accounting for the intended meaning of a formal vocabulary*

[10], and later, in 2001, Maedche and Staab [11] proposed a more artificial-intelligence-oriented definition that ontologies are described as (meta) data schemas, *providing a controlled vocabulary of concepts, each with an explicitly defined and machine processable semantics*. Ontologies, introduced in data mining for the first time in early 2000 [12], can be used in several ways: Domain Ontologies, Ontologies for Data Mining Process, and Metadata Ontologies.

Several approaches had been working on co-location mining using ontologies. [5] gave a framework on co-location rule mining interactively with the user using ontologies, which required the user to write complex and correct formulas in order to be understood by its algorithm. [13] gave an approach called OICPP to mine interesting co-location patterns using an interactive post-mining process but the output is still too large for better analyzed. In this paper, OICM can generate more accurate co-location patterns in the interactive process than OICPP because of the effective strategies for updating candidates and selecting sample co-location patterns. Moreover, with the filter process, the output of OICM is much less than that in OICPP.

3   Updating Candidates Using the Ontology

In this section, we mainly discuss how to use ontologies to measure the semantic distance between two co-location patterns, and how to update the candidates based on the measurement.

3.1   Semantic Distance Between Co-location Patterns

If feature A and feature B have some common attributes (etc. habits), they will be more likely to have the same distribution in the real world. We call that A is similar to B. Ontologies gather similar features and create a *generalized concept* or *restriction concept* to include them, as Fig. 2(b) shows, C_1 gathers A and B because A is similar to B. Thus, similar features can be inferred from ontologies.

Jaccard distance [14] is used as the distance between patterns in traditional transaction data mining: Given two patterns P_1 and P_2, the distance is defined as

$$D(P_1, P_2) = 1 - \frac{|T(P_1) \cap T(P_2)|}{|T(P_1) \cup T(P_2)|},$$ where $T(P)$ is the set of transactions which contain pattern P. In co-location data mining, there is no transaction used to calculate the similarity between two co-location patterns, but ontologies include the semantic relationships between *ontology concepts*, thus we can define the semantic distance between co-location patterns based on ontologies.

Definition 1. Given a feature f, $S(f)$ demonstrates the set of *ontology concepts* containing f directly. Given a co-location $c = \{f_1, f_2 \ldots f_n\} (n \geq 1)$, the **generalization** of c is defined as $G(c) = \bigcup_{i=1}^{n} S(f_i)$.

For example, in Fig. 2(b), for feature C, $S(C) = \{C_2, C_4\}$, for feature E, $S(E) = \{C_3, C_4\}$, then for a c-location pattern $\{C, E\}$, $G(\{C, E\}) = S(C) \cup S(E) = \{C_2, C_3, C_4\}$.

Definition 2. Given two co-location patterns c_1 and c_2, the **semantic distance** between c_1 and c_2 is defined as $\text{SD}(c_1, c_2) = 1 - \dfrac{|G(c_1) \cap G(c_2)|}{|G(c_1) \cup G(c_2)|}$, where $G(c)$ is the *generalization* of c.

For example, let $c_1 = \{A, C, D\}$, $c_2 = \{B, C, E\}$, $G(c_1) = \{C_1, C_2, C_3, C_4\}$, $G(c_2) = \{C_1, C_2, C_3, C_4\}$, $\text{SD}(c_1, c_2) = 1 - \dfrac{|G(c_1) \cap G(c_2)|}{|G(c_1) \cup G(c_2)|} = 1 - \dfrac{|\{C_1, C_2, C_3, C_4\}|}{|\{C_1, C_2, C_3, C_4\}|} = 0$.

The definition has its reasonability, for example, if it is required to study the relationships between the predators and their prey in a certain area to detect whether this area has adequate food for the predators, co-location pattern {tiger, elk} and co-location pattern {lion, antelope} express the same meaning, they can be abstracted as a higher level co-location pattern consisting of *ontology concepts* {predator, prey}. Thus, the shorter the semantic distance between two co-locations is, the more similar the two co-locations are.

3.2 Updating Candidates

Once the user feedbacks his/her preference on the provided samples, the semantic distance measure then is used to find the "similar" co-location patterns from the candidates for each sample from the user's feedback.

Definition 3. Given a co-location c and the semantic distance threshold $sdt(0 \leq sdt \leq 1)$, if there exists a co-location c' that $\text{SD}(c, c') \leq sdt$, c' is called a *similarity* of c. Particularly, if $\text{SD}(c, c') = 0$, c' is called a *hard similarity* of c.

Definition 4. If a co-location c has no *hard similarities* in the candidates or samples, c is called an *isolated pattern*.

The system first discovers all the *hard similarities* of each sample co-location pattern by the user's feedback, and then moves the interesting samples and their *hard similarities* to the result set, discards uninteresting samples and their *hard similarities* from the candidates. Furthermore, if a co-location c in candidates is an *isolated pattern*, this will cause a *conflict choice* and the interestingness choice of c depends on the minimum semantic distance calculated from its *similarities*. For example, as Fig. 2(b) shows, if 2 samples $c_1 = \{A, B, D\}$ is marked interesting and $c_2 = \{A, D, E\}$ is marked uninteresting, the *sdt* value is 0.7, for a candidate co-location $c = \{A, B, D, E\}$, $\text{SD}(c, c_1) = 1/3 < 0.7$, $\text{SD}(c, c_2) = 0 < 0.7$, c is a *similarity* of c_1 and c_2 because c is a *hard similarity* of c_2, c will be regarded as an uninteresting pattern and be discarded from the candidates. If $c = \{A, B\}$, c is an *isolated pattern*, It will cause the *conflict choice*: $\text{SD}(c, c_1) = 1/2 < 0.7$, $\text{SD}(c, c_2) = 2/3 < 0.7$, c will be regarded as an interesting co-location pattern as c_1 because c is more similar to $c_1(\text{SD}(c, c_1) < \text{SD}(c, c_2))$.

Conflict choice will only occur when $sdt > 0$ because *isolated patterns* have no *similarities* when $sdt = 0$. The *conflict choice* may cause errors because *isolated patterns* have no *hard similarities* and the interestingness decision is only based on their "nearest" *similarities*. In fact, the *isolated patterns* can be selected as the sample co-locations to avoid the possible false selection, but this may increase the times of iteration. However, the number of *isolated patterns* is few under normal conditions, so OICM can still have a high accuracy ratio. While in OICPP, any co-location pattern may be marked false interestingness because OICPP put all the *similarities* not the *hard similarities* together. Thus, the update algorithm in OICM is more effective than that in OICPP.

Algorithm 1 shows the pseudo code of the updating process. First, OICM gets the feedback on each sample co-location from the user (line 1), if the feedback is available (line 2), OICM finds the *hard similarities* of each sample and set their interestingness (lines 3–7). Then, each isolated pattern is handled (lines 8–13): OICM finds the most similar sample co-location L_s and set p's interestingness as L_s. The computational complexity is about $O(m)$, where m is the number of co-locations in the initial candidates.

Algorithm 1 Updating Candidates

Input: A set of m candidates, P
          Semantic distance threshold, *sdt*
          Relationship on *ontology concepts*, R
          List of selected sample co-location patterns, L
          List of *isolated patterns*, IL

Variables:
          Feedback of the user's selection of samples, F
          Hard similarities of co-location p in P, P_p
          The similarity having the minimum semantic distance with p in L, L_s

Procedure:
1: $F = feedback\_from\_user(L)$;
2: **if** (F != null)
3:   **foreach**(pattern p **in** L)
4:      P_p=*find_hard_similarities* (p, P, R, sdt);
5:      **if**(P_p are not empty)
6:         set each P_p's interestingness as p;
7:         $P.remove(P_p)$;
8:   **foreach**(pattern p **in** IL)
9:      L_s=*find_most_different_ pattern*(p, L, R, sdt);
10:    **if** (L_s != null)
11:      set p's interestingness as L_s;
12:      $P.remove(p)$;
13:      $IL.remove(p)$;

---

4   Strategies for Selecting Samples and Filtering Result Set

In this section, we first discuss the selection criteria and propose our samples selection method, and then present a filter to further reduce the number of co-location patterns in the result set. The importance of selecting the best samples for user feedback has been recognized by some work [15, 16]. Ideally, the system should collaborate with the user in the whole interactive process to improve the accuracy and reduce the number of interactions.

4.1 Selection Strategy

Definition 5. Given a set of co-location patterns $s = \{c_1, c_2 \ldots c_m\}$, the *information* of s is defined as $I(s) = \bigcup_{i=1}^{m} (G(c_i))$.

There are two criteria to select sample co-location patterns. The first is that the selected sample co-location patterns should not be redundant to each other. There is redundancy between two co-location patterns if they are close in pattern composition (i.e., the set of features). Since redundant co-location patterns are naturally similar (in the semantic web field) to each other, presenting redundant patterns for feedback does not maximize the learning benefit and even increases the user's burden. The second criterion is that the selected patterns should contain more *information*. For example, two co-location patterns are needed for feedback, if $c = \{A, D\}$ is selected as the first one, $c_1 = \{C, D\}$ is better than $c_2 = \{A, E\}$ for the second one because $G(c) = \{C_1, C_3\}$, $G(c_1) = \{C_2, C_3, C_4\}$, $G(c_2) = \{C_1, C_3, C_4\}$, $G(c) \cup G(c_1) = \{C_1, C_2, C_3, C_4\}$, $G(c) \cup G(c_2) = \{C_1, C_3, C_4\}$, c and c_1 have more *information* than c and c_2 have.

The two criteria for selecting sample co-location patterns require our system to select sample co-location patterns where each co-location pattern is as different as possible to the other ones. Thus, a greedy algorithm is proposed to select sample patterns efficiently. This method first chooses the co-location pattern containing the maximal *information* from the candidates. Next, the co-location pattern having the maximal *information* with the current sample co-locations is chosen as the second one, and so on.

The selecting strategy in OICPP is to find the co-location pattern c having the maximal semantic distance from the previous chosen sample pattern and $G(c)$ is not equal to the *generalization* of any selected samples as the next sample co-location pattern. It may cause a problem that the following selected samples cannot increase the *information* of the current samples. Thus, the strategy of selection in OICM is more effective than that in OICPP.

Algorithm 2 shows the pseudo code of the selection process. The candidate co-location pattern having the maximal *information* is selected as the first sample co-location pattern (line 2). The following sample co-location patterns will be chosen (lines 3–12): The candidate co-location pattern having the maximal *information* with the current samples is chosen as the following sample co-location (lines 6–11). The computational complex of Algorithm 2 is about $O(km)$, where m is the number of co-locations in the initial candidates, k is the number of co-location patterns in samples.

Algorithm 2 selecting sample co-location patterns

Input: A set of m candidates, P
        The number of sample co-location patterns, k
        Relationship on *ontology concepts*, R
Output: List of selected sample co-location patterns, L
Procedure:
1:*max_information=0*;
2: pattern $p = get\_maximal\_information(P, k, R)$
3: **do**
4:   $L.add(p)$;
5:   $P.remove(p)$;
6:   **foreach**(pattern pp **in** P)
7:     **if**(*have_same_concepts(pp, L)*)
8:       **continue**;
9:     **if**(*get_infor* $(L, pp) > max\_information$)
10:       $max\_information = get\_infor\ (L, pp)$;
11:       $p = pp$;
12: **while** $L.count >= k$

4.2 Reducing the Number of Co-location Patterns from the Result Set

If a co-location pattern is prevalent, its sub patterns are also prevalent [1]. Maximal co-location pattern is a kind of compact co-location patterns which is prevalent while any of its superset is not prevalent [3]. If {A, B, C} is prevalent, the user will know that all its subsets are also prevalent. Thus, {A, B, C} contains the prevalence of its subsets and we only need to present {A, B, C} for the user. However, in this paper, each interesting co-location pattern has its own *generalization* (semantic), so it will cause errors if our system only gets the maximal co-location patterns from the result set. Thus, in order to correctly reduce the number of result set, the system will get the maximal co-location patterns from co-location patterns having the same *generalization*. In Fig. 2(a), If the result set is {{A, B}, {A, C}, {B, C}, {A, B, C}}, the *generalization* of {A, B} is {C_1}, the maximal co-location pattern having *generalization* {C_1} is {A, B}; the *generalization* of the other 3 co-location patterns is {C_1, C_2}, the maximal co-location pattern having *generalization* {C_1, C_2} is {A, B, C}. Thus, {A, B} and {A, B, C} will be discovered as the final interesting co-location patterns.

5 Complete Algorithm

The candidates updating method and the sample selection method have been discussed. As a summary, the complete algorithm OICM for interesting co-location pattern discovering with user's interactive feedback is outlined in Algorithm 3.

Algorithm 3 Discover Interesting Co-location Patterns(OICM)

Input: A set of m candidates, P
        Number of sample patterns for feedback, k
        Semantic distance threshold, sdt
        An ontology described by an XML file, o
Variables: Relations of concepts generated by o: R
        List of selected sample co-location patterns, L
Output: Interesting Pattern List, Res
Procedure:
 1: $R=generate\_relations\_by\_ontology(o)$;
 2: $IL=get\_isolated\_patterns(P, R)$;
 3: **while** $(P.count>= 0)$
 4:   **if**$(P.count <= k)$
 5:     $L = P$;
 6:   **else**
 7:     $L= sample\_selection(P, k, R)$; //Algorithm 2
 8:   $update\_candidtes(P, sdt, R, L, IL)$; // Algorithm 1
 9: $Res=do\_filter(Res)$;
10: $output(Res)$;

OICM takes the entire collection of co-location patterns as input. The user can specify how many patterns he/she would like to judge at each round. OICM works as follows. First, the inclusion relations of *concepts* are generated from the ontology (line 1), and then the *isolated patterns* are selected from the candidates (line 2). The interactive process starts if the set of candidates is not empty (lines 3–8). If the number of candidates is no more than the number user wants to judge each round, OICM provides the entire candidates to the user (lines 4–5). Otherwise, a greedy method is used to select sample co-location patterns (line 7). After the user submits his/her choices on each sample, OICM receives the user's feedback. Once getting a *non-null* feedback, OICM travels all the co-location patterns in feedback, finds the *hard similarities* of each sample and updates the candidates (line 8). After the interactive process, all the interesting co-location patterns are stored in the result set. Thus, a filter is designed to reduce the number of co-locations in the result set further (line 9). Finally, the interesting co-location patterns will be output to the user (line 10). The computational complexity of each round is about $O(km)$, where m is the number of initial candidates, k is the number of co-location patterns in samples.

6   Experiments

In this section, we evaluate OICM with OICPP [13] using the real and synthetic data sets. All algorithms were implemented in C# and were memory-based algorithms. All the experiments were performed on a Windows 10 system with 4.0 GB memory and 3.30 GHz CPU.

6.1   Experimental Analysis on the Real Data Set

We used points of interest (POI) in Beijing for this experiment. The number of features is 16, the features are selected based on the user's interests and are mainly on the tourism, the number of instances is 90458, the range is 18 km * 18 km, the prevalence threshold is set as low as 0.1 because we want some rare co-location patterns, and the neighbor distance is 100 m.

The ontology created based on the real data is shown in Fig. 3. And the right shows the contents represented by each variable. To describe the ontology, we propose to use the web semantic representation language, OWL-DL [17]. Based on description logics, OWL-DL language permits, along with the ontological structure, to create restriction concepts using necessary and sufficient conditions over other concepts. Also, we use the Protégé [18] software to edit the ontology and validate it. The Jambalaya [19] environment was used for ontology graph exploration. Suppose the user is a traveler for the first time to Beijing, a user may want to search delicious foods around some scenic spots so he/she may consider pattern{A, B, G} interesting while another user may want to find hotels around some scenic spots so he/she may consider pattern {D, G} interesting.

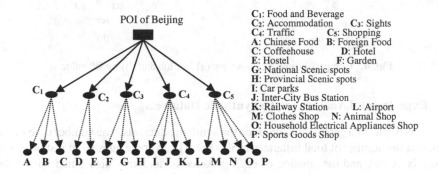

C₁: Food and Beverage
C₂: Accommodation   C₃: Sights
C₄: Traffic   C₅: Shopping
A: Chinese Food B: Foreign Food
C: Coffeehouse   D: Hotel
E: Hostel   F: Garden
G: National Scenic spots
H: Provincial Scenic spots
I: Car parks
J: Inter-City Bus Station
K: Railway Station   L: Airport
M: Clothes Shop   N: Animal Shop
O: Household Electrical Appliances Shop
P: Sports Goods Shop

Fig. 3. An ontology created based on the real dataset

In this experiment, OICM and OICPP take all the closed co-location patterns as the input. Thus, 92 closed co-location patterns are generated from the real dataset. The default k value is 5 and the default sdt value is 0.1.

The real data is mainly used to compare the accuracy of OICM with that of OICPP. In order to identify the accuracy of our algorithm, we apply all the 92 closed co-location patterns for a particular user to choose his/her interesting patterns. In order to better examine the accuracy of OICCP and OICPP, the selection on one co-location pattern and its *hard similarities* must be consistent because the selection of our algorithm is strictly under the semantic environment. In this experiment, the user selects 23 interesting co-location patterns from all the candidates.

The accuracy measure is defined as $Accuracy = \dfrac{P \cap Q}{P \cup Q}$, where P is the interesting co-location patterns selected by the user, Q is the interesting co-location patterns discovered by OICM or OICPP interactively with the user.

As Fig. 4 shows, we compare the accuracy of OICM with OICPP by k and sdt separately within 5 rounds. In Fig. 4(a), the sdt is set as 0.1, it can be concluded that with k value increases, the accuracy of the both algorithms first increases and then has a drop, and this is because a shorter k cannot supply enough information for effective update within the certain rounds while a longer k has more probability to occur the *conflict choice*. In Fig. 4(b), the k is set as 5, it can be discovered that the shorter the sdt is, the higher the accuracy is. We noticed that when $sdt = 0$, the accuracy can reach 100%, because there exists no *conflict choice* at all in both algorithms. With a bigger sdt or a higher k, any candidate co-location pattern c may cause *conflict choice* in OICCP, while in OICM only *isolated pattern* could cause *conflict choice*, which shows that OICM has higher accuracy than OICCP.

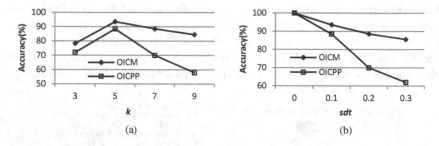

Fig. 4. Comparisons of Accuracy by (a) k value and (b) by sdt value

6.2 Experimental Analysis on the Synthetic Data Sets

We use a simple data generator to produce the synthetic data sets. The number of features is 50, and the number of total instances is 500000. The average number of instances per feature is 10000, and the position of each instance is randomly distributed within the range area 10000 * 10000. With this basic experimental setting, 1022211 prevalent co-locations are generated. The default number of samples k is 15, and the default number of semantic distance threshold sdt is 0.1. In order to get the output efficiently, OICM and OICPP will automatically simulate the user's selection process, and in each round, more than 50% of the samples will be selected interesting in order to get a large output containing plenty of information.

Efficiency of Compression of the Interaction Process. In this part, efficiency of compression of the interaction process is examined. The compression measure is defined as compression $= (n - m)/n$, where m is the selected interesting co-locations by OICM or OICPP, n is the number of co-location patterns in the candidates.

As Fig. 5 shows, the compression of the interactive process of OICM is compared with OICPP by k and sdt separately. From Fig. 5(a), it can be concluded that a bigger k value has a lower compression because when $sdt > 0$, a bigger k can more easily cause the *conflict choice* to find more co-location patterns as interesting ones. From Fig. 5(b) we can see that the compressions also drop with the std value increases because a higher std value can more easily to cause the *conflict choice*. When $sdt = 0$, there is no *conflict choice* and the compression of both algorithms are the same. Because the *conflict choice* can be more easily to cause in OICPP, the design of the interaction process in OICM is better than in OICPP.

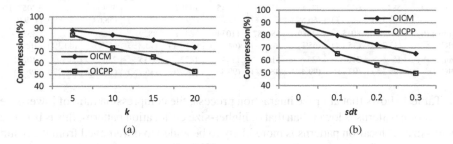

(a)   (b)

Fig. 5. Comparisons of compression (a) by k value (b) by sdt value

Efficiency of the Selecting Strategy. In this part, we mainly examine the efficiency of the times for iterations using our selection method.

As Fig. 6 shows, the efficiency of the selection strategy of OICM is compared with OICPP by k and sdt separately. From Fig. 6, we can see that the number of iterations in both algorithms decreases with k or sdt increases. In lower k or sdt, the times for inter-action process in OICPP are more than that in OICM because the selection strategy in OICPP cannot select samples with the maximal *information*. While in higher k or sdt, the times for interaction process in OICPP are less than that in OICM because the easily caused *conflict choice* makes the number of candidates less than that in OICM. Although the times for interaction in OICPP are less than that in OICM with higher k or sdt, the accuracy of the result in OICPP is much less than that in OICM.

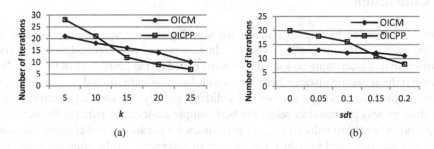

(a)   (b)

Fig. 6. Efficiency of selecting method (a) by k value (b) by sdt value

Efficiency of the Filter. This part mainly examines the compression ratio of the filter. The **size** of a co-location c is the number of features in c. Table 1 gives the number and compression ratio of the candidates, the result set and the output by OICPP and OICM separately per size. The number of sample co-location patterns k is 15 and the semantic distance threshold sdt is 0.1.

Table 1. The number of different co-location patterns per size

	2-size	3-size	4-size	5-size	6-size	7-size	8-size	9-size	10-size	Total
Candidates	869 (0)	12346 (0)	98847 (0)	423427 (0)	314569 (0)	141917 (0)	32457 (0)	907 (0)	142 (0)	1022211 (0)
Result set	587 (33.5%)	6427 (47.9%)	33143 (66.5%)	85139 (79.9%)	58245 (82.5%)	14597 (89.5%)	7233 (77.7%)	379 (58.6%)	21 (85.2%)	205771 (79.9%)
Output (by OICPP)	489 (43.7%)	8015 (35.1%)	34589 (65%)	154987 (63.4%)	110340 (64.9%)	36421 (74.3%)	8829 (73.8%)	498 (45%)	28 (80.3%)	363800 (65.4%)
Output (by OICM)	15 (98.3%)	376 (97%)	635 (99.4%)	1156 (99.7%)	1744 (99.4%)	1823 (98.7%)	4315 (86.7%)	256 (71.8%)	21 (85.2%)	10341 (99%)

Table 1 shows that after the interaction process, the compression ratio of lower-size co-location patterns is lower than that of higher-size co-location patterns, this is because lower-size co-location patterns is more likely to be added to or removed from the result set. For example, there are 5 features A, B, C, D, E having the same *ontology concept(s)*, if a co-location $c = \{A, B, D, E\}$ is marked interesting, all subsets of c will be added to the result set. While the effect of the filter is just the opposite: the compression ratio of lower-size co-location patterns is higer than that of higher-size co-location patterns, this is because a higher-size co-location pattern can represent all its subsets having the same generalization. For example, there are 5 features A, B, C, D, E having the same *ontology concept(s)*, if a co-location $c = \{A, B, C, D, E\}$ is in the result set, all subsets of c will be filtered by the filter. From Table 1, the final compression ratio can reach 99%. The filter has a good compression effect and can effectively reduce the selection burden of the user. With the same experimental settings, OICPP finally generates 363800 interesting co-location patterns. The output of OICM is much less than that of OICPP.

7   Conclusion

In many co-location mining approaches, the algorithm or the measure is pre-designed and the user accepts the results passively. In this paper, we proposed an interactive approach to find interesting co-location patterns based on the preference of the user. We measured the semantic distance between two co-location patterns and gave an efficient update method to reduce the number of candidates rapidly. Besides, an effective selection strategy was proposed to select the best sample co-location patterns to maximize the selection benefit to reduce the number of times for iteration. Furthermore, a reasonable filter was designed to reduce the number of interesting co-location patterns. The high compression of OICM can help the user to do the decision better. The high accuracy of OICM can increase the reliable of decision-making. The experiments showed the effectiveness and efficiency of our algorithm.

Acknowledgements. This work was supported in part by grants (No. 61472346, No.61262069, No. 61662086) from the National Natural Science Foundation of China, by grants (No. 2016FA026, No. 2015FB149, and No. 2015FB114) from the Science Foundation of Yunnan Province and by the Spectrum Sensing and borderlands Security Key Laboratory of Universities in Yunnan (C6165903).

References

1. Huang, Y., Shekhar, S., Xiong, H.: Discovering co-location patterns from spatial data sets: a general approach. IEEE Trans. Knowl. Data Eng. (TKDE) **16**(12), 1472–1485 (2004)
2. Yoo, J.S., Bow, M.: Mining top-k closed co-location patterns. In: IEEE International Conference on Spatial Data Mining and Geographical Knowledge Services, pp. 100–105 (2011)
3. Wang, L., Zhou, L., Lu, J., et al.: An order-clique-based approach for mining maximal co-locations. Inf. Sci. **179**(2009), 3370–3382 (2009)
4. Xin, D., Shen, X., Mei, Q., et al.: Discovering interesting patterns through user's interactive feedback. In: ACM SIGKDD International Conference on Knowledge Discovery and Data Mining, pp. 773–778 (2006)
5. Bao, X., Wang, L., Fang, Y.: OSCRM: a framework of ontology-based spatial co-location rule mining. J. Comput. Res. Dev. **52**(Suppl.), 74–80 (2015)
6. Marinica, C., Guillet, F.: Knowledge-based interactive postmining of association rules using ontologies. IEEE Trans. Knowl. Data Eng. (TKDE) **22**(6), 784–797 (2010)
7. Yoo, J.S., Shekhar, S.: A partial join approach for mining co-location patterns. In: Annual ACM International Workshop on Geographic Information Systems, pp. 241–249 (2004)
8. Yoo, J.S., Shekhar, S., Celik, M.: A join-less approach for co-location pattern mining: a summary of results. In: IEEE International Conference on Data Mining, pp. 813–816 (2005)
9. Wang, L., Bao, Y., Lu, Z.: Efficient discovery of spatial co-location patterns using the iCPI-tree. Open Inf. Syst. J. **3**(2), 69–80 (2009)
10. Gruber, T.R.: A translation approach to portable ontology specifications. Knowl. Acquis. **5**(2), 199–220 (1993)
11. Guarino, N.: Formal ontology in information systems. In: International Conference Formal Ontology in Information Systems, pp. 3–15 (1998)
12. Maedche, A., Stabb, S.: Ontology learning for the semantic web. IEEE Intell. Syst. **16**(2), 72–79 (2001)
13. Bao, X., Wang, L., Chen, H.: Ontology-based interactive post-mining of interesting co-location patterns. In: Li, F., Shim, K., Zheng, K., Liu, G. (eds.) APWeb 2016. LNCS, vol. 9932, pp. 406–409. Springer, Heidelberg (2016). doi:10.1007/978-3-319-45817-5_35
14. Jain, A., Dubes, R.: Algorithms for Clustering Data. Prentice Hall, Upper Saddle River (1988)
15. Yu, H.: SVM selective sampling for ranking with application to data retrieval. In: ACM International Conference on Knowledge Discovery in Databases, pp. 354–363 (2005)
16. Shen, X., Zhai, C.: Active feedback in ad hoc information retrieval. In: Annual International ACM SIGIR Conference on Research and Development in Information Retrieval, pp. 59–66 (2005)
17. Horrocks, I., Patel-Schneider, P.F.: A proposal for an OWL rules language. In: International Conference World Wide Web, pp. 723–731 (2004)
18. Grosso, W.E., Eriksson, H., Fergerson, R.W., Gennari, J.H., Tu, S.W., Musen, M.A.: Knowledge modeling at the millennium. In: Workshop Knowledge Acquisition, Modeling and Management, pp. 16–21 (1999)
19. Storey, M.A., Noy, N.F., Musen, M., Best, C., Fergerson, R., Ernst, N.: Jambalaya: an interactive environment for exploring ontologies. In: International Conference Intelligent User Interfaces, p. 239 (2002)

LFLogging: A Latch-Free Logging Scheme for PCM-Based Big Data Management Systems

Wenqiang Wang[1,2], Peiquan Jin[1,2(✉)], Shouhong Wan[1,3], and Lihua Yue[1,3]

[1] School of Computer Science and Technology,
University of Science and Technology of China, Hefei 230027, China
jpq@ustc.edu.cn
[2] Science and Technology on Electronic Information Control Laboratory,
Chengdu 610036, China
[3] Key Laboratory of Electromagnetic Space Information, Chinese Academy of Sciences,
Hefei 230027, China

Abstract. Big data introduces new challenges to database systems because of its big-volume and big-velocity properties. Specially, the big velocity, i.e., data arrives very fast, requires that database systems have to provide efficient solutions to process continuously-arriving queries. However, traditional disk-based DBMSs have a large overhead in maintaining database consistency. This is mainly due to the logging, locking, and latching mechanisms inside traditional DBMSs. In this paper, we aim to reduce the logging overheads for DBMSs by using new kinds of storage media such as PCM. Particularly, we propose a latch-free logging scheme named LFLogging. It uses PCM for both updating and transaction logging in disk-based DBMSs. Different from the traditional approaches where latches contention and complex logging schemes like WAL, LFLogging provides high performance by reducing latches and explicit logging. We conduct trace-driven experiments on the TPC-C benchmark to measure the performance of our proposal. The results show that LFLogging achieves up to 4~5X improvement in system throughput than existing approaches including WAL and PCMLogging.

Keywords: Phase-change memory · Logging · Concurrency · Latch-free

1 Introduction

Big Data has merged in recent years as a new paradigm providing abundant data and improvement for various researches and applications. It also introduces news challenges to database systems with hardware trends (e.g., multi-cores and large main memory) and application requirements (e.g., millions of transactions per second).

Traditional database systems are driven by the assumption that disk I/O is the main bottleneck. Most database systems use the write-in-place mode to perform update operations. However, such method may cause serious latches contention, especially in multi-core processors system [1].

© Springer International Publishing AG 2017
Z. Bao et al. (Eds.): DASFAA 2017 Workshops, LNCS 10179, pp. 90–102, 2017.
DOI: 10.1007/978-3-319-55705-2_7

For a transaction in update-in-place database systems, two kinds of locks namely a regular lock and a latch need be applied before accessing a record in the database. Regular locks, at the level of transaction, are used to guarantee the independence of transactions in concurrent environments. Latches are used to protect data structures in database pages from being accessed by different transaction threads at the same time. Different from regular locks that can be applied in a record level, latches are usually applied in the granularity of database page.

When a confliction on data accessing happens, locks and latches trigger different operations. A lock confliction may let the transaction itself sleep until another transaction holding the target transaction lock awakes it when it commits/aborts, or may end the transaction and restart it according to different concurrency control protocols. For a latch confliction, transaction threads will enter endless loop until other latches are released, saving extra overheads without context switching. In the order of applying, the regular lock should be acquired first, then to consider latches.

Under the record-level transaction concurrency control, there is such a situation like Fig. 1 where several transactions request read/write operations to different records in the same database page. They are not conflicting in regular locks but they cannot access the slot at the same time because of the latch contention. Even when a transaction maintaining a latch lock has very short run-time, e.g., several microseconds, the number of the latch conflicts increases rapidly in multi-core systems, which has been considered as the main bottleneck for database systems in the near future [2].

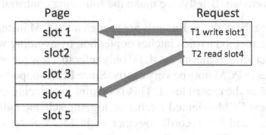

Fig. 1. Latch conflicts

Another issue is about the logging scheme in traditional DBMSs. WAL (write ahead logging) is a common mechanism that has been extensively adopted by database systems to ensure transaction atomicity and durability [3]. Under the WAL logging scheme, transactions need to write both dirty data and log records when performing write operations. Every time a transaction commit/abort or a dirty page replacement occurs, WAL flushes log records to persistent storage instead of flushing dirty data. Over decades, WAL has been performing well due to the large speed gap between DRAM and disk, as well as the volatility of DRAM. However, many researches reveal that the explicit logging scheme occupies significant overheads across database systems [4]. Although various simplifications can be admitted to improve the logging performance and reliability, such as removing synchronous flushing and group commit [5], this issue has not be solved successfully.

Upon above observations, we note that the emerging non-volatile memory (NVM) technologies may bring a new solution to cut the logging overhead [6]. Recently, byte-addressable, non-volatile memory (NVM) technologies, such as phase-change-memory (PCM), spin-transfer torque RAM (STT-RAM) and Memristor provide high performance, high scalability, and low power consumption [7–9]. Among the proposed NVMs, PCM is considered to be one of the most promising ones. PCM's read/write latency is similar with DRAM and its non-volatility and byte-addressability make it an alternative for logging. Table 1 shows a comparison between PCM and several current storage technologies in density, read/write latency, and endurance.

Table 1. Comparison of different storage technologies [10]

Parameter	DRAM	NAND flush	Hard disk	PCM
Density	1X	4X	N/A	2-4X
Read latency (granularity)	20–50 ns (64B)	~25ηs (4 KB)	~5 ms (512B)	~50 ns (64B)
Write latency (granularity)	20–50 ns (64B)	~500ηs (4 KB)	~5 ms (512B)	~1ηs (64B)
Endurance	N/A	10^4–10^5	∞	10^6–10^8

In this paper, we propose a novel method by adapting the write-out-place scheme and integrating log into the cached updates in PCM so as to eliminate latch conflicts and complex logging overhead. Briefly, we make the following contributions in this paper:

- We use PCM to cache updates. Records are written into PCM instead of into DRAM. Such an updating method avoids latches contention by avoiding write-in-place.
- We note that most write requests in OLTP only refer to a few records in pages. Thus, caching full pages in PCM may be very costly. Since PCM supports byte addressable, we manage PCM in the record level. This is helpful in reducing extra I/Os.
- We propose a new PCM-oriented latch-free logging scheme called LFLogging. It maintains updates and log records together in PCM so as to save extra logging management and simplify transaction process mode.
- We develop a trace-driven simulator to evaluate the performance of LFLogging on the TPC-C benchmark and compare our proposal with existing approaches like WAL and PCMLogging. The results suggest the efficiency of our proposal.

The rest of the paper is organized as follows. Section 2 discusses the related work. Section 3 introduces overall design of LFLogging and discusses the operation in details. Section 4 presents the results of our experiment and confirms the performance improvement. Finally, Sect. 5 concludes this paper and discusses future directions.

2   Background and Related Work

PCM is a kind of non-volatile semiconductor memories and is a promising candidate for the storage and the main memory [8, 9]. The basic unit of PCM, called a PCM cell, uses the phase change material to store a bit by switching between an amorphous state and a crystalline state with electrical pulses. Writing a PCM cell includes two operations: SET, which requires wild pulse and low current to crystallize the phase change material, and RESET, which is controlled by high-power pulse to make the material amorphous. Reading a PCM cell is done by sensing the resistivity of phase change material, which requires very low power. As such, PCM, being non-volatile and bit-addressable, also bears the advantages of having low idle power and low read latency. However, the long SET operation increases the write latency, and the PCM cell can only sustain a limited number of writes, between 106 and 108 times in general [16]. Therefore, frequently writing to PCM will not only deteriorate the I/O performance but also shorten the lifetime of PCM.

Gihwan et al. proposed SQLite/PPL (per page logging) scheme to optimize SQLite with PCM [18]. SQLite/PPL can avoid the costly journaling of SQLite without giving up atomicity and durability of transactions. In addition, it simplifies the commit and abort procedure since it prevents any uncommitted changes from propagated to data pages in flash memory. Different from our logging scheme, SQLite/PPL focuses on reducing write amplification to flush. On the other hand, SQLite/PPL is designed for SQLite where the concurrency level is low and there is at most one writer transaction is allowed to access the same database at any moment. Arulraj et al. proposed WBL (write-behind logging) to improve the runtime and recovery performance of DBMS [19]. WBL only logs what parts of the database have changed rather than how it is changed. It reduces the amount of data in logs and supports nearly instantaneously recovery from system failures. However, it is still limited by traditional logging scheme and explicit logging and concurrency problems cannot be solved successfully.

The work much close to our study is PCMLogging [11] which uses PCM to optimize logging and transaction recovery performance. Under the hybrid memory architecture where the main memory is consisted by DRAM and PCM, the authors considered PCM as dirty data caching. Like shadow paging technology, PCMLogging uses a dirty page table to manage different version of the page by adapting an out-of-place update scheme in PCM. Every time transaction commits or page replacements occur, PCMLogging flushes dirty page to PCM rather than disk. The low write latency of PCM reduces flushing time and the non-volatility ensures transaction can be recovered from failure. PCMLogging exploits non-volatility and low read/write latency of PCM, almost removes explicit logging overhead.

However, PCMLogging has following drawbacks. First, it does not change the way of updating in DRAM, latch conflicts are still serious when transactions execute in high levels of concurrency. Second, with the shrinking the access granularity from page to record on PCM, the size of auxiliary data structures in DARM, such as Mapping Table and Dirty Page Table, grows increasingly. This brings extra space costs for buffering capacity of main memory and time cost for searching on these structures.

3 Design and Implementation

In this section, we present the detailed design and implementation of our latch-free logging scheme called LFLogging. Our logging scheme is based on the record-level of concurrency control. It is meaningless at page-level since there is no latch contention at coarse-gained granularity. We consider the storage architecture using hybrid main memory, as shown in Fig. 2, due to the restriction of limited endurance and price which makes PCM still unsuitable to replace DRAM entirely.

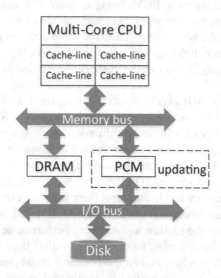

Fig. 2. The hybrid memory system architecture

3.1 Overview

The latch contention may occur is based on the assumption that a transaction performs update operation in original page. We can eliminate such contention by removing update-in-place in page. The basic idea of LFLogging is to put all updates in PCM and integrate updates with logs. The whole system is almost latch-free and explicit logging is removed. Based on the observations that most write requests in OLTP are small, we use fine-grained granularity to manage PCM space to reduce the write traffic on PCM. To support high access performance, we maintain the following data structures in PCM/DRAM.

- FreeSlotBitmap. As we use record-level granularity to manage PCM, we can split PCM space in units. A record can occupy one or more units. So we use FreeSlot-Bitmap, a bitmap maintained in PCM, to track free space in PCM, providing high performance to malloc and free PCM space.
- Mapping Table. A mapping table, maintained in DRAM, is used to map LPID (logical page IDs) to physical PCM addresses. At the beginning, the LPID is mapped to original physical page in DRAM, with updating in PCM, we use CAS instruction [17] to compare and swap the pointer with the physical address of the updates. We can access record list quickly by consulting Mapping Table.

- Record List. We organize the whole updates in record list, all updates belongs to the same page are on the same list mapped with LPID. Such structure supports efficient write and consolidation operation because both two operations require finding all records belongs to the same page.

Mapping Table and Record List are demonstrated in Fig. 3. When a transaction requests write operation, we only change pointer in mapping table with CAS which ensures the atomicity after mallocing space and writing record complete. The whole process is almost latch-free, and database systems can serve the read/write request in the same page for any transactions as long as there is no regular lock conflict.

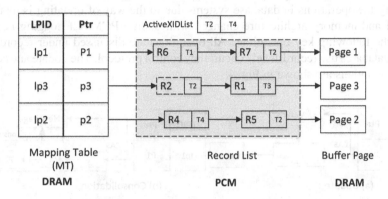

Fig. 3. The mapping table and record list

When searching a record, traversing the record list in PCM is unavoidable (if present). The search stops at the first occurrence of the search key in list. If the record containing the key represents an insert or update, the search succeeds and returns the record. If the record represents a delete, the search fails. If the record list does not contain the target record, the search would continue on original physical page in DRAM. Occasionally, we consolidate pages with records wrote by committed transaction. Consolidations reduce memory space for PCM and improve search improvement.

To support transaction recovery, some data structures, contains transaction information like XID (transaction id) of in-progress transaction and PID (page id) of dirty page, should be maintained in PCM in case of system failures. These structures are followed.

- ActiveTxList. This list is maintained in PCM to record XID of all in-process transaction that performed write operation in PCM. All records in PCM should record the XID of transaction that wrote it. The XID of the transaction must be written to ActiveTxList before writing the record to PCM to ensure the atomicity and it wouldn't be removed until the transaction commit/abort.
- Transaction Table (TT). This table is used to track all records a transaction updated during its in-progress. It, contains all the page ids that a transaction updates, is used to find all update records quickly when a transaction is aborted or recovery. These tracks will not be removed until the transaction commit/abort.

Totally speaking, LFLogging eliminates latch conflict and reduces explicit logging overhead with fine-grained management for PCM. Records are organized in record list in PCM and it is suitable for extending page level to record level without much space cost and searching cost.

3.2 Operations

Here, we will describe the LFLogging operation in details. As we write all data in PCM, durability is ensured by the nature of PCM once write operation achieved. We do not need any additional flush operation when a transaction is to commit. However, we need to modify the operations in database systems due to the way of updating is totally be changed and memory architecture changed with adding PCM. LFLogging needs to handle the following key operations. All operations are discussed under record-level and the databases use record-level concurrency control protocol. The update and consolidation operations are shown in Fig. 4.

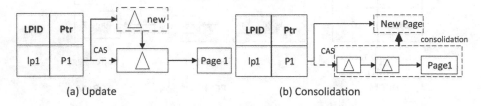

(a) Update   (b) Consolidation

Fig. 4. Update and consolidation operation

Update. We never update record in place (i.e., modify its memory contents). Rather all updates are performed by writing a record to PCM, suitable for delete and insert operation with different record mark. Such implementation allows us to incrementally update page state in a latch-free manner. We first write a new record that (physically) points to the latest version record of the page, and then we use the atomic compare and swap (CAS) instruction to replace the current p1 with the address of new record.

Read. A little more complex than traditional database, we need first search the record list in PCM to get the latest version of the record, if not hit, and then we will search DRAM or HDD to find the original version. It may cause read amplification but it can be compensated by PCM hit and appropriate consolidation operation.

Consolidation. We cache the small writes in record list in PCM to reduce unnecessary write traffic to HDD. However, search performance eventually degrades if record list grows too long. To combat this, as shown in Fig. 5, we occasionally do group update by creating a new page containing all record list based on original page, then use CAS to compare and swap pointer in Mapping Table with physical address of the new page. The consolidation is triggered if a thread, during a page search, notices the length of a record list has exceeded a system threshold. The thread performs consolidation after finishing its read operation.

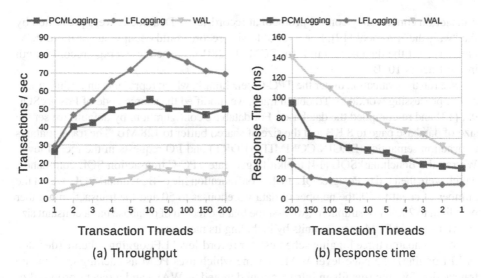

Fig. 5. The overall performance

Commit. As all updates have been persistent once write operation finished, we just need to clean the Transaction Table and ActiveXIDList. A subtle issue is hidden here is we need to make sure all writes including cacheline in CPU flush to PCM, otherwise we cannot recovery from failure without explicit log.

Abort. When a transaction is aborted, we need to invalidate the records wrote by the transaction. First, we get all dirty page ids by consulting the Transaction Table. Then we traverse the record list to delete the record wrote by the transaction. After that, we clean the Transaction Table and ActiveXIDList.

Recovery. A recovery process is invoked when the system restarts after a failure. First, we need rebuilt Mapping Table by scanning PCM. Then we need traverse record lists by consulting the Transaction Table, all records whose XID is still in ActiveXIDList would be discard.

Compared with traditional memory architecture, we add a PCM between DRAM and HDD, so a de-staging process is required when PCM is close to full or bandwidth of HDD is idle. This operation almost makes no influence to transaction response time as it is asynchronous. We simply adopt high-low watermark algorithm. Every time we flush data, we need perform consolidation in DRAM, and then flush it to HDD.

4 Experiments

4.1 Experimental Setup

We developed a trace-driven simulator to test our logging scheme against with WAL and PCMLogging. We adopt a transaction processing mode on top of simulated DRAM. In this transaction processing model, we employed the strict two-phase locking as transaction concurrency control protocol at record-level. Deadlocks were prevented by the "wait-die" protocol [12]. For a variable-size record could occupy one or more PCM units. We set the data access unit for PCM at 128B because the average record length in our trace is 108B.

We ran the evaluation under the TPC-C benchmark, which represents an on-line transaction processing workload. To obtain trace, we modified the source code of PostgreSQL 9.2 [13] and re-compiled the databases. For database configuration, by default, we set the size of database page to 8 KB and the size of shared buffer to 128 MB. The trace contains transaction semantics (BEGIN/COMMIT/ABORT) and I/O requests in the record level. We ran the BenchmarkSQL [14] toolkit to generate TPC-C transaction SQL commands, which were then fed to PostgreSQL. In the configuration file of BenchmarkSQL, we set the number of client to 50, the number of data warehouses to 50 and the transaction number per client to 2000. For simplicity, we assume that the record of database has a constant size and fits into one PCM or more units by including its metadata.

We compared three logging schemes: our record-level LFLogging scheme (denoted as LFLogging), the page-level WAL scheme which uses PCM as caching of data and transaction logging (details in introduction, denoted as WAL) and a recent proposal of record-level PCMLogging scheme (denote as PCMLogging). For WAL, we simply used PCM to run as disk like the way in traditional database with lower I/O latency. For DRAM-based simulation experiments, we configured to read/write latency the same as the IBM DNES-309170 hard disk without write cache. For the de-staging process, we adopt high-low watermark [15] with simply setting low threshold to 60% and high threshold to 90%. Above three schemes, the data archived in PCM can be flushed out to external disks asynchronously.

All of the simulation experiments are conducted on a desktop computer running Ubuntu 14.04 with an Intel i5-4590 CPU. By default, we simulate a database of 5.7 GB. The size of PCM is set to 128 MB which is the same as DRAM. We run the simulator for 10 min to warm up the system. The system performance is measured by the same number of transactions (i.e., 100000 transactions).

Table 2. The default parameter settings

Parameter	Default setting
DRAM read/write latency	20 ns/20 ns
HDD read/write latency	8.05 ms/8.20 ms
PCM read/write latency	20 ns/1000 ns
Logical page size	8 KB
PCM unit size	128B
TPC-C database size	5.7 GB
TPC-C client number/warehouse number	50/50
TPC-C transaction number per client	2000
Total Transaction Number	100000
Main memory (DRAM) size	128 MB
PCM size	128 MB

To get the performance of database under different concurrent workload, we set the number of transaction threads by varying 1 to 200. The number of transaction threads here represents the maximum concurrency of the database. The default parameter settings of our experiment are listed in Table 2 in details.

4.2 Experimental Results

We report the overall performance comparison of LFLogging with WAL and PCMLogging. We plot the transaction throughput and response time of the three schemes in Fig. 5 respectively, by varying the number of transaction threads from 1 to 200. We make the following observations of the experiment results. First, all three schemes have the worst performance at the number of thread at 1. This is mainly because there is no concurrency in database. Second, with the increasing of the thread number, the throughput also grows. This can be explained that concurrency level increased with thread number. The throughput may have slightly down because the overhead of lock contention including regular lock and latch contention neutralized the benefits of the increasing concurrency level. Third, our LFLogging has the best performance among all the three schemes. This is because that LFLogging is latch-free and improves the overall concurrency of database. The throughput improvement of LFLogging over PCMLogging increases about 47% on average and increases by 4~5X compared with WAL. A similar trend is observed for response time.

We generate two different trace under the same setting of PostgreSQL and BenchmarkSQL, then fed them to simulator. We plot the performance of our logging scheme, as shown in Fig. 6, by varying record list length thresholds which represents the maximum length of a record list before we trigger page consolidation operation. For small record lengths transaction thread need do consolidation operation frequently which reduces our benefit of write-out-place mode, and the worst, degrade to write-in-place with setting thresholds at 1. For big lengths, we need increase read I/O to PCM as we need to search the latest version of record. This read amplification would cut the overall system performance.

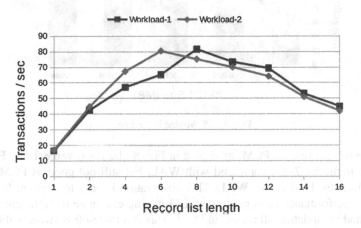

Fig. 6. Threshold of record list length

Thus, this is a trade-off between searching cost and consolidation cost. Under our trace, we obtain the best transaction throughput at the threshold of 6 for workload-1, but 8 for workload-2, which means the best threshold may be affected by different workloads.

We observe that repeat writes are rare under the benchmark of TPC-C, so the length of record list will not grow too fast and we get the best performance with a relatively big threshold. The threshold may have great difference under other benchmark like TPC-A.

We evaluate the impact of the size of PCM for three logging scheme by varying the PCM size from 32 MB to 256 MB where the size of DRAM is set at 256 MB. We set transaction threads to 50. As shown in Fig. 7, all three schemes make improvement under larger PCM. We conduct following observation. First, when the size of PCM was set at 32 MB, PCMLogging got the highest transaction throughput; this is mainly because the small PCM cannot fully exploit our scheme because of frequent de-staging operation. Our scheme, updates all writes on PCM, would exhaust small PCM space quickly, and this nearly turns asynchronous de-staging operation to synchronous ones, affecting system implementation seriously. Second, the improvement of LFLogging over WAL and PCMLogging is the highest with the increasing size of PCM. LFLogging outperforms WAL by nearly 5X in terms of throughput at 256 MB of PCM and by 47.6% compared with PCMLogging. This can be explained as followed, LFLogging works best among them as database concurrency level increased with big main memory, the main bottleneck of the system becomes restriction of concurrency from long latency of disk I/O with high hit rate. LFLogging is latch free and take the best advantage of such system architecture.

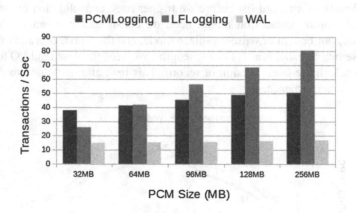

Fig. 7. PCM size influence

For the write traffic on PCM, as shown in Fig. 8, the improvement of LFLogging reduce the traffic by 7~8X compared with WAL, but still not good as PCMLogging which reduces by 11~12X for WAL. This observation is easy to explain. In order to obtain high performance of concurrency, LFLogging eliminate the efficient write-in-place method by updating all record in PCM. This is a trade-off between validation of writes and the write traffic on PCM. Compared with WAL, LFLogging and

PCMLogging that both use fine-grained granularity to manage PCM while WAL still adopt page-lever management, saving much unnecessary extra page writes on PCM as most writes in OLTP is small. But undeniable, write traffic on PCM is a big problem as it may decrease its lifetime.

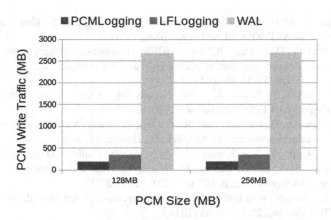

Fig. 8. Write traffic on PCM

5   Conclusions and Future Work

In this paper, we present a latch-free logging scheme for traditional database systems using PCM as updating device. It is clear that computer will be dominated by new system architecture (i.e. multi-core and new storage technologies) and high external require-ment. Under such background, traditional disk-based database systems will not be restricted by low I/O latency, and logging and locking will be the main bottleneck. Recently, the breakthrough of NVM brings opportunity to solve the problem together. Our proposed LFLogging gave outstanding performance in evaluation and gave us much confidence in our design, and we will continue to solve these problems with our design.

LFLogging still remains some problems. The first problem is how to set the threshold of record list to get the best performance under various workloads. We just used an empirical setting in our experiment and it is not suitable under other benchmark like TPC-A. We will continue to study and propose an adaptive solution to set the threshold. Second, our scheme has relatively high write traffic. This may reduce the lifetime of PCM. We will focus on this issue and find better solutions in future work.

Acknowledgements. This work is partially supported by the National Science Foundation of China under the grant numbers 61472376 and 61672479, the Fundamental Research Funds for the Central Universities, and a fund from the Science and Technology on Electronic Information Control Laboratory.

References

1. Larson, P.A., Blanas, S., Diaconu, C., Freedman, C., Patel, J.M., Zwilling, M.: High-performace concurrency control mechanisms for main-memory databases. PVLDB 5(4), 298–309 (2011)
2. Harizopoulos, S., Abadi, D., Madden, S., et al.: OLTP through the looking glass, and what we found there. In: SIGMOD, pp. 981–992 (2008)
3. Mohan, C., Haderle, D., Lindsay, B., et al.: ARIES: a transaction recovery method supporting fine-granularity locking and partial rollbacks using write-ahead logging. ACM Trans. Database Syst. (TODS) 17(1), 94–162 (1992)
4. Johnson, R., Pandis, I., Stoica, R., Athanassoulis, M., Ailamaki, A.: Aether: a scalable approach to logging. PVLDB 3(1), 681–692 (2010)
5. Helland, P., Sammer, H., Lyon, J., Carr, R., Garrett, P., Reuter, A.: Group commit timers and high volume transaction systems. In: Gawlick, D., Haynie, M., Reuter, A. (eds.) HPTS 1987. LNCS, vol. 359, pp. 301–329. Springer, Heidelberg (1989). doi:10.1007/3-540-51085-0_52
6. Fang, R., Hsiao, H.-I., He. B., Mohan, C., Wang, Y.: High-performance database logging using storage-class memory. In: ICDE, pp. 1221–1231 (2011)
7. Kawahara, T.: Scalable spin-transger torque Ram technology for normally-off computing. IEEE Des. Test Comput. 28(1), 52–63 (2011)
8. Wu, Z., Jin, P., Yue, L.: Efficient space management and wear leveling for PCM-based storage systems. In: ICA3PP, pp. 784–798 (2015)
9. Chen, K., Jin, P., Yue, L.: Efficient buffer management for PCM-enhanced hybrid memory architecture. In: APWeb, pp. 29–40 (2015)
10. Chen, S., Gibbons, P.B., Nath, S.: Rethinking database algorithms for phase-change memory. In: CIDR, pp. 21–31 (2011)
11. Gao, S., Xu, J., Härder, T., He, B., Choi, B., Hu, H.: PCMLogging: optimizing transaction logging and recovery performance with PCM. IEEE Trans. Knowl. Data Eng. 27(12), 3332–3346 (2015)
12. Soisalon-Soininen, E., Ylönen, T.: Partial strictness in two-phase locking. In: Gottlob, G., Vardi, M.Y. (eds.) ICDT 1995. LNCS, vol. 893, pp. 139–147. Springer, Heidelberg (1995). doi:10.1007/3-540-58907-4_12
13. Postgresql: Open source object-relational database system. http://www.postgresql.org/
14. BenchmarkSQL. http://www.sourceforge.net/projects/benchmarksql
15. Nam, Y.J., Park, C.: An adaptive high-low water mark destage algorithm for cached RAID5. In: PRDC, pp. 177–184 (2002)
16. Lee, B., Zhou, P., Yang, J., et al.: Phase-change technology and the future of main memory. IEEE Micro 30(1), 143 (2010)
17. Levandoski, J.J., Sengupta, S.: The BW-tree: a latch-free b-tree for log-structured flash storage. IEEE Data Eng. Bull. 36(2), 56–62 (2013)
18. Oh, G., Kim, S., Lee, S., et al.: Sqlite optimization with phase change memory for mobile applications. PVLDB 8(12), 1454–1465 (2015)
19. Arulraj, J., Perron, M., Pavlo, A.: Write-behind logging. PVLDB 10(4), 337–348 (2016)

RTMatch: Real-Time Location Prediction Based on Trajectory Pattern Matching

Dong Zhenjiang[1,2], Deng Jia[3], Jiang Xiaohui[3], and Wang Yongli[3(✉)]

[1] Department of Computer Science and Engineering,
Shanghai Jiao Tong University, Shanghai, China
dong.zhenjiang@zte.com.cn
[2] Cloud and IT Institute, ZTE Corporation, Nanjing, China
[3] Department of Computer Science and Engineering,
Nanjing University of Science and Technology, Nanjing 210094, China
yongliwang@njust.edu.cn

Abstract. Due to the universality of mobile devices, such as GPS and the devices of location-based services, there is a growing number of mobile trajectory data. This provides the opportunities for innovation of analyzing trajectory and extracting information. We proposed a new method to predict next location of moving object - RTMatch. The main idea of the method is to store and query the trajectory frequency pattern (named T-pattern) of moving objects by designing a data structure - RTPT (Real Time Pattern Tree) and HT (Hash Table) contains the spatio-temporal information, and then find a best matched path on the tree (the best T-pattern matches the trajectory to be predicted). The RTMatch can provide real-time analysis during the on the fly processing. Experiments on the actual data prove that our method is more accuracy and efficiency than some existing methods.

Keywords: Trajectory pattern · Spatio-temporal data mining · Real time · Location prediction

1 Introduction

In recent years, the study of spatio-temporal trajectory data is getting more and more attention in the academic domain and the industry domain. There are so many mobile devices with wireless communication, such as blue-tooth, Wi-Fi, GPRS, etc., which can provide the location through the Global Position System (GPS). The moving people or cars will leave trajectory data through their mobile devices. Researchers can observe the data after collecting them. And devices equipped with GPS can record trajectory data, which contains time-stamp, latitude and longitude, named spatio-temporal trajectory data.

In this paper, we aim at predicting the future location of moving objects by the analysis of spatio-temporal trajectory data. Location information of a moving object has wide applications, such as mobile e-commerce (location-based services), human activities POI finding, traffic coordination and management (traffic congestion forecasting) and so on. On one hand, due to the uncertainty of conditions and the instability

Z. Bao et al. (Eds.): DASFAA 2017 Workshops, LNCS 10179, pp. 103–117, 2017.
DOI: 10.1007/978-3-319-55705-2_8

of Global Position System, we can't know the position information completely. On the other hand, several application scenarios, such as traffic management, require forward looking to achieve better management results. So it is necessary to predict the future position though the trajectory pattern mining method.

Some of the developed methods predict next location directly from the history trajectory of a single object [1]. While some using the trajectory of all the objects in a certain region based on a hypothesis that individuals follow a similar path [2, 8]. However, these existing methods have various limitations, such as low precision or over fitting.

First of all, in this paper, we focus on spatio-temporal group pattern mining of all objects in a certain region, store it using real-time trajectory pattern tree, then we predict the future location of moving objects. Secondly, because of the uncertainty of personal wishes, traffic conditions, road conditions, etc., there is no absolute similar or identical in individual paths, it will reduce the accuracy of prediction to analysis from a static angle, we propose a dynamic update store method.

Figure 1 shows two moving object trajectory frequent patterns (T - pattern, see Definition 4) named P1, P2, and a moving object trajectory named B to be predicted, which contains six time-stamped trajectory points, we want to predict the next location of B by calculating matching degree between trajectory to be predicted and trajectory pattern. If we analyze it statically, P2 matches B mostly, the prediction results will also be closest to the P2. However, if we analyze it dynamically in the time-stamp increasing order, and update the new mining frequent patterns to the storage data structure, we get the trajectory pattern named RTP. As a result, P1 matches B mostly in the former three time-stamp and P2 in the latter three time-stamp, the prediction results

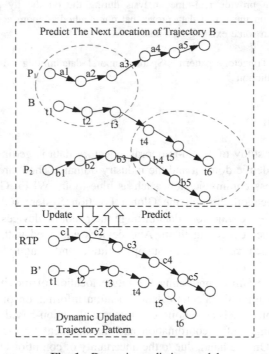

Fig. 1. Dynamic prediction model

named B' is closer to B. The example above has made it clear that real-time and dynamic analysis of trajectory pattern benefits to the improvement of the prediction accuracy providing a rich granular and more accurate pattern.

We propose a new location prediction method, named RTMatch, contains off-line training module and on-line prediction module. In the former module, we build a storage structure including RTPT (Real Time Pattern Tree) and HT (Hash Table), which not only pay attention to the historical moving objects trajectory, but also provide functions of analyzing trajectory pattern dynamically as well as updating the storage data structure. In the latter one, we propose a calculation method to compute path-matching degree, and then find the best path. Showed in Sect. 3.

The rest of our paper was as follows: we review the related work in Sect. 2. We propose a spatio-temporal trajectory pattern mining method in Sect. 3, and we describe the details to build real-time store structure of trajectory pattern named RTPR and HT in Sect. 4. In Sect. 5, we describe the location prediction method. Finally, we present the evaluation results in Sect. 6 and discuss our conclusion in Sect. 7.

2 Related Work

In recent years, there are some researches about position prediction of moving trajectory, which mainly can be divided into two types according to the style of prediction: historical trajectory modeling and forecasting based on Markov models [4, 5, 13, 14], and trajectory prediction based on historical trajectory frequent pattern mining [2]. Figueiredo et al. [7] use unstable, transient, temporal diversified trajectory sequence to build a nonparametric model and make a prediction on geographic location, product recommendations, etc. Feng et al. [9] consider the location correlation of the user's interests and travel sequence. Some researchers predict the location of users through user interest point mining and user travel sequence detecting by the HITS-based model [10]. Chen et al. [14] use the prefix tree to express historical movement patterns, and then find the part, which match the current trajectory mostly by iterating through the tree. At last, they calculate matching degree, and return the trajectory path with the best matching value as the prediction result.

Apriori and PrefixSpan are two classic algorithms about trajectory frequent pattern mining. There are some related researches about use them to predict location. Morzy et al. use improved Apriori algorithm [11] and improved PrefixSpan algorithm [12] to discover historic frequent items from trajectory, and generate the prediction rules. All matching method mentioned above are based on supporting degree.

Monreale et al. [2] put forward WhereNext method to predict the next location of the moving objects according to prediction method, which extracts motor pattern behavior from the historical trajectory. The motor pattern behavior covers three different movement patterns, including location, travel time and frequency of user access.

The existing work only consider historical trajectory information and ignore the real-time information, as a result, the prediction effect is not so good. In this paper, we put forward a new method named RTMatch to predict the future location of the moving object using the storage structure RTPT and HT, which can be updafunction has different usageted dynamically and can provide dynamic analysis of trajectory pattern according to historical information as well as real-time information. At different supporting degree, compared to the classic method WhereNext [2], the average accuracy of RTMatch increases about 15% and the average coverage of RTMatch increases about 50%. Due to the real-time updating property of RTPT, the matching probability is increased, which improves the speed to find the optimum matching path, and reduces the running time to predict the location. At the same time, it decreases the information storing in the process of matching degree calculation, and reduces the space occupation.

3   Spatio-Temporal Trajectory Frequent Pattern Mining

The trajectory of moving object consists of the position sequence with time-stamp. We can obtain trajectory of all the objects in a particular area through wireless mobile devices, such as GPS or Big Dipper. And the location is denoted with cartesian coordinates.

The framework of RTMatch method to predict next location we proposed is described in Fig. 2.

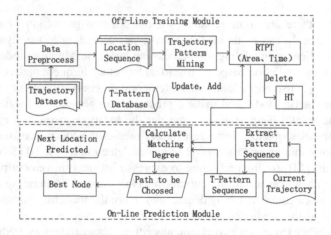

Fig. 2. The framework of RTMatch

To make it easy, RTMatch's results will be achieved through the steps below. Firstly, we choose the trajectories in specific area at specific time through querying spatio-temporal information. Secondly, we use the best working trajectory pattern mining algorithm to get T-pattern set. And then, we build storage structure as well as update it in real time. Finally, we propose the prediction algorithm to predict the next location of moving object.

3.1   Spatio-Temporal Trajectory Sequence

The location sequence with time-stamp can be called spatio-temporal trajectory sequence, and there are some related definitions below.

Definition 1. Spatio-temporal Trajectory Points Sequence: It is a triple sequence that $T = \{I_0, \ldots, I_i, \ldots, I_n\}$, in which $I_i = (x_i, y_i, t_i)$ indicates two-dimensional location coordinates, $t_i(i = 0\ldots n)$ is the timestamp of the point. Each (x_i, y_i, t_i) indicates that at time t_i, the moving object is at the location (x_i, y_i).

Definition 2. Spatio-temporal Trajectory Edges Sequence: The Spatio-temporal trajectory edges sequence E is transform among trajectory points, which is also a triples sequence. $E = \{s_1, \ldots, s_i, \ldots, s_n\}$, in which $s_i = \langle(x_{i-1}, y_{i-1}, t_{i-1}), (x_i, y_i, t_i)\rangle$ is a edge of the space, (x_i, y_i, t_i) indicates two-dimensional location coordinates, $t_i(i = 0\ldots n)$ is the timestamp of the point.

3.2 Spatio-Temporal Trajectory Frequent Pattern Mining

The spatio-temporal trajectory frequent pattern means the sequence that found from spatio-temporal trajectory set. We can use it to finish the analysis and prediction towards moving object.

The research result can be applied to social networking recommend [13], tourism [9] and city traffic management [6], etc.

Definition 3. Trajectory Frequent Pattern: In the database, each trajectory is a set of tracing points. The emergence frequency of a certain pattern refers to the number of trajectory including that pattern. The frequency of a pattern means the number of trajectory containing that pattern. If the ratio of frequency of a pattern and total trajectory number is greater than the specified minimum support specified in advance, then the pattern regarded as the trajectory frequent pattern.

In this paper we improved the T-Pattern mining algorithm proposed by Giannotti et al. [3]. The algorithm aims at mining frequent sequence that made up of ROI (Region of Interest), which extracts frequent sequence with time annotation in dense space region. The each extracted T-pattern is a brief description of frequent action on both time and space.

Definition 4. T-pattern: A two-tuples of Sequence and annotation (S, A), in which, $S = \langle R_0, R_1, \ldots R_n\rangle$, is region sequences, $A = a_1, \ldots a_n \in R_+^k$, is time annotation towards S, (a_i is arithmetic number, indicating time interval from region R_0 to region R_1). A T-pattern can also be expressed directly as $(S, A) = R_0 \xrightarrow{a_1} R_1 \xrightarrow{a_2} \ldots \xrightarrow{a_n} R_n$.

We extra time interval between R_0 and R_1 as time annotation. As a result, time annotation set of sequence is time interval set in fact. In this paper, we put forward a new trajectory pattern mining model, described in Fig. 3, based on Giannotti et al. [1]

There are some rules Inferred from T-pattern $A \xrightarrow{a_1} B \xrightarrow{a_2} C \xrightarrow{a_3} D \xrightarrow{a_4} E$:

$$A \xrightarrow{a_1} B \xrightarrow{a_2} C \xrightarrow{a_3} D \xrightarrow{a_4} E \tag{1}$$

$$A \xrightarrow{a_1} B \xrightarrow{a_2} C \xrightarrow{a_3} D \xrightarrow{a_4} E \tag{2}$$

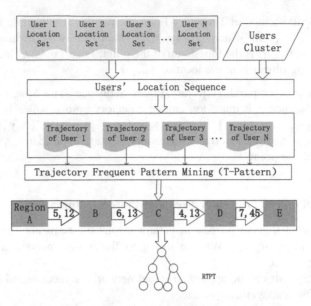

Fig. 3. Flow path of trajectory pattern mining

According to the first rule, the next location of $(x_0, y_0, t_0) \rightarrow (x_1, y_1, t_1) \rightarrow (x_2, y_2, t_2)$ is region D. But as for the second one, the next location of $(x_0, y_0, t_0) \rightarrow (x_1, y_1, t_1)$ is region C.

In order to avoid some unnecessary rules and make the prediction more efficient, we propose Real Time Pattern Tree (for short, RTPT) to store the real-time patterns and Hash Table (for short, HT) to store the updated pattern nodes.

4   Real-Time Storage Structure of T-PATTERN

4.1   The Introduction of RTPT

RTPT: A structure tree made up of a triple. RTPT $= (N, E, Root)$, in which, N is a set of nodes, E is a set of edges with time interval mark, $Root \in N$ is a virtual root node.

Each node $v \in N$, except the *Root*, contains {Region, Support, Child Node}:

Region: Indicate a region of T-pattern;
Support: The support degree of T-pattern;
Child Node: A child node list of node v.

Each edge $e \in E$, connects parent node and child node, whose interval form is $[time_{min}, time_{max}]$. What's more, the edge connecting root and its child is an empty sign, whose interval is $\emptyset$.

4.2   RTPT Update

In order to improve the efficiency and accuracy of prediction, we update RTPT instantly, and store the latest and most frequent T-pattern. The following rules are put forward.

Rule 1 RTPT Insert: At each input T-pattern, expressed as *tp*, we look for paths in the RTPT consistent with its longest prefix, and then add a branch for the rest of elements. If the pattern tree is prefix of *tp*, then *tp* will be attached to the tree.

The creation process of RTPT is to start from the Root node and to perform the rule 1.

Path means belt edge sequence starting from the root node. For example, the path of node *c* on RTPT, $P(c, RTPT)$ can be expressed as:

$$P(c, RTPT) = (Root, a, \emptyset), (a, b, interval_1), (b, c, interval_2).$$

In which, $interval_1$ is the time interval from region of node *a* to region of node *b*.

As shown in Fig. 4, when we update T-pattern $\langle O, E \rangle \langle (12, 82), B \rangle, \langle (3, 12), C \rangle$, we can get the node *E* in the second layer.

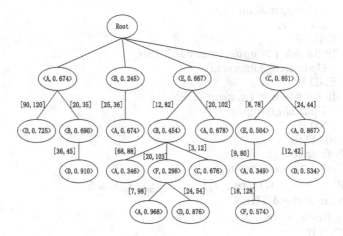

Fig. 4. An example of RTPT

Rule 2 RTPT Update: We simplify T-pattern *tp* to a sequence pair $\langle r, i \rangle$, *r* indicates region, *i* indicates time interval. We set threshold δ as minimum support for pattern, the forerunner of $\langle r_1, i_1 \rangle$ in *tp* is $\langle r_0, i_0 \rangle$. As for $node_{r_1}$ in RTPT whose region is r_1, the parent node is $node_P$. After recursive traversal of $node'_P$s child node, if there exists a path $P = (r_1, r_2, i_1), (r_2, r_3, i_2), \ldots, (r_{n-1}, r_n, i_{n-1}) \supset$

$\langle r_1, i_1 \rangle, \langle r_2, i_2 \rangle, \ldots, \langle r_m, i_m \rangle$, as well as $supp_p < \delta$, then update the region of $node_P$ as r_0, time interval as i_0.

Rule 3 RTPT Update: For each pair of *region* and *time* $\langle r, i \rangle$ in *tp*, let's set $r.supp = n$. Then get the child $\langle r', i' \rangle$. If $\langle i' < i \rangle$ then update to *i*, if $\langle n' > n \rangle$ then update to *n*.

Rule 4 RTPT Delete: As one of the situation of Rule 2. In Rule 2, when $\langle r_1, i_1 \rangle$. in *tp* has no forerunner, we delete $node_P$ from RTPT, and insert into HT according to Rule 5.

The RTPT updating method is described as Algorithm 1: Execute insert rule towards nodes and edges (Line 2–10), execute update rule towards nodes (Line 18–21), execute update rule towards nodes (Line 11–16), execute delete rule towards nodes (Line 23–25).

Algorithm 1: RTPTUpdate(T-patterns(tps))

Input: A Set of T-patterns $Tset$

Output : Updated $RTPT$

1) FOREACH tp in $Tset$ DO
2)   node = Root($RTPT$);
3)   FOREACH $\langle r, \ i \rangle$ in tp DO
4)   (edge, n) = findchild(r);
5)   IF $\nexists n \ || \ edge.interval \not\supseteq i$ THEN
6) v = new Node(r);
7) v.support =tp.supp;
8)   node.appendChild(v,i);
9)   node = v;
10)   END IF
11)   IF $\exists n$ && $i > egde.interval$ THEN
12)   Update(edge.interval,i);
13)   END IF
14)   IF $\exists n$ && $n < tp.supp$ THEN
15)   Update(n,tp.supp);
16)   END IF
17)   END FOR
18)   P=findpath(r_1);
19)   Node$_p$=Node(r_1).parent;
20)   IF P $\supset \langle r_1, \ i_1 \rangle, \ \langle r_2, \ i_2 \rangle, \ ..., \ \langle r_m, \ i_m \rangle$ &&
        Node$_p$.supp<δ THEN
21)   Node$_p \leftarrow \langle r_0, \ i_0 \rangle$;
22)   END IF
23)   IF P $\supset \langle r_1, \ i_1 \rangle, \ \langle r_2, \ i_2 \rangle, \ ..., \ \langle r_m, \ i_m \rangle$ &&
        Node$_p$.supp<δ && $\nexists \langle r_0, \ i_0 \rangle$ THEN
24)   Delete(Node$_p$);
25)   END IF
26) END FOR
27) return $RTPT$;

The algorithm complexity analysis is as following:

RTPT Update algorithm contains the building and updating of RTPT, is a top-down process. The time complexity is $O(n^2)$, depending on the size of trajectory pattern set and number of nodes.

4.3 HT Introduction and Update

As shown in Fig. 5, in HT structure, each key value corresponds a deleted node; the key is followed by a link list of region and time interval. According to Sect. 4.2,

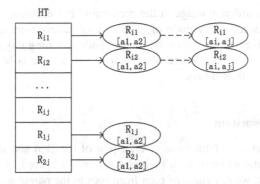

Fig. 5. An example of HT

RTPT retains the nodes and edges of all T-pattern with appropriate support. And the nodes whose support degree is less than the minimum support will be inserted into HT.

Rule 5 HT Insert: The time interval of deleted node of RTPT $Node_{(r,i)}$ should be put to the end of link list, if there existed a node with region r, otherwise, we build a new node in HT, and insert $Node_{(r,i)}$ to the list.

In order to maintain each T-pattern, we build a hash table. After all, support degree isn't everything. When we delete the minimum support node, our starting point is to improve the RTPT storage efficiency. As to the presented strategy in the next section, deleted nodes may also have high compatibility, and be elected to the chosen path.

5   Location Prediction

5.1   Prediction Strategy

In this section, we will introduce how to predict location of moving object using RTPT. The main idea of our forecasting method is to find the best path on the tree (the best T-pattern matching the current trajectory). For the given trajectory, we calculate the best match degree of the chosen path, all child nodes of the best node is likely to be the next position. Before calculating the match degree of the whole path, we calculate compatibility of each node on the path, called accurate value.

Total value of a path (Path Value) is based on the exact value of each path node. We define three different scoring functions below:

Definition 5 Scoring Function: There is a trajectory t, a path $\langle P = p_1, p_2, \ldots, p_n \rangle$, and the accurate value $pScore_k$ of each $p_k \in P$, we give the functions below:

$$\text{Trajectory path Average Value} = \frac{\sum_1^n pScore_k}{n}$$

$$\text{Trajectory path Total Value} = \sum_1^n pScore_k$$

$$\text{Trajectory path Maximum Value} = \max\{pScore_1, pScore_2, \ldots pScore_n\}$$

Each function has different usage in the prediction. For example, "Average" get the average distance between every node of trajectory or pattern tree, and summarize the similarity. "Total" give the priority of the longest path-matching trajectory based on the concept of depth. If matching degree of trajectory and a node is high, the priority of "Maximum" is higher than others.

5.2   Prediction Algorithm

Suppose that a trajectory T (simply as $\langle l, t \rangle$, a pair of location and time) and a PT are given, we calculate the value of each path about T on PT according to the function. If no exact value exists, we put value of path from root to the parent node as candidate of trajectory prediction. After the whole tree is traversed, we choose candidates with best value. We return regions have contact with candidates as prediction, there may be some results. Prediction method is described as Algorithm 2.

Algorithm 2: Prediction(RTPT,   Trajectory)

1) Input: RTPT,   A Trajectory T
2) Output: prediction
3) FOREACH $\langle l,   t \rangle$ in T DO
4)   P = findpath($\langle l,   t \rangle$);
5) $score_p$=Compute(P);
6)   IF $\exists score_p$ THEN
7)   candidate=$score_p$;
8)   END IF
9)   ELSE THEN
10)   P'=Path: Root $\rightarrow$ $node_l$. parent;
11) $score_{p'}$=Compute(P');
12)   candidate=$score_{p'}$;
13) END ELSE
14) predictions=Best(candidate);
15) END FOR
16) Return predictions;

We can calculate the scores as well as compare them, so we can choose the best candidate as the prediction. The DPT mentioned above can update the pattern dynamically, so we can get the more effective prediction. We can calculate score according to the definition of "Score Function", and we can improve the threshold, then discard more un-satisfiable path. The algorithm complexity analysis is as following:

In Prediction algorithm, the process of inputting RTPT and Trajectory, looking for the best match path for each trajectory point, is actually a process of traversal the tree. The time and space complexity are $O(nm)$, where n is the number of location that T contains, and m is the node number of RTPT.

6 Experiment Results

In this section, we conduct a series of experiments to evaluate the performance of our proposed location prediction method RTMatch. We use a real dataset T-Drive Trajectory of Geo-Life, which Contains 10,357 taxi GPS trajectory in Beijing during a week. Each trajectory in the dataset contains the information of number, time-stamps, latitude and longitude. It is typical spatio-temporal dataset. In experiments, we select nearly 2,000 trajectories on each Monday as the training set, and select 1,000 trajectories on each Tuesday as the training set.

The experiments were conducted using Matlab 2014 on an Intel(R) Core(TM) i3-4130 3.40 GHZ computer with 4.00 GB memory and 500 GB hard disk. We present our results and discussions.

6.1 Evaluation on Effectiveness of RTMatch

First of all, we use two evaluation measurement in our experiment, where A_+ indicate the number of correct predictions, A_- indicate the incorrect ones, and R indicate the total number of trajectories.

$$\text{Measurement 1 : Accuracy} = \frac{A_+}{A_+ + A_-}$$

$$\text{Measurement 2 : Coverage} = \frac{A_+ + A_-}{R}$$

In this section, we compare WhereNext method and proposed method RTMatch, calculate the prediction accuracy and coverage under different minimum support. The former one, model the historical trajectory of moving objects in a particular area, gets average prediction accuracy of 56.8% and coverage of 37.5%. While the latter, update prediction model dynamically based on historical trajectory modeling, gets average prediction accuracy 72% and the average coverage of 92.1%, obviously, higher than the former one.

As shown in Fig. 6(a), we conduct the experiment about WhereNext [2], and certificate the result that the accuracy is about 50%. The accuracy of RTMatch we proposed based on real-time model, increased 15.2%. It indicates that trajectory model will change in some way because of subjective or objective factors, using historical trajectory mode invariably is inadvisable, dynamic updating will reach a more reasonable result.

In Fig. 6(b), in the terms of prediction coverage, the RTMatch method we proposed increased by 54.6% than WhereNext. It is not only more accurate in prediction, but also process those trajectories can't be predicted.

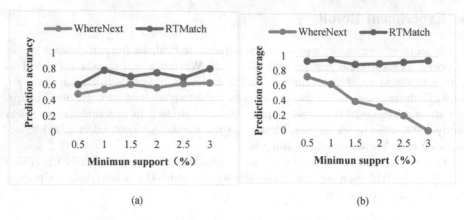

(a)   (b)

Fig. 6. Evaluation on effectiveness of RTMatch method

6.2 Evaluation on Efficiency of RTMatch

In Fig. 7, it shows out the results of efficiency evaluation of RTMatch compared to Not RealTime method. Under all of the minimum support, the execution time of RTMatch is 0.7 s shorter than the Not Real Time one. That's to say, after building a storing model, we can predict location of moving objects more quickly when we use the real-time prediction method. It is of great practical significance in intelligent traffic management application.

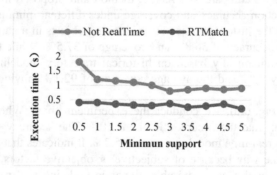

Fig. 7. Evaluation on efficiency of RTMatch method

6.3 Evaluation on Efficiency of RTPT and HT

In this part, we evaluate the efficiency of our prediction model RTPT and HT using different size of dataset. Firstly, we choose a larger dataset for training and testing, as shown in Fig. 8(a), when the dataset increases, the training time makes exponential growth, which is necessary in order to make a more precise location prediction.

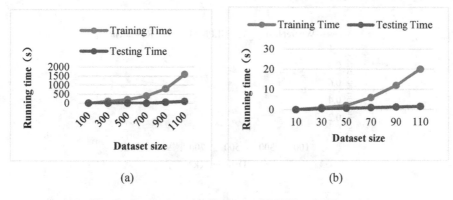

Fig. 8. Evaluation on efficiency of RTPT and HT model

Relatively speaking, the testing time just makes linear growth as the dataset increases, and the speed is slow. In Fig. 8(b), the change can be observed clearly. We can draw a conclusion from the figures that the more demand the more cost.

6.4 Evaluation on Prediction Accuracy of RTMatch

In this section we compare the prediction accuracy of three methods, RTMatch, WhereNext [2] and NLPMM [15], using the comparable setting parameters. The method NLPMM trains Markov model according to the historical track information in which rules change from 1 to N. And predicts the next location of the moving objects, where N is the number can be set by users. When N = 3, the prediction accuracy of this method reaches the highest value.

In this experiment, the relevant parameters are set as follows: for RTMatch and WhereNext, the support is 10. For NLPMM, the N is set to 3. Figure 9 shows the prediction accuracy and coverage of the three methods. As shown in Fig. 9a, the

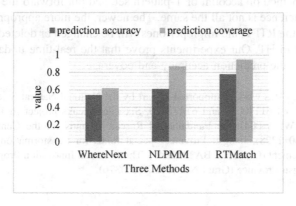

Fig. 9a. Comparing prediction accuracy and prediction coverage

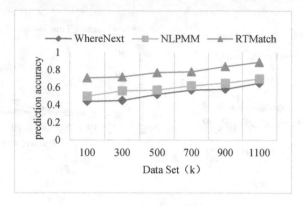

Fig. 9b. Comparing prediction accuracy with the different size of data set

RTMatch method proposed in this paper has the best prediction effect. And the prediction accuracy is about 24% higher than WhereNext, 17% higher than NLPMM. As shown in Fig. 9b, the prediction accuracy of the three methods increases with the increasing of the size of data set, and the prediction accuracy of RTMatch method is higher than that of the other two methods over the different data sets.

7 Conclusion

In this paper, we proposed a predicting next location method for moving object named RTMatch based on trajectory pattern mining. The main idea of the method is to construct the storage structure RTPT and HT which can store and update trajectory pattern information of moving objects in real time, and to generate the prediction strategy through calculating best matching path. RTMatch method uses the proposed algorithm T-pattern for mining frequent patterns, the object of pattern mining is all the moving object trajectory in a particular area, regardless of the semantic, habits, unexpected situation etc. When constructing RTPT, we choose the best mode through the calculation method on account of T-pattern set, and put forward the updating rules for prediction reference is not all the same. The newer, the more appropriate T-patterns will be added to the RTPT, the improper ones will be changed or deleted. Deleted ones should be stored to HT. Our experiments prove that the real-time updating RTMatch method improves the prediction accuracy and coverage.

Acknowledgment. This work is supported in part by the National Natural Science Foundation of China under Grant 61170035 and 61272420, Six talent peaks project in Jiangsu Province (Grant No. 2014 WLW-004), the Fundamental Research Funds for the Central Universities (Grant No. 30916011328), Jiangsu Province special funds for transformation of science and technology achievement (Grant No. BA2013047). The Research Innovation Program for College Graduates of Jiangsu Province (Grant No. KYLX15 0376).

References

1. Niedermayer, J., Zufle, A., Emrich, T., Renz, M., Mamouliso, N., Chen, L., Kriegel, H.: Probabilistic nearest neighbor queries on uncertain moving object trajectories. Proc. VLDB Endow. **7**(3), 205–216 (2014)
2. Monreale, A., Pinelli, F., Trasarti, R., Giannotti, F., et al.: WhereNext: a location predictor on trajectory pattern mining. In: ACM SIGKDD International Conference on Knowledge Discovery and Data Mining, Paris, France, 28 June–July 2009, pp. 637–646 (2009)
3. Giannotti, F., Nanni, M., Pinelli, F., Pedreschi, D.: Trajectory pattern mining, KDD 2007, pp. 330–339 (2007)
4. Yang, J., Xu, J., Xu, M., et al.: Predicting next location using a variable order Markov model. In: ACM SIGSPATIAL International Workshop on Geostreaming, pp. 37–42. ACM (2014)
5. Nodari, A., Nurminen, J., Siekkinen, M.: Energy-efficient position tracking via trajectory modeling. In: Mobiarch 2015 Proceedings of the International Workshop on Mobility in the Evolving Internet Architecture, pp. 33–38 (2015)
6. Ma, S., Zheng, Y., Wolfson, O.: Real-time city-scale taxi ridesharing. IEEE Trans. Knowl. Data Eng. **99**(2014), 2334313 (2015)
7. Figueiredo, F., Ribeiro, B., Almeida, J., et al.: TribeFlow: mining & predicting user trajectories. In: International Conference on World Wide Web, pp. 695–706 (2015)
8. Ying, J.J.C., Lee, W.C., Weng, T.C.: Semantic trajectory mining for location prediction. In: Proceedings of the 19th ACM SIGSPATIAL International Conference on Advances in Geographic Information Systems, ACM (2011)
9. Feng, S., Li, X., Zeng, Y., Cong, G., Chee, Y.M., Yuan, Q.: Personalized ranking metric embedding for next new POI recommendation. In: Proceedings of the IJCAI (2015)
10. Zheng, Y., Zhang, L., Xie, X., et al.: Mining interesting locations and travel sequences from GPS trajectories. In: International Conference on World Wide Web, WWW April 2009, Madrid, Spain, pp. 791–800 (2009)
11. Morzy, M.: Prediction of moving object location based on frequent trajectories. In: Levi, A., Savaş, E., Yenigün, H., Balcısoy, S., Saygın, Y. (eds.) ISCIS 2006. LNCS, vol. 4263, pp. 583–592. Springer, Heidelberg (2006). doi:10.1007/11902140_62
12. Morzy, M.: Mining frequent trajectories of moving objects for location prediction. In: Perner, P. (ed.) MLDM 2007. LNCS (LNAI), vol. 4571, pp. 667–680. Springer, Heidelberg (2007). doi:10.1007/978-3-540-73499-4_50
13. Chaney, A.J., Blei, D.M., Eliassi-Rad, T.: A probabilistic model for using social networks in personalized item recommendation. In: Proceedings of the RecSys (2015)
14. Chen, S., Xu, J., Joachims, T.: Multi-space probabilistic sequence modeling. In: Proceedings of the KDD (2013)
15. Chen, M., Liu, Y., Yu, X.: NLPMM: a next location predictor with Markov modeling. In: Tseng, V.S., Ho, T.B., Zhou, Z.-H., Chen, A.L.P., Kao, H.-Y. (eds.) PAKDD 2014. LNCS (LNAI), vol. 8444, pp. 186–197. Springer, Heidelberg (2014). doi:10.1007/978-3-319-06605-9_16

Online Formation of Large Tree-Structured Team

Cheng Ding, Fan Xia(✉), Gopakumar, Weining Qian, and Aoying Zhou

ECNU-RMU-Infosys Data Science Joint Lab,
School of Data Science and Engineering,
East China Normal University, Shanghai, China
{fxia,wnqian,ayzhou}@sei.ecnu.edu.cn, cding@ecnu.cn,
Gopakumar@infosys.com

Abstract. Software projects are often divided into different components and groups of individuals are assigned to various parts of the project. The matching of modular components of the project with right set of individuals is a fundamental challenge in both commercial and open source software projects. However, most of the extant studies on team formation have only considered the problem of creating flat teams, i.e., teams without communities and central authorities. In this paper, we study the problem of forming a hierarchically structured team. We use tree structure to model both teams and task specifications and introduce the notion of sub-team. Next, we define local density to minimize communication costs in sub-teams. Then, two algorithms are proposed to address this team formation problem in bottom up and top down manners. Furthermore, sub-teams are pre-computed and indexed to facilitate online formation of large teams. Results of experiments with a large dataset suggest that the index based algorithm can achieve both good effectiveness and excellent efficiency.

1 Introduction

In enterprises, complex projects with large teams may suffer from the scaling issues. As Brucks [7] pointed out, the coordination cost may scale quadratically with the team size. Therefore, teams with sub-teams are more efficient for large projects as most communications will be confined among sub-teams and coordination is required across sub-teams for effective performance. In the design of softwares, design principles such as separation of concerns [53] are often applied. It utilizes the divide and conquer approach to partition a software into components and each component is tasked by a team of individuals. "Conway's Law" [9] suggests that this rule is suitable for most intellectual activities building the whole from diverse parts. The matching between modularized artifacts of products and individuals is practiced in open source projects such as PostgreSQL, Python and Perl as well [6]. The modularity computed for graphs based on emails among open source developers indicates the existence of strong community structure with "boundary spanning" or "gatekeeping" roles, which are akin to leaders in traditional project teams who coordinate different teams.

Z. Bao et al. (Eds.): DASFAA 2017 Workshops, LNCS 10179, pp. 118–132, 2017.
DOI: 10.1007/978-3-319-55705-2_9

The problem of staffing project teams with right individuals has received substantial scholarly attention. The assignment problem [1] in operations research, which is a closely related area, emphasizes assigning tasks to qualified agents with minimal overall assignment cost. However, those studies only consider the cost of selected agents in objective functions. With the availability of information regarding collaboration history or social network of individuals, an effective team can be composed by optimizing the communication cost among team members. Several criteria focusing on the sub-graph consisting of team members or their path distance have been proposed. For instance, Lappas et al. [12] focused on minimizing the diameter of the sub-graph or the weight of the minimum spanning tree. Likewise, Rangapuram et al. [13] utilized density of the sub-graph to optimize the average communication cost between members.

However, most of such studies are focused on the problem of solving flat teams, i.e., teams without sub-groups and central authorities. The problem of forming hierarchically structured teams become more salient as software projects are increasingly designed in a modular fashion. In this paper, both teams and task specifications are modeled as the tree structure. The sub-team led by a person is defined as the group of people that are directly managed by her. Furthermore, we define local density as an objective function that emphasizes the communication cost in sub-teams similar to the analyses of communities in open source projects. Then, two heuristic algorithms are proposed to solve the problem in top down and bottom up manners. The sub-team index is designed to improve the performance to facilitate online formation of large teams. Through a series of experiments we demonstrate that our algorithms achieve good balance between efficiency and effectiveness as compared to the existing methods. Main contributions of this paper are the following:

1. The tree structure is used to model both the team and task specification. Based on these notations, the local density is defined to assess and optimize the communication cost in sub-teams.
2. Two heuristic algorithms are proposed to solve the tree structured team formation problem in top down and bottom up manners. We also present the analysis of their complexity.
3. Based on the performance analysis, we propose the sub-team index, which indexes pre-computed sub-teams. It can alleviate expensive sub-team construction in bottom up team formation algorithm.
4. Extensive experiments conducted on real project assignment data of a large company highlights superiority of our algorithms over the existing ones.

The remaining part of this paper is organized as follows. We provide an overview of related research in Sect. 2. In Sect. 3, we define the model of team and task specification. Then the team formation problem is presented. We introduce two heuristic algorithms and analyze their performance in Sect. 4. The sub-team index is introduced in Sect. 5. The details of experiments conducted on a real project assignment dataset is presented in Sect. 6 and conclusions are presented in the final section.

2 Related Work

The problem of team formation has been extensively studied by operations research scholars [5,8,14,15]. Most of them have examined various characteristics of people that influence the success of a team. For instance, Baykasoglu et al. [5] identified four skills such as oral communication skills, technical expertise, problem solving ability, and decision-making skills to determine the suitability of people to project teams. The fuzzy model proposed by Yaakob and Kawata [16] is often used to represent the endorsement level of those skills. Wi et al. [15] have proposed a random walk like approach to match skills of people with those of their neighbors in the social network. Their objective function consider the competence and connections of team members and managers. Genetic algorithms or simulated annealing algorithms are used to solve such mathematical optimization problems.

Job scheduling is another closely related research area. Given a set of tasks, the goal of job scheduling is to assign tasks to a set of machines such that the load of each machine is minimized. It has been proven to be NP-hard and its online version has been studied by Azar [3] and Azar et al. [4]. In a similar vein, Anagnostopoulos et al. [2] have examined online team formation where tasks arrive in a sequential manner. Their algorithms sought to form teams that have skills required by tasks while imposing minimal load on any expert across all tasks and bounding the coordination cost by a given value. Diameter cost and Steiner tree cost are used as the coordination cost and approximation algorithms are proposed respectively.

The work of database scholars [1] in this field are similar to studies in operations research area. Here, both people and team profile are defined as a vector of skills. Several functions are defined to compute the team profile from individual skill vectors. Further, several objective functions, such as least skill dominant, are defined based on team profiles. The problem is proved to be NP-hard and approximation algorithms with provable guarantees are provided.

Social network research scholars have also made significant contributions to address the team formation problem. Studies in this stream of research model teams as networks wherein members are nodes and connections represent different types of relationships between them. Various criteria are proposed to assess the communication cost of teams. Lappas et al. [12] first used the diameter and cost of minimum spanning tree to assess the communication cost. Kargar and An [11] subsequently proposed two other distance based criteria such as sum distance and leader distance that are based on the shortest path distance between nodes of the graph. To our knowledge, leader distance is the only measurement that explicitly considers the structure of a team. However, it is merely a summation of the shortest distance between the leader and other members for each required skills. Their optimization problem is proved to be a NP-Hard. A greedy algorithm with an approximation guarantee of two is provided.

However, distance based criterion are not suitable for situations when people are not connected or far from each other in the social network. Gajewar and Sarma [10] and Rangapuram et al. [13] have used density of the sub-graph to evaluate the effectiveness of a team. It can generate more compact teams even though some people can be disconnected. Rangapuram et al. [13] present a realistic definition of team formation with constraints such as budgets, team size and distance among team members. Their problem formulation leads to a generalized version of densest sub-graph problem with cardinality constraints. Algorithms based on the relaxation of the problem is provided with an approximation guarantee.

3 Problem Definition

We use $S = \{s_1, s_2, \ldots, s_r\}$ to denote a set of r skills. Then each person n is associated with a skill vector p_n. $p_{n,i}$, the ith element of vector p_n, indicates whether person n is endorsed for skill s_i. Without ambiguity, we also use p_n to denote person n. Set H_n represents the work history of person n. Each item $<t_i, T_j>$ stored in H_n means person n worked in project team T_j at time t_i. We provide colleague graph in Definition 1 to represent the relationship among people based on their work history (Table 1).

Table 1. Notations used in the paper

Notation	Description
s_i	The ith skill
p_n	Skill vector of person n or simply person n
H_n	Work history of person n
T_i	Tree-structured team of project i
J_i	Tree-structured team specification for project i

Definition 1 (Colleague graph). *Colleague graph is modeled as an undirected weighted graph $G(V, E)$. Each node v_n in V corresponds to person n, represented by skill vector p_n. An edge $e_{m,n}$ is created between two nodes v_m and v_n if p_m and p_n worked together in the same project at the same time. Its distance $d_{m,n}$ is defined as $1 - \frac{|H_m \cap H_n|}{|H_m| + |H_n|}$. $|H_m \cap H_n|$ is the total time in which p_m and p_n worked together in same projects.*

The team defined in previous studies is typically flat, which means people have equal positions in the team. However, teams that operate in organizations are often hierarchical. Thus, we define the team T_i of project i as a tree with node set V and edge set E. The node set V is a subset of whole people and each edge $e_{n,m}$ in E indicates person p_m is managed by p_n. Furthermore, the sub-team led by p_n in project i is denoted by $T_i(p_n) = \{p_m | p_m = p_n \vee e_{n,m} \in E\}$. It consists of p_n and people that are directly managed by p_n.

Example 1. Figure 1 demonstrates an example of a tree-structured team. In this example, p_1, p_4 and p_5 constitute the group developing user interface with HTML

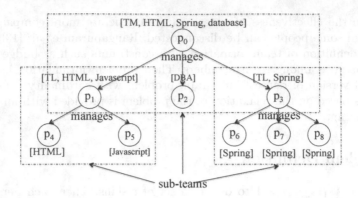

Fig. 1. Tree structured project team and corresponding task specification

and Javascript. p_3, p_6, p_7 and p_8 constitute the group developing the backend with Spring. p_2 is a database administrator while p_0 is the project manager. It is important to consider this team structure when forming an efficient team. For example, people in the UI group should be familiar with each other while familiarity with every other member in the backend group is not that critical.

In correspondence, we define *team specification* J_i for a new project i as a tree with node set V and edge set E. Each node v_n in V is a skill vector specifying skills required by the position. For each edge $e_{n,m}$, the people p_{v_n} assigned to position v_n manages p_{v_m} assigned to v_m. We define the communication cost of a sub-team led by p_n as the sum distance of the sub-team. It is denoted as $Sum(T_i(p_n))$ and its definition is given in Eq. 1, where $dist_G(p_m, p_k)$ returns the distance between p_m and p_k in the colleague graph.

$$Sum(T_i(p_n)) = \sum_{p_m, p_k \in T_i(p_n)} dist_G(p_m, p_k) \tag{1}$$

Taking the team in Fig. 1 as an example, the overall team is efficient if sub-teams $T(p_0)$, $T(p_1)$ and $T(p_3)$ are dense. Accordingly, we define *local density* in Definition 2 as our score function to measure the communication cost of a team. It accounts for edges in sub-teams to prefer a team with dense sub-teams.

Definition 2 (Local density). *Given a tree-structured team T, its local density, denoted by $Dens_{loc}(T)$, is defined by the following equation:*

$$Dens_{loc}(T) = \frac{\sum_{p_n \in V} Sum(T(p_n))}{\sum_{p_n \in V} |T(p_n)| * (|T(p_n)| - 1)} \tag{2}$$

Example 2. Assume the communication cost between people in the team shown in Fig. 1 equals one. When computing the local density for this team, we need to compute the sum distance $Sum(T(p_n))$ of sub-team led by each person p_n in the team. For each sub-team $T(p_n)$ containing only one person, e.g., $T(p_2)$ and $T(p_4)$, its sum distance $Sum(T(p_n))$ and $|T(p_n)| * (|T(p_n)| - 1)$ both equal zero. For the person p_3, its sub-team communication cost $Sum(T(p_3))$ equals 6 (summation of communication cost between any two people in $T(p_3)$) and $|T(p_3)| * (|T(p_3)| - 1)$ equals 12. Thus, the local density of this team is $(3 + 6 + 6)/(6 + 12 + 12) = 0.5$.

Based on the above definition, the team formation problem is formulated as Eq. 3. The target is to select a project team T which minimizes the local density while guaranteeing team members are competent for positions they are assigned to. The comparison $p_{v_n} > v_n$ requires each element of p_{v_n} is greater than corresponding part of v_n.

$$\text{minimize } Dens_{loc}(T)$$
$$\text{subject to:} \tag{3}$$
$$p_{v_n} > v_n, \forall v_n \in J.V$$

For teams with only two levels, our team formation problem is equivalent to the team formation without a leader as defined in [11]. It has been proved that the problem of forming a team without a leader is NP-hard. Thus, our problem which is more generic is also a NP-hard problem. In the following sections, we propose two heuristic algorithms to solve this problem and the sub-team index to accelerate the performance.

4 Methods

There are two straightforward intuitions of constructing a tree structured team. We can first select sub-team leaders and then choose sub-team members led by them. Correspondingly we propose the Top Down Team Formation (*TDTF*) algorithm based on this heuristic. The other way is to select sub-team members first and then assign sub-team leaders. Based on this intuition, the Bottom Up Team Formation (*BPTF*) algorithm is proposed. Those two algorithms are introduced in this section.

4.1 Top Down Team Formation (TDTF) Algorithm

The *TDTF* algorithm is presented as Algorithm 1. It solves the team formation problem in a top down manner. It takes a social network G and a tree-structured task specification J as parameters. The constructed team is returned as a set T, consisting of <position, person> pairs.

Suppose $J.v_r$ denotes the root node of specification J. The algorithm starts by constructing the sub-team rooted at $J.v_r$. It invokes procedure $Cands$ in line 5 to find out candidates fulfilling the skill requirements. In our implementation an in-memory inverted index is build with skills as keywords and people as documents. $Cands$ queries the inverted index to find competent people. For each qualified candidate p_n of the root position, the algorithm (lines 5–10) invokes Finding Followers (*FF*) procedure to construct the best sub-team led by p_n. The *FF* procedure is introduced in the following paragraph. At the end of the loop, the sub-team with smallest communication cost is chosen and merged into

---

Algorithm 1. Top down team formation algorithm TDTF

    Input   : Social network G and task specification J
    Output : Position assignment set T

1 **begin**
2      $queue \leftarrow Queue(), T \leftarrow \emptyset$;
3      $bestDist \leftarrow \infty$;
4      $bestSubTeam \leftarrow \emptyset$;
5   **for** $p_n \in Cands(J.v_r)$ **do**
6         $bestSubTeam \leftarrow \emptyset$;
7         $< team, dist > \leftarrow FF(G, T, J.v_r, p_n)$;
8   **if** $dist < bestDist$ **then** //choose the team with smallest cost
9            $bestSubTeam.add(< J.v_r, p_n >)$; //assign p_n to $J.v_r$;
10            $bestSubTeam.union(team)$;//merge sub-team member assignments;

11      $T.union(bestSubTeam)$;
12   **for** $v_i \in J.v_r.children$ **do**
13         $queue.add(v_i)$; //append v_i to the tail of $queue$;
14   **while** $queue \neq \emptyset$ **do**
15         $v_i \leftarrow queue.pop()$;//fetch the data from the head of $queue$;
16         $< team, dis > \leftarrow FF(G, T, v_i, T.getMember(v_i))$;
17         $T.union(team)$;
18   **for** $v_j \in v_i.children$ **do**
19            $Q.add(< v_j, T.getMember(v_j) >)$;

20   RETURN T;

---

current assignment set T (line 11). Then, the algorithm starts a breadth first traverse over the task specification tree by first appending children of $J.v_r$ to $queue$. As the position leading a sub-team is always computed first, we can use $T.getMember(v_i)$ to obtain the person assigned to v_i. Thus, there is no need to iterate through competent candidates of v_i. We only need to invoke FF to construct the best sub-team for v_i given $T.getMember(v_i)$ in line 16.

The FF Algorithm presented as Algorithm 2 selects members for the sub-team led by a given person. Given the position v_i and person p_{v_i} of the sub-team leader, the algorithm iterates though each child position v_j of v_i. For each competent candidate of v_j, the person who is closest to p_{v_i} in the colleague graph is chosen (line 5). It is added to the position assignment set $bestTeam$ and the $sumDistance$ is also updated. After all team members are selected, FF returns $bestTeam$ together with its $sumDistance$.

Example 3. In this example we describe how $TDTF$ constructs the project team for the task specification in Fig. 1. The loop at lines 5–10 first assign members to positions v_0, v_1, v_2 and v_3. For each candidate p_n of v_0, the algorithm invokes FF to construct sub-team $T(p_n)$. At line 11, the best-SubTeam $\{<v_0, p_0>, \ldots, <v_3, p_3>\}$ is computed and merged into T. After lines 12 and 13 are executed, $queue$ consists of v_1, v_2, v_3. In the first

---

Algorithm 2. Finding followers procedure FF

Input   : Colleague graph G, current position v_i and person p_{v_i} assigned to v_i
Output : Best sub-team $bestTeam$
1 **begin**
2 | $bestTeam \leftarrow \emptyset$;
3 | $sumDistance \leftarrow 0$;
4 | **for** $v_j \in v_i.children$ **do**
5 | | $p_{v_j} \leftarrow closestCand(G, v_j, p_{v_i}, \emptyset)$;
6 | | $bestTeam.add(< v_j, p_{v_j} >)$;
7 | | $sumDiatance \leftarrow sumDistance + dist_G(p_{v_i}, p_{v_j})$;
8 | RETURN $< bestTeam, sumDistance >$;

---

iteration of the loop scoped by line 14, v_1 is first popped from *queue*. Then $FF(G, \{<v_0, p_0>, \ldots, <v_3, p_3>\}, v_1, p_1)$ is invoked. p_4 and p_5 will be selected for v_1' child positions v_4 and v_5. At last *queue* is updated with those child positions and becomes $\{v_2, v_3, v_4, v_5\}$.

4.2   Bottom Up Team Formation (BPTF) Algorithm

In contrast to the top down algorithm, Bottom Up Team Formation (BPTF) algorithm constructs the project team from the lowest level of the task specification tree to the root position. It is presented as Algorithm 3, which has the same input parameters and output type as *TDTF*.

The *BPTF* algorithm forms the team by constructing sub-teams without the leader from bottom to top. The code in lines 5 to 10 executes a breadth first traverse of the task specification tree and stores siblings into *stack*. Then the second loop pops each sibling set from the stack and invokes Find Siblings (*FS*) procedure to choose best suitable members for positions in the sibling set. The procedure *FS* behaves similar to algorithms finding flat teams with an exception that people assigned to children of each position affect the choice of current position. At last the result of *FS* is merged into the position assignment set T.

The *FS* algorithm, presented as Algorithm 4, traverses each position v_i in B in line 4. For each candidate p_n, the loop in lines 5 to 15 selects competent people for remaining positions such that the sum of distance from them to p_n is minimized. It first initializes *team* with p_n assigned to v_i. For each remaining position v_j, it invokes procedure *closestCand* to choose skilled people whose distance to p_n and selected children of v_j are smallest. Then, the assignment pair $<v_j, p_m>$ is appended to *team* and the sum of distance *sumDist* is updated by the distance between p_m and p_n. Finally, it chooses the best solution by comparing *sumDist* with *bestSumDistance*.

Example 4. In this example we describe how *BPTF* constructs the project team for the task specification in Fig. 1. After the breadth first traverse indicated by lines 5–10, the data stored in *stack* is $\{v_4, v_5\}$, $\{v_6, v_7, v_8\}$, $\{v_1, v_2, v_3\}$,

---

Algorithm 3. Bottom up team formation algorithm BPTF

Input   : Social network G, task specification J
Output : Position assignment set T
1 **begin**
2 | $queue \leftarrow Queue(), stack \leftarrow Stack(), T \leftarrow \emptyset$;
3 | $queue.add(J.v_r)$; //begin breadth first traverse of tree J;
4 | $stack.push(\{J.v_r\})$; //store sibling sets in $stack$;
5 | **while** $queue \neq \emptyset$ **do**
6 | | $v_i \leftarrow queue.pop()$;
7 | | **if** $v_i.children \neq \emptyset$ **then**
8 | | | $stack.push(v_i.children)$;
9 | | | **for** $v_j \in v_i.children$ **do**
10 | | | | $queue.add(v_j)$;

11 | **while** $stack \neq \emptyset$ **do** //begin post-order like traverse of tree J
12 | | $sibling \leftarrow stack.pop()$;
13 | | $subteam \leftarrow FS(G, sibling, T)$;
14 | | $T.union(subteam)$;
15 | RETURN T;

$\{v_0\}$. In the first iteration of the loop scoped by line 12, $\{v_4, v_5\}$ is on the top of $stack$ and is popped first. Then, FS(G, $\{v_4, v_5\}$, $\emptyset$) is invoked and returns $\{<v_4, p_4>, <v_5, p_5>\}$. When the loop comes to the third iteration, $\{v_1, v_2, v_3\}$ is on the top of $stack$ and the arguments passed to FS are G, $\{v_1, v_2, v_3\}$ and $\{<v_4, p_4>, \ldots, <v_8, p_8>\}$. Thus child positions are always selected first and influence the choice of current sibling positions.

4.3 Performance Analysis

To analyze the complexity of those two algorithms, we introduce symbols used in the analysis. Suppose s_m is the skill that has the maximum number of people endorsed to it. Then M_s is used to represent the number of people endorsed to s_m. We use $|J|$ to denote the number of nodes in $J.V$ and M_d as the maximum degree of those nodes.

First, we analyze the performance of $TDTF$. The algorithm traverses the specification tree in pre-order. For the root position of the task tree, it needs to invoke FF to construct sub-team for each candidate. The size of the candidates is bounded by M_s. For the remaining positions, FF is only invoked once as team members are already fixed. In algorithm FF, the loop at line 4 visits each child of current position v_i, whose size is also bounded by M_d. For each child position, $closestCand$ is invoked to select the proper member from people fulfilling skills of v_j. Thus, the complexity of FF is $O(M_s * M_d)$ and the overall complexity of $TDTF$ is $O(M_s^2 * M_d + M_s * M_d * |J|)$.

Then, we analyze the performance of $BPTF$ as well. The breadth first traversal in $BPTF$ visits each node in the specification tree and for each node FS

Algorithm 4. Find siblings procedure FS

Input   : Social network G, sibling set B, task specification J and current
          position assigment set T
Output : Position assigment set T_B for B
1 **begin**
2 | $T_s \leftarrow \emptyset$;
3 | $bestSumDistance \leftarrow MAX$;
4 | **for** $v_i \in B$ **do**
5 | | **for** $p_n \in Cands(v_i)$ **do**
6 | | | $team \leftarrow \{< v_i, p_n >\}$;
7 | | | $sumDist = 0$
8 | | | **for** $v_j \in B$ and $i \neq j$ **do**
9 | | | | $children \leftarrow T.getMembers(v_j.children)$;
10 | | | | $p_m \leftarrow closestCand(G, v_j, p_n, children)$;
11 | | | | $team.add(< v_j, p_m >)$;
12 | | | | $sumDist \leftarrow sumDist + dist_G(p_m, p_n)$;
13 | | | **if** $sumDist < bestSumDistance$ **then**
14 | | | | $bestSumDistance \leftarrow sumDist$;
15 | | | | $T_B \leftarrow team$
16 | RETURN T;

---

is invoked. The algorithm FS scans positions in the sibling set B, whose size is bounded by M_d, in loops at lines 4 and 8. For each position, lines 5 and 10 iterate through all people having the desired skills, whose size is bounded by M_s. Thus the complexity of $BPTF$ is $O(|J| * (M_d * M_s)^2)$.

5   Sub-team Index

According to the above performance analyses, a large fraction of time is required for constructing sub-teams. Thus, we compute the skill bags, e.g., {Spring, Spring, Spring} for p_6, p_7 and p_8 in Fig. 1, required for team members of sub-teams in project history of our dataset. We compute the number of occurrence of those skill bags and the frequency of those occurrences. Figure 2 shows the result, which complies to the power distribution. By pre-computing and indexing sub-teams whose skill bags happens more than once, $BPTF$ may avoid expensive sub-team construction and accelerate the performance.

Inverted index [17] is a common index used to support text search over documents. It maintains a directory that maps a term to a posing list which is a list of documents containing the term. We use it to implement the sub-team index. The data to be indexed are sub-teams without the team leader. The skill pairs consists of skills required by sub-teams are treated as terms. We construct the sub-team index in the following steps. For each past project team T_i, we generate corresponding task specification J_i. Then we use $BPTF$ algorithm to

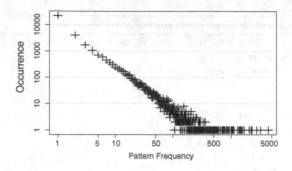

Fig. 2. Statistic of skills required for members in sub-teams

construct new team $T_{i'}$ for task J_i based on currently available people. For each sub-team $T_{i'}(p_n)$ led by p_n in $T_{i'}$, we obtain the skill set S_n which contains all skills required for people in set $T_{i'}(p_n)/p_n$. For any two skills s_j and s_k belonging to S_n, we use the skill pair $<s_j, s_k>$ as a term of the inverted index and insert the sub-team into corresponding posting list. We also compute the communication cost for each sub-team, which is the summation of communication cost between any two team members. The posting list is sorted in ascending order of the communication cost.

Example 5. Assume $p_0, p_1, \ldots, p_8$ are team members that algorithm *BPTF* selects from current available people for the task specification in Fig. 1. The data to be indexed are three sibling sets $\{p_1, p_2, p_3\}$, $\{p_4, p_5\}$ and $\{p_6, p_7, p_8\}$. For the sibling set $\{p_1, p_2, p_3\}$ led by p_0, we obtain the skill set S_0, which consists of TL, HTML, Javascript, DBA and Spring. Skill pairs such as $<$TL, HTML$>$ are enumerated from S_0. Then the sibling set $\{p_1, p_2, p_3\}$ is inserted into the posting lists corresponding to skill pairs like $<$TL, HTML$>$.

Given the sub-team index, *FS* used in algorithm *BPTF* is modified. Similarly we can compute the set of skills that are required for members in siblings B. Then we enumerate the skill pairs and use them to query the sub-team index to find out sub-teams which contain all skill pairs. As posting lists are sorted in ascending order of communication cost, the sub-teams are also returned in the same order. For each sub-team returned, we verify whether it fulfills the specification of siblings B, i.e., it has competent team members assigned to positions in B. Currently, we stop the verification process until k qualified sub-teams are obtained from the index. For those k sub-teams, we update their corresponding communication cost by adding the cost between team members and their children. Finally, the one with the smallest communication cost is returned. We call the bottom up algorithm *BPTF* using the sub-team index as the index-based algorithm.

6   Experimental Study

In this section, we evaluate our proposed algorithms on a real-life enterprise dataset and present experimental results on their effectiveness and efficiency. Apart from the three proposed algorithms, we also implement *best leader* [11], *best sum distance* [11] and *rare first* [12], which solve variants of the flat team formation problem. They are used as the baseline solutions in experiments.

6.1   Experiment Setups

We use the project assignment data of a large multinational IT services and consulting organization to conduct the experiments. The data consists of employee profile, employee skill and project assignment records. The employee profile dataset contains basic information, such as employee id and job level, of 140,000 employees. The skill records in the employee skill dataset store the skills of each employee. The team formation tasks used in experiments are constructed based on the project assignment dataset as the following. For each record like "p_i reports to p_j using skill s" in the dataset, we create corresponding positions for those two employees, specify their skill requirements and create a directed edge between them. When constructing the colleague graph, we use the rule that p_i works with p_j if one of them reports to the other or they both reports to the same manager during same time interval.

To study the effectiveness of these approaches, we compare them in terms of our proposed local density and two other measurements, i.e., global density and percentage of existing edges. The global density equals the sum of distances between each pair of team members divided by the number of pairs for every two members. The percentage of existing edges equals the percentage of father-child and sibling pairs in the generated team that are reachable within two hops in the colleague graph.

6.2   Experimental Results

To study how different approaches perform against different kinds of teams, we compute the average of proposed measurements and execution time in terms of two kinds of project team complexity. Currently team size and *depth × breadth* are used to reflect the complexity of a team. Team size simply equals the number of people in the team. The depth of a team equals the maximum length of paths from root to leaf nodes. The breadth equals the number of nodes in the level of the tree which has the maximum number of nodes.

First, we study how the global density of formed teams changes along with complexity of teams. Figures 3(a) and 4(a) report the result of global density for different team sizes and *depth × breadth*. In terms of global density, Fig. 3(a) suggests that best sum distance always performs best. This is because it enumerates all possible combination of team members to find the one with optimal global distance. However, bottom up algorithm *BPTF* and index-based algorithm beat the remaining ones when team size and *depth × breadth* is greater than 5 and

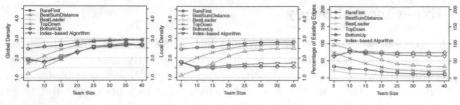

(a) team size vs. global dis- (b) team size vs. local dis-   (c) team size vs. density
tance   tance

Fig. 3. The Effectiveness of approaches under different team sizes

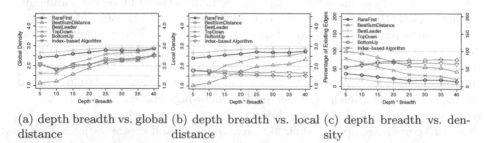

(a) depth breadth vs. global (b) depth breadth vs. local (c) depth breadth vs. den-
distance   distance   sity

Fig. 4. The Effectiveness of approaches under the defined team complexity

15 respectively. Besides, the gap between them and best sum distance nearly diminishes when team size and depth * breadth become large. Those results imply that optimizing local density also tends to reduce global density, which is especially obvious for large complex teams.

Second, Figs. 3(b) and 4(b) show the result of local density for the two tree complexity measurements. From the results, we have four observations: first, *TDTF* performs better than *BPTF* and indexed-based algorithm when the team size is smaller than 5 or depth * breadth is smaller than 10; second, local density of best sum distance is smallest when the team size is smaller than 10 or depth * breadth is smaller than 15; third, *BPTF* and indexed based algorithm work best when complexity becomes larger. Besides, their local density is not sensitive to the team complexity while the performance of other approaches degrades for large team complexity; finally, index-based algorithm only causes minor decrease in local density in comparison with bottom up algorithm *BPTF*.

By comparing the three proposed approaches, we notice that the top down heuristic is more suitable for constructing small simple teams. In such case, most of those teams are just trees with depth of two. Selecting the proper team leader may be the better choice. It is also not surprising that *best sum distance* works best as global density equals local density for those trees. However, a project team may consist of more sub-teams when it becomes larger and more complex. The performance of *best sum distance* shows optimizing global density can not lead to teams with small local density, but not vice versa as shown in results

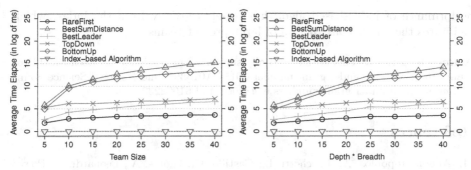

(a) team size vs. runtime (log scaled)   (b) depth breadth vs. runtime (log scaled)

Fig. 5. Results of efficiency experiments

of global density. At last, the local density of formed teams only incurs minor penalty even though only top-k candidates returned by the sub-team index are used to construct members of sub-teams.

At last the average percentage of existing edges is reported in Figs. 3(c) and 4(c). The results show this measurement of *BPTF* and index-based algorithm first raises and then keeps steady when team size or *depth * breadth* increases. It explains why local density of those two approaches decreases in Figs. 3(b) and 4(b). On the contrast, this measurement of all remaining approaches drops when teams become larger and complexer. It means local densities are sacrificed by approaches like *best sum distance* to optimize global density.

The log-scaled average time elapse of all approaches is presented in Fig. 5(a) and (b). The results show *BPTF* and *best sum distance* get the largest time elapse and don't scale quite well with two team complexities. Their effectiveness benefits from the extensive enumeration of teams or sub-teams, but their performance also incurs great penalty. The computation complexity makes them unsuitable for real-time formation of large tree-structured team. On the other hand, the execution latency of index-based algorithm is rather small and is not sensitive to two team complexities. The pre-computed sub-teams greatly alleviate the construction cost although only choosing optimal sub-team from top-k results leads to little loss in effectiveness.

7   Conclusion

In this paper we study the problem of forming a tree-structured team. Such kind of teams is quite ubiquitous in projects that are either running in companies or open source. Two heuristic approaches are proposed to solve the problem in top-down and bottom-up manners respectively. Furthermore, index is built on pre-computed sub-teams to accelerate performance of *BPTF*. Experiments on the real-life enterprise projects dataset show both *BPTF* and index-based algorithm can construct teams with small local density. The efficiency results reveal that performance of index-based algorithm benefits greatly from the index, which facilitates the online construction of large project teams.

Acknowledgment. This work is partially supported by National Hightech R&D Program (863 Program) under grant number 2015AA015307, and National Science Foundation of China under grant numbers 61432006 and 61672232.

References

1. Anagnostopoulos, A., Becchetti, L., Castillo, C., Gionis, A., Leonardi, S.: Power in unity: forming teams in large-scale community systems. In: CIKM
2. Anagnostopoulos, A., Becchetti, L., Castillo, C., Gionis, A., Leonardi, S.: Online team formation in social networks. In: WWW, pp. 839–848. ACM (2012)
3. Azar, Y.: On-line load balancing. In: Online Algorithms
4. Azar, Y., Naor, J., Rom, R.: The competitiveness of on-line assignments. J. Algorithms **18**(2), 221–237 (1995)
5. Baykasoglu, A., Dereli, T., Das, S.: Project team selection using fuzzy optimization approach. Int. J. Cybern. Syst. **38**(2)
6. Bird, C., Pattison, D., D'Souza, R., Filkov, V., Devanbu, P.: Latent social structure in open source projects. In: SIGSOFT, pp. 24–35. ACM (2008)
7. Brooks, F.P.: The Mythical Man-month, vol. 1995. Addison-Wesley, Reading (1975)
8. Chen, S.-J.G., Lin, L.: Modeling team member characteristics for the formation of a multifunctional team in concurrent engineering. IEEE Trans. Eng. Manag. **51**(2), 111–124 (2004)
9. Conway, M.E.: How do committees invent. Datamation **14**(4), 28–31 (1968)
10. Gajewar, A., Sarma, A.D.: Multi-skill collaborative teams based on densest subgraphs. SIAM
11. Kargar, M., An, A.: Discovering top-k teams of experts with/without a leader in social networks. In: CIKM, pp. 985–994. ACM (2011)
12. Lappas, T., Liu, K., Terzi, E.: Finding a team of experts in social networks. In: SIGKDD, pp. 467–476. ACM (2009)
13. Rangapuram, S.S., Bühler, T., Hein, M.: Towards realistic team formation in social networks based on densest subgraphs. In: WWW, pp. 1077–1088 (2013)
14. Strnad, D., Guid, N.: A fuzzy-genetic decision support system for project team formation. Appl. Soft Comput. **10**(4), 1178–1187 (2010)
15. Wi, H., Oh, S., Mun, J., Jung, M.: A team formation model based on knowledge and collaboration. Expert Syst. Appl. **36**(5), 9121–9134 (2009)
16. Yaakob, S.B., Kawata, S.: Workers' placement in an industrial environment. Fuzzy Sets Syst. **106**(3), 289–297 (1999)
17. Zobel, J., Moffat, A.: Inverted files for text search engines. ACM Comput. Surv. (CSUR) **38**(2), 6 (2006)

Cell-Based DBSCAN Algorithm Using Minimum Bounding Rectangle Criteria

Tatsuhiro Sakai[1,2]($\boxtimes$), Keiichi Tamura[1], and Hajime Kitakami[1]

[1] Graduate School of Information Sciences, Hiroshima City University,
Hiroshima, Japan
{ktamura,kitakami}@hiroshima-cu.ac.jp
[2] JSPS, Tokyo, Japan
da65003@e.hiroshima-cu.ac.jp

Abstract. The density-based spatial clustering of applications with noise (DBSCAN) algorithm has been well studied in database domains for clustering multi-dimensional data to extract arbitrary shape clusters. Recently, with the growing interest in big data and increasing diversification of data, the typical size and volume of databases have increased and data have increasingly become high-dimensional. Therefore, a large number of speed-up techniques for DBSCAN algorithms including exact and approximate approaches have been proposed. The fastest DBSCAN algorithm is the cell-based algorithm, which divides the whole data set into small cells. In this paper, we propose a novel exact version cell-based DBSCAN algorithm using minimum bounding rectangle (MBR) criteria. The connecting cells step is the most time-consuming step of the cell-based algorithm. The proposed algorithm can process the connecting cells step at high speed by using MBR criteria. We implemented the proposed cell-based DBSCAN algorithm and show that it outperforms the conventional one in high dimensions.

Keywords: DBSCAN · Density-based clustering · Cell-based DBSCAN algorithm · Minimum bounding rectangle

1 Introduction

With the growing interest in big data, speed-up techniques for data clustering have attracted much attention from data mining researchers. In general, because the data size and number of dimensions are large, it is difficult to set the number of clusters in advance; therefore, general partition-based clustering techniques such as k-means and k-medoids are not an appropriate clustering strategy. A density-based clustering algorithm is one of the simplest but most robust clustering techniques and can extract an arbitrary number of clusters automatically. In density-based clustering, clusters are extracted as high-density regions in which the data is dense.

T. Sakai—JSPS Research Fellow.

© Springer International Publishing AG 2017
Z. Bao et al. (Eds.): DASFAA 2017 Workshops, LNCS 10179, pp. 133–144, 2017.
DOI: 10.1007/978-3-319-55705-2_10

The density-based spatial clustering of applications with noise (DBSCAN) algorithm was first introduced by Ester et al. [2,9], and it applies a density-based concept of spatial clusters. In DBSCAN, clusters are recognized by analyzing the density of data points. Areas with a high density of data points are clusters, whereas areas with a low density are not. The key concept of the DBSCAN algorithm is that for each data point in a cluster, the neighborhood with a user-defined radius has to contain at least a minimum number of points; i.e., the density in the neighborhood must exceed the predefined threshold $MinPts$.

Since the DBSCAN algorithm was first proposed, a large number of speed-up techniques including approximate algorithms and algorithms using index structures have been proposed. The computation time still requires $O(n^2)$ for an exact algorithm on a non-indexed data set because DBSCAN consists of two computationally expensive procedures: range queries and label propagation. Progress on this problem stopped for many years; however, Gunawan recently released a breakthrough technique called the cell-based DBSCAN algorithm [4]. The cell-based DBSCAN algorithm decreases the number of range queries and does not require any label propagation. Gan et al. then proposed both an improved algorithm and an approximate version of cell-based DBSCAN [3].

In this study, we focus on the exact version of the cell-based DBSCAN algorithm. In [3], an approximate method based on the cell-based algorithm, which is called ρ-approximate DBSCAN, showed good performance and the authors claim that their approximate algorithm can extract as same clusters as exact algorithms in almost all cases. However, it is still necessary to speed up the exact version of the cell-based algorithm. As a result of our pre-analysis using benchmark data sets, the most time-consuming processing step of the exact version cell-based algorithm is the search for all connecting cells. Therefore, we propose a novel exact version cell-based algorithm that speeds up this step.

In the cell-based DBSCAN algorithm, the whole data set is divided into cells, where each cell is a $\epsilon/\sqrt{2} \times \epsilon/\sqrt{2}$ square if the data set is in two-dimensional space. If the number of data points in a cell is larger than $MinPts$, which is the threshold value for detecting density, all the data points in the cell are determined to be dense data points automatically. In the connecting cells step, two neighboring cells are connected if and only if an arbitrary pair of data points in these cells are within ϵ. This process requires $O(n^2)$ in the worst case. The proposed algorithm uses minimum bounding rectangle (MBR) criteria to speed-up this checking. If two neighboring cells satisfies conditions of MBR criteria, the proposed algorithm can process this checking with $O(d)$ (d is dimension). We implemented the proposed cell-based DBSCAN algorithm, and evaluated it on benchmark data sets. The results show that the proposed algorithm outperforms the conventional one in high dimensions.

The rest of this paper is organized as follows. In Sect. 2, related work is reviewed. In Sect. 3, the definitions of the density-based spatial clustering algorithm are presented. In Sect. 4, we propose a novel cell-based DBSCAN algorithm. In Sect. 5, we report the experimental results, and in Sect. 6, we conclude the paper.

2   Related Work

Recently, the typical size and volume of databases have increased and data has become increasingly high-dimensional. Therefore, since the DBSCAN algorithm was first proposed, a large number of speed-up techniques for DBSCAN have been proposed, including approximate algorithms [1,6,10]. The approximate algorithms can extract the same clusters as exact algorithms in almost all cases. In this study, we consider the exact version of the DBSCAN algorithm.

Subsequently, a grid-based DBSCAN algorithm that divides the whole data set into grids was proposed [7,10]. The main motivation of the grid-based algorithms is to calculate range queries at high speed by calculating range queries for each grid; however, grid-based algorithms have a time-consuming label propagation process. An exact DBSCAN algorithm using index structures such as the kd-tree was also proposed in [8]. This algorithm requires a huge amount of memory to construct an index for huge volumes of data.

The fastest of DBSCAN algorithm is the cell-based algorithm. Gunawan developed a cell-based algorithm in two dimensions [4]. The cell-based algorithm does not require any label propagation processes because it divides and connects cells. Gan et al. then proposed both an exact and an approximate version of the cell-based algorithm in multiple dimensions [3]. However, as a result of our pre-analysis, the exact version using the bichromatic blosest pair (BCP) consumes a large amount of time to determine all the connecting cells. The proposed novel algorithm can reduce the processing time of this step using MBR criteria.

3   DBSCAN

In this section, the definitions used in DBSCAN are briefly reviewed.

3.1   Definitions

Let P be a set of data points in d-dimensional space. In DBSCAN, the ϵ-neighborhood of a data point is defined as the data points in the neighborhood of a user-defined given radius ϵ.

Definition 1 (ϵ-neighborhood $DPN_\epsilon(p)$). The ϵ-neighborhood of data point p, denoted by $DPN_\epsilon(p)$, is defined as

$$DPN_\epsilon(p) = \{q \in P | dist(p,q) \le \epsilon\}, \tag{1}$$

where the function $dist$ returns the Euclidean distance between data point p and data point q.

Definition 2 (Core data point, Border data point). Data point p is called a core data point if there is at least the minimum number of data points, $MinPts$, in the ϵ-neighborhood $DPN_\epsilon(p)$ ($|DPN_\epsilon(p)| \ge MinPts$). Otherwise, ($|DPN_\epsilon(p)| < MinPts$), p is called a border data point.

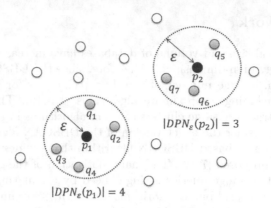

Fig. 1. Example of Definition 2.

An example of this definition is shown on Fig. 1. Suppose that $MinPts$ is set to four. A data point p_1 is a core data point, because there are four data point in $DPN_\epsilon(p_1)$. In contrast, a data point p_2 is a border data point because the number of data point in $DPN_\epsilon(p_2)$ is less than $MinPts$.

Definition 3 (Density-based reachable). Suppose that there is a data point sequence $(p_1, p_2, \cdots, p_n)$. Data point p_n is density-based reachable from p_1 if there is a sequence such that

(1) $p_1, p_2, \cdots, p_{n-1}$ are core points;
(2) p_{i+1} is the ϵ-neighborhood of p_i ($p_{i+1} \in DPN_\epsilon(p_i)$).

A density-based cluster consists of two types of data: core data points, which are mutually density-based reachable, and border data points, which are in the ϵ-neighborhood of the core data point. A density-based cluster is defined as follows.

Definition 4 (Density-based cluster). A density-based cluster (DC) in a data point set satisfies the following restrictions:

(1) $\forall p,\ q \in P$, if and only if $p \in DC$ and q is density-based reachable from p, and q is also in DC.
(2) $\forall p,\ q,\ o \in DC$, p and q are density-based reachable from o.

4 Proposed Algorithm

In this section, we propose a novel cell-based DBSCAN algorithm using MBR criteria.

4.1   Overview

The proposed algorithm is divided into four main steps, which are cell division, core data point determination, connecting cells using MBR criteria and determining border data points or noise. The outline of these four main steps are as follow:

- The cell division step performs dividing the whole data set into cells. The proposed algorithm processes clustering based on the divided cells.
- The core data point determination step performs detemining whether each data point is a core or non-core data point. The proposed algorithm can determine whether each data point satisfy Definition 2 at high speed by using the cells.
- The connecting cells step performs connecting neighborhood cells to form clusters. This step is processed at high speed by using MBR criteria.
- In the determining border data points or noise step, all non-core data points are determined to be a border data point or noise. When all steps is complete, all data points are determined as belonging to a cluster or noise.

4.2   Cell Division

In the first step, the proposed algorithm divides the whole data set into small cells, where each cell is $\epsilon/\sqrt{d} \times \epsilon/\sqrt{d}$ square. This cell size is chosen because it is easy to determine whether the cell is dense or not. Then, the algorithm assigns each data point to the cells.

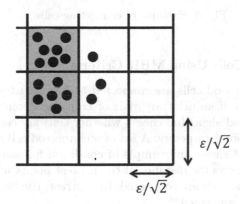

Fig. 2. Example of core data point determination.

4.3   Core Data Point Determination

This step detemines whether each data point is a core or non-core data point. If the number of data points in a cell is lager than $MinPts$, which is the threshold value used to detect density, all data points in the cell are determined to be core data points automatically. That is, if the number of data points in a cell is lager than $MinPts$, all data points in the cell are determined to be core data points. Figure 2 shows an example of cell division and identification of the core data points. In Fig. 2, in which $MinPts = 5$, there are two cells that exceed $MinPts$, as indicated by gray shading. All data points in these cells are determined to be core data points. A data point in the other cells is determined as a core or non-core point by calculating the distance between it and the data points in the neighborhood cells. The neighborhood cells of a cell are the cells within a distance of ϵ from the first cell. If there is at least $MinPts$ in its ϵ-neighborhood, the data point is determined to be a core data point.

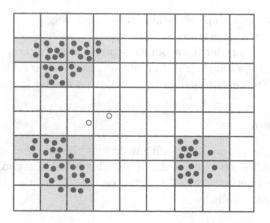

Fig. 3. Example of connecting cells.

4.4   Connecting Cells Using MBR Criteria

In this step, neighborhood cells are connected to form clusters. These cells are connected if and only if an arbitrary pair of core data points in two cells are within ϵ. The proposed algorithm checks whether cells are connected or not for all cells containing core data points. A set of a connected cell forms a cluster, as shown Fig. 3. Figure 4 shows an example of condition for connecting cells. The upper side of Fig. 4 shows the distance of the nearest points in two cells. On the left side of Fig. 4, these cells are connected. In contrast, the two cells on the right side of Fig. 4 are not connected.

This process requires $O(n^2)$ in the worst case because if there is not pair of core data points, this process must calculate the distance between all core data points. Hence, the proposed algorithm uses MBR criteria [5] to speed up this checking. The proposed algorithm calculates the $MINMINDIST$ and $MINMAXDIST$ between MBRs. Let d be the dimension and $MBR(S, T)$ be the MBR of the cell; then, S and T are the maximum and minimum for each dimension: $S = (s_1, s_2, \cdots, s_d)$ and $T = (t_1, t_2, \cdots, t_d)$. The calculating of

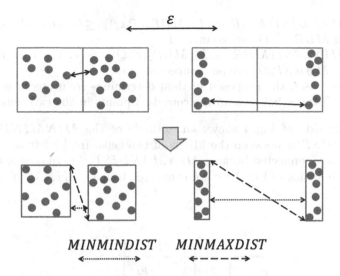

MINMINDIST   MINMAXDIST

Fig. 4. Example of connecting cells using MBR criteria.

$MINMINDIST$ requires $O(d)$. The $MINMINDIST$ between $MBR_1(S,T)$ and $MBR_2(P,Q)$ is defined as follows:

$$MINMINDIST(MBR_1(S,T), MBR_2(P,Q)) = \sqrt{\sum_{i=1}^{d} y_i^2}$$

$$y_i = \begin{cases} p_i - t_i, & if\ p_i > t_i \\ s_i - q_i, & if\ s_i > q_i \\ 0, & otherwise \end{cases} \tag{2}$$

The order time for calculating $MINMAXDIST$ is $O(2^d)$ on $d \geq 4$. That is, higher dimensions increase the calculation time of $MINMAXDIST$. The proposed method uses an upper bound of $MINMAXDIST$, which is $O(d)$ to calculate. The upper bound of $MINMAXDIST$ between $MBR_1(S,T)$ and $MBR_2(P,Q)$ is defined as follows:

$$MINMAXDIST(MBR_1(S,T), MBR_2(P,Q)) =$$

$$\sqrt{\min_{1 \leq j \leq d} \left\{ x_j^2 + \sum_{i=1,i \neq j}^{d} y_i^2 \right\}} \tag{3}$$

where, $x_j = \min\{|s_j - p_j|, |s_j - q_j|, |t_j - p_j|, |t_j - q_j|\}$ and $y_i = \max\{|s_i - q_i|, |t_i - p_i|\}$.

In the proposed algorithm, the connecting cells conditions using the $MINMINDIST$ and $MINMAXDIST$ between the MBRs of two cells are as follows:

- If $MINMAXDIST(MBR_1(S,T), MBR_2(P,Q)) \leq \epsilon$, the cells of MBR_1 (S,T) and $MBR_2(P,Q)$ are connected.
- If $MINMINDIST(MBR_1(S,T), MBR_2(P,Q)) > \epsilon$, the cells of MBR_1 (S,T) and $MBR_2(P,Q)$ are not connected.
- In all other cases, the proposed method determines connectedness by calculating the distance between all the core data points in the two cells.

The lower side of Fig. 4 shows an example of the $MINMINDIST$ and $MINMAXDIST$ of between the MBRs of two cells. In the left side of Fig. 4, the two cells are connected because $MINMAXDIST \leq \epsilon$. In contrast, the two cells on the right side of Fig. 4 are not connected because $MINMINDIST > \epsilon$.

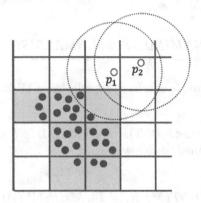

Fig. 5. Example of determining border data points or noise.

4.5   Determining Border Data Points or Noise

Finally, the proposed method determines border data points or noise. All non-core data points are determined to be a border data point or noise. If there is at least one core data point in the ϵ-neighborhood of a data point, it belongs to the cluster of the closest core data point. In contrast, a data point does not belong to a cluster if there is not a core data point in its ϵ-neighborhood. Figure 5 shows an example of determining border data points or noise. There is a core data point in the ϵ-neighborhood of p_1. That is, p_1 belongs to the cluster of the core data point. In contrast, there is not a core data point in the ϵ-neighborhood of p_2. In this case, p_2 does not belong to a cluster and become noise. When all steps is complete, all data points are determined as belonging to a cluster or noise.

4.6   Algorithm

The overall algorithm is as follows:

(1) The algorithm receives a set of data points P in d-dimensional space, and parameters ϵ and $MinPts$.

(2) The algorithm divides the whole data set into cells, where each cell is $\epsilon/\sqrt{d} \times \epsilon/\sqrt{d}$ square.

(3) Each data point are assigned to the cells.

(4) The algorithm calculates a neighborhood cells for each cell included a data point.

(5) The algorithm detemines whether each data point is a core or non-core data point.

(6) The algorithm connects the cells using MBR criteria for each cell. A set of connected cell forms a cluster.

(7) The algorithm determines border data points or noise for each data point not assigned to a cluster.

(8) The algorithm returns a set of clusters.

5   Experiments

To evaluate the proposed algorithm, we conducted an experiment using synthetic datasets. We implemented the proposed cell-based DBSCAN algorithm using MBR criteria and conducted the experiment on a PC with an Intel Xeon E5-1270 v2 @3.5 GHz CPU and 32 GB of memory. In the experiment, we used synthetic datasets generated by the data generator scheme of [3]. We used datasets of dimension $d = 3, 5, 7$, and 9 and varied the number of data points from 1,000,000 to 10,000,000. Moreover, the parameters were set to $\epsilon = 5,000$ and $MinPts = 100$, which are same parameters used [3].

In the experiments, we measured the processing time of the proposed cell-based DBSCAN algorithm using MBR criteria (CDBSCAN_MBR). We compared the proposed algorithm with the simplest cell-based DBSCAN algorithm (CDBSCAN) extended to higher dimensions [4] and the fastest exact cell-based DBSCAN algorithm (CDBSCAN_BCP) [3]. We used the binary file of CDB-SCAN_BCP provided by the authors. Figure 6 shows the processing time for each dimension. For $d = 3$, $d = 5$ and $d = 7$, CDBSCAN is the slowest of the three algorithms. For $d = 3$ and $d = 5$, CDBSCAN_BCP is faster than CDB-SCAN and CDBSCAN_MBR. However, for $d = 7$ and $d = 9$, CDBSCAN_MBR is faster than CDBSCAN_BCP. For $d = 9$, the difference between the processing time of CDBSCAN_BCP and CDBSCAN_MBR for 5,000,000 data points is about 900 seconds. These results show that CDBSCAN_MBR is faster than CDBSCAN_BCP in higher dimensions.

Table 1 shows the number of times that the MBR criteria conditions were used to determine connecting cells in CDBSCAN_MBR. At low dimensions, the number of times the MBR criteria conditions were met is low. However, at high dimensions, the MBR criteria conditions were often used. These results show that the proposed algorithm is an effective approach at high dimensions.

Tables 2, 3, and 4 show the processing time for 5,000,000 data points for each step. The connecting cells step consumes the most time for all three algorithms, as shown these tables. Therefore, if an algorithm can decrease the processing

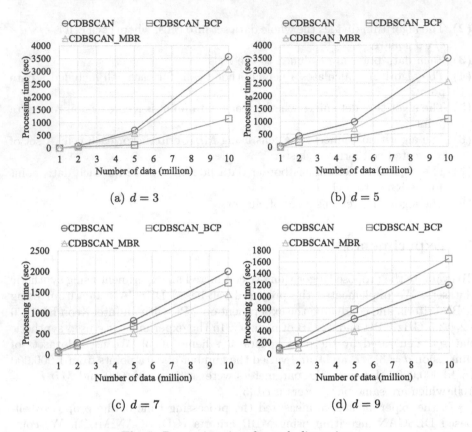

Fig. 6. Processing time for each dimension.

Table 1. The number of times (rate) that applied MBR criteria conditions in the connecting cells step in CDBSCAN_MBR.

d	$MINMINDIST$	$MINMAXDIST$	Other cases	Total
3	398 (0.12)	1524 (0.44)	1536 (0.44)	3458
5	23262 (0.20)	46985 (0.41)	45657 (0.39)	115904
7	219819 (0.25)	385310 (0.44)	278164 (0.31)	883293
9	946982 (0.27)	1636529 (0.48)	860849 (0.25)	3444360

time of this step, the entire processing time can be speed up. CDBSCAN_MBR is faster at calculating this step than CDBSCAN_BCP for $d = 7$ and $d = 9$. As a result, the overall processing time of CDBSCAN_MBR is faster than that of CDBSCAN_BCP for $d = 7$ and $d = 9$.

Table 2. The processing time for each step on CDBSCAN.

d	Processing time (sec)			
	Cell division	Core data point determination	Connecting cells	Determining border data points or noize
3	2.23	0.18	669.69	0.02
5	3.26	3.56	971.27	0.01
7	4.45	17.64	787.17	0.01
9	5.50	47.38	552.39	0.01

Table 3. The processing time for each step on CDBSCAN_BCP.

d	Processing time (sec)			
	Cell division	Core data point determination	Connecting cells	Determining border data points or noize
3	0.42	0.06	106.77	0.02
5	0.47	0.40	368.43	0.01
7	0.56	3.55	680.98	0.01
9	0.65	15.03	760.12	0.01

Table 4. The processing time for each step on CDBSCAN_MBR.

d	Processing time (sec)			
	Cell division	Core data point determination	Connecting cells	Determining border data points or noize
3	2.09	0.19	560.03	0.02
5	3.31	3.52	719.06	0.01
7	4.46	17.29	485.32	0.01
9	5.60	46.77	333.95	0.01

6   Conclusion

This paper proposed a novel exact cell-based DBSCAN algorithm. The proposed algorithm decreases the number of range queries and does not require any label propagation processes because it divides and connects cells instead. Moreover, the proposed algorithm utilizes MBR criteria to speed-up the processing of connecting cells. The results of the experiments showed that the proposed algorithm outperforms the conventional exact cell-based DBSCAN algorithm at high dimensions. In our future work, we intend to develop a faster algorithm than the proposed algorithm at low dimensions. Moreover, we intend to conduct experiments using real datasets.

Acknowledgments. This work was supported by JSPS KAKENHI Grant Number JP16J05403 and JP26330139, and the MIC/SCOPE #162308002.

References

1. Borah, B., Bhattacharyya, D.K.: An improved sampling-based DBSCAN for large spatial databases. In: 2004 Proceedings of International Conference on Intelligent Sensing and Information Processing, pp. 92–96 (2004)
2. Ester, M., Kriegel, H.P., Sander, J., Xu, X.: A density-based algorithm for discovering clusters in large spatial databases with noise. In: Proceedings of the Second International Conference on Knowledge Discovery and Data Mining KDD-1996, pp. 226–231 (1996)
3. Gan, J., Tao, Y.: DBSCAN revisited: mis-claim, un-fixability, and approximation. In: Proceedings of the 2015 ACM SIGMOD International Conference on Management of Data SIGMOD 2015, pp. 519–530 (2015)
4. Gunawan, A.: A faster algorithm for DBSCAN. Master's thesis, Technische University Eindhoven (2013)
5. Liria, A.L.C.: Algorithms for processing of spatial queries using R-trees. The closest pairs query and its application on spatial databases. Ph.D. thesis, Department of Languages and Computation, University of Almeria (2002)
6. Liu, B.: A fast density-based clustering algorithm for large databases. In: 2006 International Conference on Machine Learning and Cybernetics, pp. 996–1000 (2006)
7. Mahran, S., Mahar, K.: Using grid for accelerating density-based clustering. In: 2008 8th IEEE International Conference on Computer and Information Technology, pp. 35–40 (2008)
8. Mai, S.T., Assent, I., Storgaard, M.: AnyDBC: an efficient anytime density-based clustering algorithm for very large complex datasets. In: Proceedings of the 22nd ACM SIGKDD International Conference on Knowledge Discovery and Data Mining KDD 2016, pp. 1025–1034 (2016)
9. Sander, J., Ester, M., Kriegel, H.P., Xu, X.: Density-based clustering in spatial databases: the algorithm GDBscan and its applications. Data Min. Knowl. Discov. **2**(2), 169–194 (1998)
10. Tsai, C.F., Wu, C.T.: GF-DBSCAN: a new efficient and effective data clustering technique for large databases. In: Proceedings of the 9th WSEAS International Conference on Multimedia Systems Signal Processing MUSP 2009, pp. 231–236 (2009)

Time-Aware and Topic-Based Reviewer Assignment

Hongwei Peng, Haojie Hu, Keqiang Wang, and Xiaoling Wang[(✉)]

Shanghai Key Laboratory of Trustworthy Computing,
MOE International Joint Lab of Trustworthy Software,
East China Normal University, Shanghai, China
penghongwei_phw@163.com, ideashhj@126.com,
sei.wkq2008@gmail.com, xlwang@sei.ecnu.edu.cn

Abstract. Peer review has become the most widely-used mechanism to judge the quality of submitted papers at academic conferences or journals. However, a challenging task in peer review is to assign papers to appropriate reviewers. Both the research directions of reviewers and topics of submitted papers are often multifaceted. Besides, reviewers' research direction may change over time and their published papers closer to current time reflect their current research direction better. Hence in this paper, we present a time-aware and topic-based reviewer assignment model. We first crawl papers published by reviewers over years from web, and then build a time-aware reviewers' personal profile using topic model to represent the expertise of reviewers. Then the relevant degree between reviewer and submitted paper is calculated through the similarity measure. In addition, by considering statistical characteristics such as TF-IDF of the papers, the matching degree between reviewer and submitted paper is further improved. At the same time, we also consider the quality of all past reviews to measure the reviewers' present reviews. Extensive experiments on a real-world dataset demonstrate the effectiveness of the proposed method.

Keywords: Reviewer assignment · Topic model · Expert retrieval · Time aware

1 Introduction

Peer review has become the most common way to judge the quality of the submitted papers at academic conferences or journals. However, assigning all submitted papers to appropriate reviewers by conference organizers and journal editors manually is truly a time-consuming task [2]. Conferences or journals require both reviewers and submitted papers to fill in the research directions so that conference organizers or journal editors assign papers to reviewers based on their research directions. Recently, another more common way of assigning papers is to allow reviewers to bid papers they are interested in, which reduces the workload of conference organizers and journal editors and takes into account

© Springer International Publishing AG 2017
Z. Bao et al. (Eds.): DASFAA 2017 Workshops, LNCS 10179, pp. 145–157, 2017.
DOI: 10.1007/978-3-319-55705-2_11

the reviewers' preferences. Apparently, there are many problems with these manual assignments. For the first solution, since the reviewers' research directions may change over time while their research directions do not be regularly updated, it may cause a bias of assignment. Even though the information of reviewers is updated in time, it is also very difficult to assign the papers to right reviewers only based on key words by conference organizers or journal editors, who are not familiar with all research directions. In bidding, reviewers may not have enough time to go through the entire submission list hence their bidding is unable to accurately reflect their interest preferences [1]. Besides, reviewers are only able to read abstracts and titles of papers instead of the complete papers which may affect their judgments. To conclude all these, an automatic reviewer assignment system based on the limitations of manual assignment is an urgent need.

Topic model is an effective method to model the research directions of reviewers and the submitted papers. Maryam *et al.* [8] capture the feature that reviewers and submitted papers may have multiple research directions. They use mixture language models to model research directions of reviewers and submitted papers. But they do not consider the temporal impacts on reviewers' research directions. Li *et al.* [9] present a reviewer assignment system which takes into account both reviewers' preferences and the relevance of reviewers and submitted papers. Though they consider the impacts of time, they do not combine time with the reviewers' research direction.

Therefore, in this paper, we present a time-aware and topic-based reviewer assignment model, which not only considers the referent degree between the research directions of reviewers and submitted papers, but also captures the changes of research directions of reviewers as time goes on. Firstly, we use topic model such as Latent Semantic Allocation (LDA [10]) to get the topic distribution of reviewers' published papers. Then we integrate temporal impact and build a time-aware reviewers' personal profile. Above process can be completed offline. Next, we measure the matching degree through calculating the similarity between the time-aware reviewers' topic distribution and submitted papers' topic distribution. In addition, by combining the statistical characteristics of the text (*e.g.,* TF-IDF), the matching degree between reviewer and submitted paper is further improved. We also capture the influence of reviewers' review quality, which can ensure that conference organizers or journal editors assign a paper to a relevant reviewer who has good review quality at the same time. At last, we demonstrate the effectiveness of our proposed model with a real-world dataset obtained from a journal which used our system. The experimental results show that our proposed model can assign the paper to proper reviewers.

This paper is organized as follows. Section 2 describes the related work. Then we propose a time-aware and topic-based reviewer assignment model in Sect. 3. Section 4 demonstrates the effectiveness of the proposed model by experiments. Conclusion and future work are followed in Sect. 5.

2 Related Work

In this section, we review the related work. Generally, reviewer assignment problem could be divided into two categories: retrieval-based and assignment-based methods.

Retrieval-based method regards reviewer assignment as an information retrieval problem which is solved by retrieval techniques. In retrieval-based method, a submitted paper is a query, and a reviewer is a document (the papers published by the reviewer). Therefore, finding the most appropriate reviewer becomes a document top N ranking problem. The earliest researchers who studied reviewer assignment problem are Dumais and Nielsen [3]. They match papers to reviewers by using Latent Semantic Indexing (LSI) trained on the abstracts of reviewers' published papers. Basu et al. [4] extract reviewers' published papers from web via search engines and model reviewer assignment problem by using Vector Space Model. Hettich and Pazzani [5] present a recommendation system for recommending panels of reviewers for NSF grant applications. They use TF-IDF weighted vector space to calculate the matching degree between reviewers and papers. In addition to these information retrieval techniques, topic discovery model is also an effective technique to solve reviewer assignment problem. Wei and Croft [6] propose a document retrieval model based on LDA within the language modeling framework. David and Andrew [7] propose a novel topic model, Author-Persona-Topic (APT). APT model uses some features to capture the relevance between a paper and a reviewer. The underlying assumption of this model is authors might write about different topic areas; so they learn the topical components as well as divide each author's papers into several "personas" by learning topic model. Similar to [7], Maryam et al. [8] believe that there are more than one research direction in both reviewers and papers. So they use a mixture language model to model multiple research directions so that the reviewers should not only have the required expertise to review a paper but also can cover all the research directions of a paper. Li and Watanabe et al. [9] design a reviewer assignment system combining a preference and topic based approach, which takes into account both reviewers' preferences and the relevance of reviewers and submitted papers to measure final matching degree between reviewers and submitted papers.

Another approach is assignment-based method [11–19]. The main idea of this method is to maximize total matching degree of reviewer assignment in the premise of given workload of each reviewer and the number of reviewers which each submitted paper should be reviewed. The method generally constructs the reviewers set and the submitted papers set into a weighed bipartite graph. The weight of the graph's edge is denoted by the matching degree between reviewers and papers. Then we can use the bipartite graph to find the maximum reviewer assignment scheme that satisfies the constraints. It can obtain a better reviewer assignment scheme, but some interdisciplinary papers may be reviewed only by some reviewers with narrow expertise [20]. The topic-based approach is a good solution to solve this problem. At the same time, by limiting the workload of each reviewer and the number of reviewers which each paper should be reviewed by, it can guarantee the fairness of the assignment. Therefore, the model presented in this paper is exactly based on topic model.

3   Framework of Reviewer Assignment

Although there have been many studies that focus on reviewer assignment problem, most of the researches are still focus on bidding. Considering that the research directions of both reviewers and submitted papers may be multifaceted, it is a good way to use topic model to build profile of reviewers and submitted papers. Noting that the research directions of reviewers may change over time. Therefore in this paper, we propose a time-aware and topic-based reviewer assignment model.

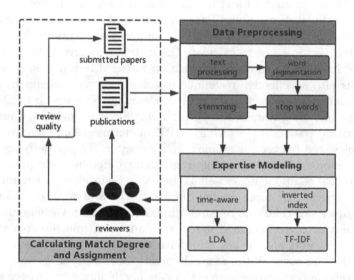

Fig. 1. Framework of proposed reviewer assignment system

Figure 1 shows our system framework. It consists of three parts: Data Preprocessing, Expertise Modeling, and Calculating Matching Degree and Assignment. Data Preprocessing focuses on processing reviewers' publications and submitted papers into proper data form which is available in the next phase. It mainly contains text processing, word segmentation, filtering out the stop words and stemming. Expertise Modeling focuses on modeling the profile for reviewers and submitted papers. We measure the matching degree between reviewers and submitted papers from two aspects of both topic model (*e.g.*, LDA) and text statistical feature (*e.g.*, TF-IDF). In building reviewers' personal profile, we consider the influence of time on the research directions of reviewers. The first two parts are completed offline, and online calculating is only in the third part. Calculating Matching Degree and Assignment is divided into two parts. In the part of Calculating Matching Degree, we consider the influence of review quality on reviewer assignment problem. Firstly, we obtain the degree of expertise matching between reviewers and submitted papers by the similarity calculation. Then we get the final matching degree of them by combining review quality of reviewers. The part of Assignment is to make the best assignment under the constraint conditions. Table 1 summarizes the frequently used notations in this paper.

Table 1. The frequently used notations in this paper

Symbol	Description
R	The set of reviewers $\{r_1, r_2, ..., r_{|R|}\}$
D	The set of papers $\{d_1, d_2, ..., d_{|D|}\}$
X	A $|D| \times |R|$ assignment matrix
K	The number of topics
$\mathbf{p}_j$	The topic distribution of paper d_j
$\mathbf{q}_i$	The topic distribution of reviewer r_i
δ_p	Paper Size Constraint
δ_r	Reviewer Workload Constraint
S_{ij}	Review score of paper d_j Given by reviewer r_i
T_{ij}	Time score of reviewer r_i for reviewing paper d_j
f_n^i	The n-th review quality of reviewer r_i
$\hat{f}_n^i$	Reviewer r_i's overall quality of the first n-th reviews

3.1 Data Preprocessing

Data preprocessing is to transfer reviewers' publications and submitted papers into text form which can be processed for modeling expertise in the next stage. We extract titles, abstracts and texts of papers to model the profile of reviewers and papers. Because those papers are all in pdf format, we first convert their format into text and removed the irrelevant information. By those ways of word segmentation, filtering out the stop words and stemming, we obtain a standardized text form finally. Since the title and abstract are the most important parts in a paper, we extract papers' titles and abstracts and copied N copies (*e.g.*, $N = 20$) to the original text to increase their exposure, and we do this after text processing.

3.2 Expertise Modeling

The process of Expertise Modeling is to establish profile of reviewers and submitted papers. Research directions of reviewers and papers often involve in many aspects, and topic model is exactly a good approach by using topic distribution to characterize these research directions. Here, we use LDA model to generate topic distribution for submitted papers and reviewers' published papers.

Topic distribution of submitted papers can be obtained directly through LDA model. For the topic distribution of reviewers, we need to consider how to get topic distribution of reviewers by using topic distribution of those reviewers' submitted papers. A natural idea is to add all topic distributions of papers published by the reviewer to indicate the topic distribution of the reviewer. Then the probability of reviewer r_i belonging to the k-th topic can be expressed as:

$$q_i[k] = \frac{\sum_{j=1}^{|D_i|} p_j^i[k]}{\sum_h^K \sum_{j=1}^{|D_i|} p_j^i[h]} \tag{1}$$

where $|D_i|$ denotes the number of reviewer r_i's papers. $p_j^i[k]$ denotes the probability that the paper d_j of reviewer r_i belongs to the k-th topic. Denominator is a normalization factor that guarantees the sum of all topics' probabilities is 1.

However, it does not consider the influence of time on research directions of reviewers. With continuous updating and progress of specialized domain knowledge, reviewers' research directions may change over time. For example, an expert's early research direction was database theory some years ago, but now he may be more interested in data mining and machine learning in today's big data age. Therefore, the research directions of reviewers are affected by time. We use publications of reviewers to represent their expertise, so those papers whose published dates are closer to the current time reflect reviewers' current research directions better. Thus we can characteristic the probability of reviewer r_i on the k-th topic by following formula:

$$q_i[k] = \frac{\sum_{j=1}^{|D_i|} \alpha^{t_{now}-t_{ij}} \cdot p_j^i[k]}{\sum_h^K \sum_{j=1}^{|D_i|} \alpha^{t_{now}-t_{ij}} \cdot p_j^i[h]} \tag{2}$$

where t_{now} represents current year, and t_{ij} denotes paper d_j's published date of reviewer r_i. We represent impact of time on reviewers' topics by $\alpha^{t_{now}-t_{ij}}$, and control the influence degree of papers' published date on expertise direction by α.

The process of building profile for reviewers is offline. Next, we can obtain the degree of expertise matching between reviewer r_i and submitted paper d_j based on topic model by similarity calculation (*e.g.*, cosine similarity):

$$degree\_topic_{ij} = \frac{\mathbf{q}_i \cdot \mathbf{p}_j}{|\mathbf{q}_i| \cdot |\mathbf{p}_j|} \tag{3}$$

In order to further improve matching degree between reviewers and submitted papers, we consider to combine other information retrieval techniques. Paper [5] verifies the effectiveness of TF-IDF technique on reviewer assignment problem. As a kind of statistical characteristics in the language model, TF-IDF can express texts' characteristics well by using statistical characteristics of vocabularies as a feature set. Assuming that the TF-IDFs of reviewer r_i's publications and submitted paper d_j are represented as $tfidf_i$ and $tfidf_j$, respectively, the matching degree between reviewer r_i and submitted paper d_j based on statistical characteristics can also be obtained by calculating cosine similarity:

$$degree\_tfidf_{ij} = \frac{tfidf_i \cdot tfidf_j}{|tfidf_i| \cdot |tfidf_j|} \tag{4}$$

We establish inverted index for collected papers of reviewers. By calculating those collected papers' TF-IDFs in advance, we save the time required to calculate matching degrees. Therefore, combining with matching degrees based on topic model and statistical characteristics, we obtained expertise matching degree between a reviewer r_i and the submitted paper d_j:

$$degree\_expertise_{ij} = \beta \cdot degree\_topic_{ij} + (1 - \beta) \cdot degree\_tfidf_{ij} \tag{5}$$

β can be used to control the proportion of expertise matches based on topic model and text features in entire match.

3.3   Calculating Matching Degree and Assignment

This part consists of two parts: Calculating Match Degree and Assignment. Calculating Match Degree part calculates final matching degree between reviewers and submitted papers. Assignment part is to make a reasonable reviewer assignment on the premise of satisfying constraints.

Calculating Matching Degree. Intuitively, reviewers should be related to the assigned papers and give high quality reviews. If a reviewer submits his review comment on deadline, he maybe just want to complete the task and write the review comment in a hurry. As a consequence, it is possible that the reviewer may not have a thorough understanding of this paper and the quality of review comment is low. Therefore, for those reviewers who have poor review quality, even though their expertise matching degrees are high, it is not a good way to assign the submitted papers to these kinds of reviewers or give a high weight of these kinds of reviewers.

In general, a reviewer has a review for each review paper. A part of review is reviewer's rating for the review paper, which indicates the reviewer's admissive degree to the paper. In addition, a rating can be given to the submitted date of review comment by conferences or journals which is called time score. These two indicators directly affect the quality of review comment, which is shown as follows:

(1) Review quality is primarily determined by review score of reviewers. We assume that most reviewers' review scores could represent true review score of the paper. So when a reviewer's review score is significantly different from others', we think his review quality is relatively poor. An intuitive approach is to evaluate review quality by calculating the difference between a reviewer's review score and the average of all the review scores. The smaller the difference is, the better review quality is.

(2) In general, we think the quality of a review comment whose submission time is close to deadline is relatively poor, since this situation may imply

that reviewers give the review score hastily in order to complete the task. Therefore, we can quantize the time score as a weighting factor T, and T can obtain three difference values in three different stages: in normal time, in 1–2 days before the due date, and beyond the due date. The higher the value of T is, the better the review quality is.

Assuming $\hat{f}_n^i$ denotes reviewer r_i's overall quality of the first n reviews. And d_j denotes the $(n+1)$-th review paper. S_{ij} denotes the review score of paper d_j given by reviewer r_i. M denotes the number of reviewers assigned to paper d_j. The review time score of reviewer r_i from conference or journal is T_{ij}. Considering both time score and review score, we get the $(n+1)$-th review quality of review r_i:

$$f_{n+1}^i = e^{-\frac{|S_{ij} - \frac{\sum_{k=1}^M S_{ik}}{M}|}{T_{ij}}} \tag{6}$$

The greater the value of the review score difference $|S_{ij} - \frac{\sum_{k=1}^M S_{ik}}{M}|$ or the smaller the value of the time score T_{ij} is, the greater the value of f_{n+1}^i is. Moreover, we guarantee that the value of f_{n+1}^i is between 0 and 1 by an exponential function. Therefore, the $(n+1)$-th overall review quality of reviewer r_i is:

$$\hat{f}_{n+1}^i = \frac{\hat{f}_n^i \cdot n + f_{n+1}^i}{n+1} \tag{7}$$

Reviewer's review quality can be calculated offline. Once a new review comment arrives, the updating is also fast. In this way, the final matching degree between reviewer r_i and submitted paper d_j can be obtained by combining expertise matching degree with reviewer's review quality:

$$degree_{ij} = \gamma \cdot degree\_expertise_{ij} + (1 - \gamma) \cdot \hat{f}_n^i \tag{8}$$

where γ indicates the weights of both expertise matching degree and reviewer's review quality in the final matching degree.

Assignment. Reviewer assignment problem not only requires high matching degree between a reviewer and a submitted paper, but also requires a reasonable assignment scheme. Reviewer assignment scheme can be defined as achieving the maximum matching degree under the constraint conditions. The constraints mainly include two conditions:

- **Paper Size Constraint** δ_p: the number of reviewers assigned to each paper.
- **Reviewer Workload Constraint** δ_r: the maximum papers' number that each reviewer could review.

We use an assignment matrix $X \subseteq |D| \times |R|$ to denote papers' assignment condition:

$$X_{ij} = \begin{cases} 1 \text{ if } d_j \text{ is assinged to } r_i \\ 0 \text{ otehrwise} \end{cases} \tag{9}$$

After defining a weight function A, we can define reviewer assignment problem as follows:

$$max \quad \sum_{r_i \in R} \sum_{d_j \in D} A(i,j) \times X_{ij}$$

$$where \begin{cases} \forall d_j \in D : \sum_{i=0}^{|R|} X_{ij} = \delta_p \\ \forall r_i \in R : \sum_{j=0}^{|D|} X_{ij} \le \delta_r \\ \forall d_j \in D , \ \forall r_i \in R : X_{ij} \in \{0,1\} \end{cases} \quad (10)$$

Reviewer assignment problem is actually an NP-hard problem. In this paper we use greedy algorithm to get a local optimal assignment scheme. There are many studies having been done. Several studies have been done about this research aspect. The simplest greedy algorithm can achieve a $\frac{1}{3}$-approximation ratio compared to the exact solution. The latest research [20] has approximation guarantee of factor $\frac{1}{2}$ to get a near-optimal solution. Because the main focus of this paper is to improve the matching degree between reviewers and submitted papers, we only pick up a simple greedy algorithm in assignment scheme.

4   Experiment

In this section, we do some experiments to evaluate the performance of the model on real-world journal system. We develop a reviewer assignment system and apply it to FCS (Frontiers of Computer Science) journal. In order to prepare experiment data, we crawled 190 experts' publications from the Internet. These experts are in the field of computer and are all reviewers of FCS. We crawled at least 10 papers for each expert, a total of 3642 papers. We evaluate our model's performance by the feedback obtained from the way assigning papers to those reviewers by using the system. The experimental settings and experimental results are described in detail below.

4.1   Experimental Settings

Here we describe the parameter settings in the experiment. Time weighted factor α is set to 0.7. Expertise weighted factor β is set to 0.8. Match degree factor γ is set to 0.8. Our experimental results shows that under these parameters, our system can achieve great performance. In addition, topic K value of 100 can reflect the number of research directions of computer science in the real situation.

4.2   Experimental Results

First, we test the difference between reviewers assigned by our system and the reviewers manually matched by FCS journal for a given paper. We select 10 papers published on FCS. FCS journal editors assign each paper to at least 3 reviewers, so we get 35 reviewer-paper samples. However, since some reviewers do not exist in our system, there are only 23 samples being actually effective. We assign the best five matched reviewers to each paper. The experimental results show that there are 8 samples that are in the manual assignment set of FCS, which is to say the hit rate is 34.8%.

We are not surprised by this result. Because manual matching cannot capture the feature that reviewers and submitted papers may have multiple research directions, it is normal that the hit rate of reviewers assigned by our system is not high. Actually, by studying these 10 papers carefully, we find the research directions of both reviewers assigned by our system and test papers are more consistent, while there are some inconsistencies in the research directions of both reviewers matched manually and test papers.

In order to verify the matching degree between papers assigned through our system and reviewers, we do a further experiment. We pick up a five-level relevance scheme to represent the matching degree between reviewers and papers, which is expressed as follows:

- **Perfect Relevant** is expressed in 5 points. This indicates that the research directions of the review paper are exactly in line with the current research directions of the reviewer.
- **Very Relevant** is expressed in 4 points. This indicates that the research directions of the review paper are basically consistent with the current research directions of the reviewer.
- **Relevant** is expressed in 3 points. This indicates the research directions of the review paper and the current research directions of the reviewer are only in a same big research direction, but specific research directions of them are different. For example, the research direction of a paper is Software Verification and Testing, while the reviewer's research direction is Software platform. Though the research directions of them are both Software, their relevant degree is not strong.
- **Slightly Relevant** is expressed in 2 points. This indicates the research directions of the review paper and the research directions of the reviewer only have a little correlation, and are just similar in the big research direction.
- **Irrelevant** is expressed in 1 point. This indicates the research directions of the review paper and the reviewer have no correlation.

We select 21 representative papers (they cover 15 research directions) and match 5 most suitable reviewers to each paper through our system. Then we assign the papers to those reviewers and let them rate those papers according to the relevant degree we specify. Because some reviewers do not give the feedback to us, we finally get 29 reviewers' records. The score records are demonstrated in Table 2.

You can see that 55% of reviewers believe that the papers recommended for them are perfect relevant with their current research directions, and 31% believe that the papers recommended to them are very relevant with their current research directions. Those show the model we propose is effective. Obviously, we should notice that there are 4 reviewers who think that the papers recommended

Table 2. The number of relevance at different levels

Relevance	5	4	3	2	1
Numbers	16	9	0	3	1

for them are slightly relevant or even irrelevant to their research directions. The main reason for the low relevant degree is that the publications of our selected reviewers cannot be a good representation for their current directions, which can happen in the following cases:

1. Those crawled papers published by those reviewers are all their early papers, not their latest papers. Hence, the papers recommended to them are not consistent with their current research directions. In fact, this also shows the influence of time on the reviewers' research directions.
2. Those papers we crawled are not wrote by those reviewers we recommended, but they are published by those experts who have the same names as reviewers we try to recommend papers to. In this case, the reviewers' profiles we obtained definitely cannot reflect those reviewers' research directions.

Finally, by observing the reviewers' topic distribution generated by our model, we can find that reviewers' research directions are indeed diverse. We randomly select five reviewers from the 190 reviewers we crawled. Table 3 shows the indexes of the five most relevant topics for the 5 reviewers. Generally, a probability value of a reviewer on a topic greater than 15% indicates that the topic is very relevant to the reviewer.

Table 3. The topic distribution of five reviewers sampled. We show the indexes of the five most relevant topics for each reviewers. The values in parentheses indicate the probability that the reviewer belongs to the topic. The bold topic indicates that the topic is very relevant to the search direction of the reviewer.

Reviewer	Five main topics of reviewer				
A	**52**(0.343)	**5**(0.221)	83(0.126)	92(0.061)	54(0.0423)
B	**55**(0.215)	**85**(0.200)	19(0.085)	35(0.060)	15(0.054)
C	**21**(0.253)	**13**(0.229)	**15**(0.154)	18(0.067)	55(0.063)
D	**74**(0.342)	**54**(0.236)	86(0.098)	88(0.079)	3(0.061)
E	**5**(0.282)	2(0.099)	55(0.098)	83(0.071)	50(0.062)

From the table, most reviewers have diverse research directions and generally have at least two research directions. Evidently, there are some reviewers whose research directions are relatively simple. For example, reviewer E's research direction is mainly topic 5. Further, by browsing E' homepage, we know topic 5 actually corresponds to the research direction of Machine Learning. While reviewer A has similar research directions with reviewer E, and his research directions have topic 52 (it corresponds to the research direction of Knowledge Discovery and Data Mining) in addition to topic 5, we can find these two topics are very relevant.

5   Conclusion

In this paper, we analyze current problems in reviewer assignment system and further propose a time-aware and topic-based reviewer assignment model. This model can capture the diverse nature of reviewers and papers' research directions and consider the influence of time on the reviewer's research direction. Meanwhile, the model can further improve the matching degree between reviewers and submitted papers by utilizing text statistical features. In addition, we model the influence of reviewers' review quality on matching degree. Thus, we ensure that through our system, conference organizers and journal editors can assign a paper to a reviewer who is relevant with the paper and of good review quality. We use a real journal' feedback data to validate the effectiveness of model we proposed. The experimental results prove that our proposed model can achieve an appropriate assignment of reviewers to papers.

A key question for future work is to make a more effective reviewer assignment scheme under the constraint conditions. Considering more informations of reviewers' published papers in future could help to improve the matching degree between papers and reviewers.

Acknowledgments. This work was supported by NSFC grants (No. 61472141 and 61532021), Shanghai Knowledge Service Platform Project (No. ZF1213), Shanghai Leading Academic Discipline Project (Project Number B412), and Shanghai Agriculture Applied Technology Development Program (Grant No. G20160201).

References

1. Kou, N.M., Mamoulis, N., Li, Y., et al.: A topic-based reviewer assignment system. Proc. VLDB Endow. **8**(12), 1852–1855 (2015)
2. Goldsmith, J., Sloan, R.H.: The AI conference paper assignment problem. In: Proceedings of AAAI Workshop on Preference Handling for Artificial Intelligence, pp. 53–57. Vancouver (2007)
3. Dumais, S.T., Nielsen, J.: Automating the assignment of submitted manuscripts to reviewers. In: 1992 Proceedings of the 15th Annual International ACM SIGIR Conference on Research, Development in Information Retrieval, pp. 233–244. ACM (2011)
4. Basu, C., Hirsh, H., Cohen, W.W., et al.: Technical paper recommendation: a study in combining multiple information sources. J. Artif. Intell. Res. (JAIR) **14**, 231–252 (2001)
5. Hettich, S., Pazzani, M.J. Mining for proposal reviewers: lessons learned at the national science foundation. In: Proceedings of the 12th ACM SIGKDD International Conference on Knowledge Discovery and Data Mining, pp. 862–871. ACM (2006)
6. Wei, X., Croft, W.B.: LDA-based document models for ad-hoc retrieval. In: 2006 Proceedings of the 29th Annual International ACM SIGIR Conference on Research, Development in Information Retrieval, pp. 178–185. ACM (2011)
7. Mimno, D., McCallum, A.: Expertise modeling for matching papers with reviewers. In: 2007 Proceedings of the 13th ACM SIGKDD International Conference on Knowledge Discovery, Data Mining, pp. 500–509. ACM (2011)

8. Karimzadehgan, M., Zhai, C.X., Belford, G.: Multi-aspect expertise matching for review assignment. In: 2008 Proceedings of the 17th ACM Conference on Information, Knowledge Management, pp. 1113–1122. ACM (2011)

9. Li, X., Watanabe, T.: Automatic paper-to-reviewer assignment, based on the matching degree of the reviewers. Procedia Comput. Sci. **22**, 633–642 (2013)

10. Blei, D.M., Ng, A.Y., Jordan, M.I.: Latent dirichlet allocation. J. Mach. Learn. Res. **3**(1), 993–1022 (2003)

11. Hartvigsen, D., Wei, J.C., Czuchlewski, R.: The conference paper-reviewer assignment problem. Decis. Sci. **30**(3), 865 (1999)

12. Benferhat, S., Lang, J.: Conference paper assignment. Int. J. Intell. Syst. **16**(10), 1183–1192 (2001)

13. Chen, M., Liu, Y., Yu, X.: NLPMM: a next location predictor with Markov modeling. In: Tseng, V.S., Ho, T.B., Zhou, Z.-H., Chen, A.L.P., Kao, H.-Y. (eds.) PAKDD 2014. LNCS (LNAI), vol. 8444, pp. 186–197. Springer, Heidelberg (2014). doi:10.1007/978-3-319-06605-9_16

14. Conry, D., Koren, Y., Ramakrishnan, N.: Recommender systems for the conference paper assignment problem. In: Proceedings of the Third ACM Conference on Recommender systems, pp. 2009: 357–360. ACM (2011)

15. Karimzadehgan, M., Zhai, C.X.: Constrained multi-aspect expertise matching for committee review assignment. In: 2009 Proceedings of the 18th ACM conference on Information, knowledge management, pp. 1697–1700. ACM (2011)

16. Garg, N., Kavitha, T., Kumar, A., et al.: Assigning papers to referees. Algorithmica **58**(1), 119–136 (2010)

17. Charlin, L., Zemel, R.S., Boutilier, C.: A framework for optimizing paper matching. arXiv preprint arXiv:1202.3706 (2012)

18. Xue, N., Hao, J.X., Jia, S.L., et al.: An interval fuzzy ontology based peer review assignment method. In: 2012 IEEE Ninth International Conference on e-Business Engineering (ICEBE), pp. 55–60. IEEE (2011)

19. Wang, F., Zhou, S., Shi, N.: Group-to-group reviewer assignment problem. Comput. Oper. Res. **40**(5), 1351–1362 (2013)

20. Kou, N.M., Hou, U.L., Mamoulis, N., et al.: Weighted coverage based reviewer assignment. In: Proceedings of the 2015 ACM SIGMOD International Conference on Management of Data, pp. 2031–2046. ACM (2011)

Adaptive Bayesian Network Structure Learning from Big Datasets

Yan Tang$^{(\boxtimes)}$, Qidong Zhang, Huaxin Liu, and Wangsong Wang

College of Computer and Information, Hohai University, Nanjing 210098, China
tangyan@hhu.edu.cn, qdzhang.hhu@gmail.com, hxliu.hhu@gmail.com,
wswang.hhu@gmail.com

Abstract. Since big data contain more comprehensive probability distributions and richer causal relationships than conventional small datasets, discovering Bayesian network (BN) structure from big datasets is becoming more and more valuable for modeling and reasoning under uncertainties in many areas. Facing big data, most of the current BN structure learning algorithms have limitations. First, learning BNs structure from big datasets is an expensive process that requires high computational cost, often ending in failure. Second, given any dataset as input, it is very difficult to choose one algorithm from numerous candidates for consistently achieving good learning accuracy. To address these issues, we introduce a novel approach called Adaptive Bayesian network Learning (ABNL). ABNL begins with an adaptive sampling process that extracts a sufficiently large data partition from any big dataset for fast structure learning. Then, ABNL feeds the data partition to different learning algorithms to obtain a collection of BN Structures. Lastly, ABNL adaptively chooses the structures and merge them into a final network structure using an ensemble method. Experimental results on four big datasets show that ABNL leads to a significantly improved performance than whole dataset learning and more accurate results than baseline algorithms.

Keywords: Bayesian network structure learning · Bayesian score · Big data sampling · Ensemble method

1 Introduction

A Bayesian network (BN) [1] is a probabilistic graphical model that represents a set of random variables and their conditional dependencies via a directed acyclic graph (DAG) and probability functions associated with the nodes in the DAG. BNs have been broadly applied to modeling and reasoning in many domains such as Health Care, Bioinformatics, Finance and Social Services [2–6]. With the increasing availability of big datasets in academia, government and business, constructing Bayesian network from big datasets is becoming more and more valuable and mission critical, especially when uncertainties are involved.

However, facing big data, most of the current BN structure learning algorithms have several limitations. First, learning BNs structure from big datasets

© Springer International Publishing AG 2017
Z. Bao et al. (Eds.): DASFAA 2017 Workshops, LNCS 10179, pp. 158–168, 2017.
DOI: 10.1007/978-3-319-55705-2_12

is an expensive process that requires expensive computational cost [7], conventional computation platforms easily end in failure when the dataset grow in size. Second, given any dataset as input, it is very difficult to choose one algorithm from numerous Bayesian network structure learning algorithms (BNSLAs) for consistently achieving good learning accuracies. For the first limitation, one solution is adopting a big data processing platform, such as the MapReduce based method proposed by Fang et al. [8] and our previous work [9]. But big data platforms are not easily available and accessible. Therefore, an alternative is sampling a small portion from the big dataset using probabilistic approximation and then learning a BN from the sampled dataset on a conventional computation platform. For solving the second limitation, in light of the ensemble methods [10], we could collect the BN structures learned by different BNSLAs and then design an ensemble method to adaptively merge the structures, consolidating them into a more correct network structure.

Therefore, we introduce a novel approach called Adaptive Bayesian network Learning (ABNL). ABNL begins with an adaptive sampling process that extracts a sufficiently large data partition from any big dataset for fast structure learning. We adopt the Markov blanket based algorithm [11] and a Bayesian score based algorithm for calculating the minimal sampling size (MSS) for learning. Then, a data partition is sampled using MSS and given to different algorithms for learning a collection of BN Structures. Lastly, ABNL adaptively chooses the structures with the top Bayesian scores and intelligently merge them into a final network structure using a weighted adjacent matrix based ensemble method. Experimental results on four big datasets show that by adaptively sampling data and intelligent merging high scored BN structures, ABNL leads to a significantly improved performance than whole dataset learning and more accurate network structure than baseline algorithms.

To summarize, the main contributions of this paper are as follows:

- This paper proposes a sampling process to obtain a small data partition from big dataset suitable for Bayesian network learning. This makes BN structure learning from bigdata feasible on conventional computation platforms.
- This paper provides new ensemble method for merging the Bayesian network structures learned by different BNSLAs. This solves the dilemma for choosing BN structure learning algorithms.
- We further evaluate ABNL on four big datasets, the experiment results validates the efficiency and accuracy of our approach.

The rest of the paper is organized as follows: Sect. 2 is related work. The proposed method including algorithms is presented in Sect. 3. After giving experimental results and discussion in Sect. 4, we conclude this work in the final section.

2   Related Works

Identifying BN structure from large dataset is known to be NP-hard [12]. Following data parallel processing approach, Fang et al. [8] propose a Map-Reduce based method for learning BN from massive datasets. Our previous work [13] adopts distributed data-parallelism techniques and scientific workflow for BN learning from big datasets to achieve better scalability and accuracy. However, little work uses data sampling to learn BN from big datasets for reducing learning complexity. Jiang et al. [14] study dataset sampling method and apply the sampled datasets to different Bayesian network classifiers to achieve better classification accuracy, they validate the effectiveness of data sampling for accurate BN learning. Our previous work [11] is an attempt following the data sampling approach. It shows that given a large DAG-faithful dataset [15], its DAG-faithful sub datasets can be sampled to approximate the learning on the whole dataset. This paper continues the work in this direction.

Over the last decades, numerous algorithms are proposed for learning Bayesian network from data, such as Hill Climbing (HC), Tabu Search (Tabu), Grow and Shrink (GS) [16], Three Phase Dependency analysis (TPDA) [15], Inter-IAMB [17] and Max-Min Hill Climbing (MMHC) [18]. However, most algorithm operates individually, few BNSLAs consider the structures learned by other algorithms and incorporate them to obtain a more accurate result. Ensemble methods [10] use learners to obtain better predictive performance than learning from any of the constituent methods. Njah and Jamoussi [19] proposed a weighted ensemble Bayesian network learning method for gene regulatory networks. Our previous work [9] achieved higher accuracy of BN structure learning through ensemble methods. In this paper, we continue to adopt ensemble methods on different structure learning algorithms for achieving better learning accuracy. Our ensemble method can be categorized as a bagging strategy based on weighted voting.

3   The Proposed Approach

3.1   Overview of the Method

ABNL consists of the following three key steps: Adaptive Data Sampling, Multiple Structure Learning and Adaptive Ensemble Learning.

Firstly, ABNL executes two greedy algorithms to adaptively obtain a minimal sampling size (MSS_D) to sample a data Partition DP for any big dataset D.

Secondly, ABNL calls a set of M Bayesian network structure learning algorithms to learn M network structures from DP.

Lastly, ABNL selects top K network structures with the best Bayesian score, and uses a weighted adjacent matrix based ensemble method to merge K network structures into a final Bayesian network.

3.2   Adaptive Data Sampling

When a dataset D is large enough and is DAG-faithful, it is reasonable to learn a BN from a partition of D for improving computation efficiency. The key is calculating the ideal sample/partition size. In our previous work [11], the notion minimal sampling size (MSS) is introduced. Given a DAG-faithful dataset D, its minimal sampling size denoted as MSS_D is the minimal size of D's sub dataset that maintains DAG-faithful. A greedy algorithm can be designed to calculate MSS_D. The algorithm starts with a small sample size, and iteratively double the sample size and evaluate the Bayesian network learned by the newly sampled data. The first greedy algorithm introduced in [11] is based on Markov blanket, it stops the iteration when the average Markov Blanket size becomes stable and returns MSS_{AMBS}. In this paper, another greedy algorithm (Algorithm 1) is designed based on Bayesian score, it stops the iteration when the Bayesian score of the learned network become stable.

Algorithm 1. CalculateMSS$_{BS}$
_____
Input:
    D: Dataset;
    ϵ: Threshold;
    $mstep$: Maximum loop steps.
Output:
    MSS_{BS}: Minimal sampling size.
1: $bestScore = -1$; $step = 0$;
2: sliceSize = 100 * number of attributes in D;
3: D_{sliced} = read the first sliceSize rows in D
4: BN_{DS} = LearnBNStructure(D_{sliced});
5: $currentScore$ = Bayesian Score of BN_{DS};
6: **while** ($currentScore > bestScore$&&$step \leq mstep$&&(($currentScore - bestScore$) $> bestScore * \epsilon$) **do**
7:    $sliceSize = sliceSize * 2$;
8:    $bestScore = currentScore$;
9:    D_{sliced} = read the first $sliceSize$ rows in D
10:   BD_{DS} = learnBNStructure(D_{sliced});
11:   $currentScore$ = Bayesian Score of BN_{DS};
12:   $step = step + 1$;
13: **end while**
14: MSS_{BS} = number of records in D_{sliced};
15: **return** MSS_{BS}
_____

Algorithm 1 starts with small sub dataset D_{sliced}. It learns the BN from D_{sliced} (Step 4) and obtains the initial Bayesian score (Line 5). In order to make D_{sliced} DAG-faithful, the loop (Line 6–13) in the algorithm doubles $sliceSize$ at each iteration, and stops when the Bayesian score of the learned structure becomes relatively stable. This indicates that the current $sliceSize$ is close to the MSS_D (Step 14). Algorithm 1 stops and then return MSS_{BS}. The MSS_D is calculated using Eq. (1).

$$MSS_D = MIN(MSS_{DS}, MSS_{AMBS}) * 2; \qquad (1)$$

3.3 Multiple Structure Learning

After obtaining MSS_D and the sampling data partition DP, ABNL feeds DP to a collection of M Bayesian network learning algorithms using Algorithm 2.

Algorithm 2. Multiple Structure Learning

Input:
    DP : sampled data partition
    Algos : The collection of Bayesian network learning algorithms
Output:
    structures: Learned network structures by *Algos*.
1: Initialize *structures* as an empty array;
2: **for** i = 1..M **do**
3:    *structures*[i] = *Algos*[i](DP);
4: **end for**
5: **return** *structures*.

Given a collection of M learning algorithms *Algos* and the data partition DP, Algorithm 2 obtains a Bayesian network structure learned by each algorithm *Algos*[i] and stores the structure in the array *structures*.

3.4 Adaptive Ensemble Learning

After obtaining learned network structures, ABNL enters the Adaptive Ensemble Learning (Algorithm 3) to produce the final BN structure from the big dataset.

Algorithm 3 first gets the Bayesian scores of the learned structures, then it adaptively selects the top K structures with the best scores (Step 1–3). This makes sure that the structures in the ensemble are the best ones. Then, it transforms each top structure into an adjacent matrix, calculates the weight of each structure and obtains the weighted adjacent matrix (step 4–6). The weight is determined by the Bayesian score of the structure (Step 5), higher score results in higher weight. In step 7, the final weighted adjacent matrix **FWAM**$[i, j]$ is obtained. **FWAM**$[i, j]$ is the collective voting results of the best structures with respect to their weights, it is calculated using Eq. (2).

$$\mathbf{FWAM} = \sum_{i=1,...,K} \mathbf{WAM}_i \tag{2}$$

Based on majority voting, if an edge exists between node i and node j in most of the structures, then **FWAM**$[i, j]$ should be larger than a threshold. The basic threshold is equal to the minimal weight value (step 8). If an entry **FWAM**$[i, j]$ is larger than $threshold * C(C > 1)$, it means there are more than C structures containing the edge $j - i$. Therefore, Algorithm 3 adds an edge between i and j in the final network (step 11). After iterating all the entries in **FWAM**, the final network structure BN_{final} is produced. For achieving satisfying results, we can choose C = (K*2)/3 to make sure that the majority of edges in the top structures appear in the final structure.

Algorithm 3. Adaptive Ensemble Learning

Input:
    DP : Data Partition ;
    K : Integer for selecting structures with top K Bayesian scores;
    structures : Learned network structures by *Algos*;
    C : Coefficient of the Threshold.
Output:
    BN_{final}: Final network structure as an Adjacent Matrix.
  1: $scores = \text{BayesianScore}(structures, DP)$;
  2: $topScores = $ Bayesian scores of BN_{Top} given DP;
  3: $BN_{Top} = $ Network structures with top K Bayesian scores ;
  4: Transform each $BN_{Top}[i]$ into an Adjacent Matrix $\mathbf{AM}_i$;
  5: $Weight_{BN_{Top}[i]} = topScores[i]/sum(\text{topScores})$;
  6: $\mathbf{WAM}_i = Weight_{BN_{Top}[i]} * \mathbf{AM}_i$
  7: Calculate $\mathbf{FWAM}$ using Equation (2);
  8: $threshold = MIN(Weight_{BN_{Top}})$;
  9: **for** Each $\mathbf{FWAM}[i,j]$ **do**
 10:    **if** $\mathbf{FWAM}[i,j] >\text{threshold}*C$ **then**
 11:        $BN_{final}[i,j] = 1$;
 12:    **end if**
 13: **end for**
 14: **return** BN_{final}.

---

4 Experiments and Discussion

4.1 Experimental Setup and Datasets

To validate the effectiveness of our proposed method, four big datasets are used to evaluate the learning accuracy and the computation efficiency of ABNL.

The experiments are run on a MacPro, with Intel(R) Core i5 2.7 GHz, 8 GB of Memory, running MAC OS X 10.11.4. Bayesian network structure learning algorithms are implemented using Bnlearn R Package [20] running on R (version 3.1.1). The learning algorithms are Hill Climbing (HC), Tabu Search (Tabu), Grow and Shrink (GS) [16], Inter-IAMB [17] and Max-Min Hill Climbing (MMHC) [18]. The threshold used in Algorithm 1 is 0.05 for calculating MSS_{BS}. Since there are five learning algorithms, in the adaptive ensemble learning, $K = 3$ and consequently, $C = 2$. BDeu score [21] is used throughout the experiment as the Bayesian score.

Table 1 lists the datasets (CSV formatted) used in our experiments. Four big datasets were generated using the data simulation module of the SamIam tool (http://reasoning.cs.ucla.edu/samiam/) from four well known Bayesian networks: Child [22], Alarm [23] and Insurance [24] and Hailfinder [25]. These known networks provide ground truths for comparing the network structures learned by ABNL with the baseline algorithms.

Table 1. Experimental datasets

Name	#Rows (million)	Size (MB)	#Nodes	Domain
Child	5	576	20	Medical
Alarm	5	994	37	Medical
Insurence	5	982	27	Finance
Haildinfer	2	846	56	Weather

4.2 Performance Results

Figure 1 shows the computation time of ABNL compared with learning on the whole big dataset using the fastest learning algorithm Hill Climbing (HC).

It is observed that ABNL achieves a significant improvement in computation time on all the big datasets compared with whole dataset learning with HC. Especially, on Child dataset, the execution time is an order of magnitude shorter than learning from the whole dataset. Most of the computation of ABNL is on the MSS_D calculation. The experimental results confirm good computation efficiency of ABNL on learning Bayesian network from big datasets.

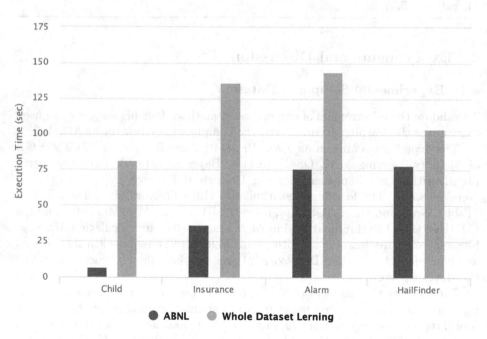

Fig. 1. Execution time of ABNL Vs whole dataset learning

4.3   Learning Accuracy Results

The key advantage of ABNL is the adaptive ensemble learning which can intelligently selects the best learning algorithms and merges their learning results into a more accurate network structure. To evaluate the effectiveness of ABNL, we compare the structures learned by ABNL, HC, Tabu and MMHC with the correct network structure on four partitioned datasets. When comparing a learned structure with the correct structure, there are three types of edges: Correct Edge, Missing Edges and Added Edge, Figs. 2, 3 and 4 show the comparison results.

Figure 2 shows the comparison results for correct edges. More correct edges indicate better structure learning accuracy. It is observed that ABNL is consistently obtaining the most number of correct edges. For Child dataset, ABNL finds all the correct edges, for insurance, Alarm and Hailfinder datasets, ABNL is capable of finding more than 80% of the correct edges.

Figure 3 shows the comparison results for missing edges. Less missing edges means more accurate network structures. It is observed that the number of missing edges learned by ABNL is the least compared with the results of three baseline algorithms. For Child dataset, HC and Tabu finds the correct network structure, the ensemble method of ABNL makes sure that these correct edges are preserved in the final structure, therefore, ABNL obtains the perfect structure without any missing edges. Furthermore, for the other three big datasets, ABNL is able to obtain a better structure by collective voting and remain unaffected by the poor individual learning results. For the added edges (Figure 4), ABNL

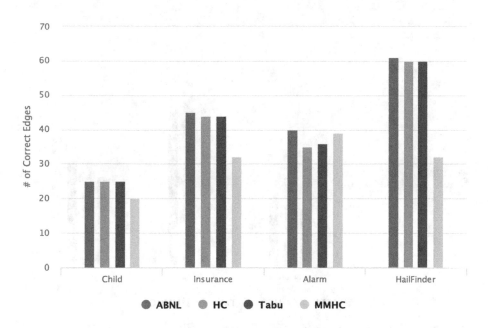

Fig. 2. Number of correct edges of ABNL Vs three baseline algorithms

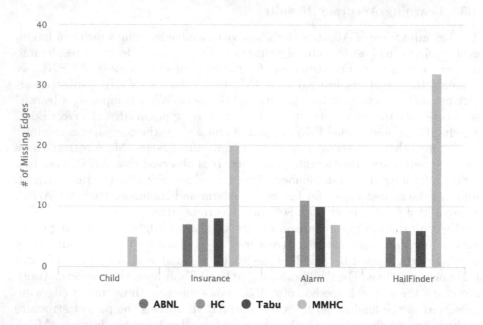

Fig. 3. Number of missing edges of ABNL Vs three baseline algorithms

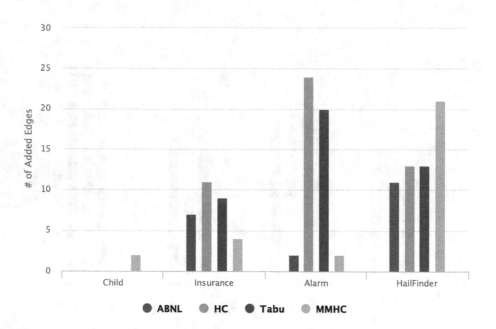

Fig. 4. Number of added edges of ABNL Vs three baseline algorithms

also outperforms the baseline algorithms by keeping the number of added edges at a minimal level.

In summary, the above experimental results confirm that ABNL can adaptively learn an accurate Bayesian network structure from big datasets.

5 Conclusion

In this paper, we have proposed an adaptive ensemble based method called ABNL for learning Bayesian network (BN) structure from big data. We demonstrate through experiments that, by adaptively sampling data and merging BN structures using the weighted adjacent matrix based ensemble method, ABNL leads to a significantly improved performance than whole dataset learning and more accurate Bayesian network structure learning than baseline algorithms. ABNL enables big data graphical modeling on conventional computation platform. It also makes algorithm selection autonomous when performing BN structure learning tasks. Our future work will focus on enriching the ensemble of learning algorithms to achieve better learning accuracy.

Acknowledgments. This work was supported by the Natural Science Foundation of Jiangsu Province, China (Grant No. BK20141420 and Grant No. BK20140857).

References

1. Ben-Gal, I.: Bayesian networks. In: Ruggeri, F., Kenett, R.S., Faltin, F.W. (eds.) Encyclopedia of Statistics in Quality and Reliability. Wiley, Hoboken (2007)
2. Yoo, C., Ramirez, L., Liuzzi, J.: Big data analysis using modern statistical and machine learning methods in medicine. Int. Neurourol. J. **18**(2), 50–57 (2014)
3. Zhang, Y., Zhang, Y., Swears, N., et al.: Modeling temporal interactions with interval temporal Bayesian networks for complex activity recognition. IEEE Trans. Pattern Anal. Mach. Intell. **35**(10), 2468–2483 (2013)
4. Njah, H., Jamoussi, S.: Weighted ensemble learning of Bayesian network for gene regulatory networks. Neurocomputing **150**(B), 404–416 (2015)
5. Yang, J., Tong, Y., Liu, X., Tan, S.: Causal inference from financial factors: continuous variable based local structure learning algorithm. In: 2014 IEEE Conference on Computational Intelligence for Financial Engineering and Economics (CIFEr), pp. 278–285. IEEE (2014)
6. Yue, K., Wu, H., Fu, X., Xu, J., Yin, Z., Liu, W.: A data-intensive approach for discovering user similarities in social behavioral interactions based on the Bayesian network. Neurocomputing **219**, 364–375 (2017)
7. Al-Jarrah, O., Yoo, P., et al.: Efficient machine learning for big data: a review. Big Data Res. **2**(3), 87–93 (2015)
8. Fang, Q., Yue, K., Fu, X.,Wu, H., Liu, W.: A mapreduce-based method for learning Bayesian network from massive data. In: Proceedings of the 15th Asia-Pacific Web Conference (APWeb 2013), pp. 697–708 (2013)
9. Tang, Y., Wang, Y., Cooper, K., Li, L.: Towards big data Bayesian network learning - an ensemble learning based approach. In: Proceedings of the IEEE International Congress on Big Data (BigData Congress), pp. 355–357 (2014)

10. Rokach, L.: Ensemble-based classifiers. Artif. Intell. Rev. **33**(1–2), 1–39 (2010)
11. Tang, Y., Xu, Z., Zhuang, Y.: Bayesian network structure learning from big data: a reservoir sampling based ensemble method. In: Gao, H., Kim, J., Sakurai, Y. (eds.) DASFAA 2016. LNCS, vol. 9645, pp. 209–222. Springer, Heidelberg (2016). doi:10.1007/978-3-319-32055-7_18
12. Chickering, D., Heckerman, D., Meek, C.: Large-sample learning of Bayesian networks is NP-hard. J. Mach. Learn. Res. **5**, 1287–1330 (2004)
13. Wang, J., Tang, Y., Nguyen, M., Altintas, I.: A scalable data science workflow approach for big data Bayesian network learning. In: Proceedings of the 2014 IEEE/ACM International Symposium on Big Data Computing (BDC 2014), pp. 16–25 (2014)
14. Jiang, L., Li, C., Cai, Z., Zhang, H.: Sampled Bayesian network classifiers for class-imbalance and cost-sensitive learning. In: Proceedings of the IEEE 25th International Conference on Tools with Artificial Intelligence (ICTAI), pp. 512–517 (2013)
15. Cheng, J., Greiner, R., Kelly, J., Bell, D., Liu, W.: Learning Bayesian networks from data: an information-theory based approach. Artif. Intell. **137**(1–2), 43–90 (2002)
16. Margaritis, D.: Learning Bayesian network model structure from data. Ph.D. thesis, Carnegie-Mellon University (2003)
17. Yaramakala, S., Margaritis, D.: Speculative Markov blanket discovery for optimal feature selection. In: Fifth IEEE International Conference on Data Mining (ICDM 2005), pp. 809–812. IEEE (2005)
18. Tsamardinos, I., Brown, L.E., Aliferis, C.F.: The max-min hill-climbing Bayesian network structure learning algorithm. Mach. Learn. **65**(1), 31–78 (2006)
19. Njah, H., Jamoussi, S.: Weighted ensemble learning of Bayesian network for gene regulatory networks. Neurocomputing **150**(PB), 404–416 (2015)
20. Scutari, M.: Learning Bayesian networks with the bnlearn R package. J. Stat. Softw. **35**(3), 1–22 (2010)
21. Heckerman, D., Geiger, D., Chickering, D.: Learning Bayesian networks: the combination of knowledge and statistical data. Mach. Learn. **20**, 197–243 (1995)
22. Spiegelhalter, D., Cowell, R.: Learning in probabilistic expert systems. In: Bayesian Statistics, vol. 4. Clarendon Press (1992)
23. Beinlich, I., Suermondt, H., Chavez, R., Cooper, G.: The alarm monitoring system: a case study with two probabilistic inference techniques for belief networks. In: Proceedings of the 2nd European Conference on Artificial Intelligence in Medicine, pp. 247–256 (1989)
24. Binder, J., Koller, D., Russell, S., Kanazawa, K.: Adaptive probabilistic networks with hidden variables. Mach. Learn. **29**(2–3), 213–244 (1997)
25. Abramson, B., Brown, J., Edwards, W., Murphy, A., Winkler, R.L.: Hailfinder: a Bayesian system for forecasting severe weather. Int. J. Forecast. **12**(1), 57–71 (1996)

A Novel Approach for Author Name Disambiguation Using Ranking Confidence

Xueqin Lin, Jia Zhu, Yong Tang$^{(\boxtimes)}$, Fen Yang, Bo Peng, and Weiling Li

School of Computer Science, South China Normal University, Guangzhou, China
{xqlin,jzhu,ytang,fyang,bpeng,weiling}@m.scnu.edu.cn

Abstract. In digital libraries, ambiguous author names may occur because of the existence of multiple authors with the same name or different name variations for the same person. In recent years, name disambiguation has become a major challenge when integrating data from multiple sources in bibliographic digital libraries. Most of the previous works solve this issue by using many attributes, such as coauthors, title of articles/publications, topics of articles, and years of publications. However, in most cases, we can only get the coauthor and title attributes. In this paper, we propose an approach which is based on Hierarchical Agglomerative Clustering (HAC) and only use the coauthor and title attributes, but can more effectively identify the disambiguation authors. The whole algorithm can divide into two stages. In the first stage, we employ a pair-wise grouping algorithm which is based on coauthors'name to group records into clusters. Then, we merge two clusters if the similarity of the article titles from two clusters reach the threshold. Here, we use three kinds of similarity algorithms such as Jaccard Similarity, Cosine Similarity and Euclidean Distance to compare the similarity between the titles of two clusters. To minimize the risk of using only one similarity metric, we design the concept of ranking confidence to measure the confidence of different similarity measusrements. The ranking confidence decides which similarity measure to use when merging clusters. In the experiments, we use PairPresicion, PairRecall and PairF1 score to evaluate our method and compare with other methods. Experimental results indicate that our method significantly outperforms the baseline methods: HAC, K-means and SACluster when only use coauthor and title attributes.

Keywords: Name disambiguation · Hierarchical agglomerative clustering · Ranking confidence · String similarity metrics

1 Introduction

Name disambiguation in digital libraries refers to the task of attributing the publications or citation records to the proper authors [15,24]. It is common that several authors share the same name or a single author has multiple names in digital library. This will make people feel inconvenient because it is difficult for

© Springer International Publishing AG 2017
Z. Bao et al. (Eds.): DASFAA 2017 Workshops, LNCS 10179, pp. 169–182, 2017.
DOI: 10.1007/978-3-319-55705-2_13

them to identify which records belong to the proper author. For example, Han et al. [7] found that the author page of "Yu Chen" contains citations authored by three individuals with the same name. Zhu et al. [25] also found that there were at least 50 different authors who called "Wei Wang" in DBLP and there were more than 400 entries under this name. In the other hand, different authors may share the same name label in multiple papers, e.g., both "Jia Zhu" and "Jiaxing Zhu" are used as "J. Zhu" in their papers. In recent years, name disambiguation has become a major challenge when integrating data from multiple sources in bibliographic digital libraries.

At present, disambiguation of homonymous names has received a growing attention with the advent of the semantic web and social networks. Successful name entity disambiguation may greatly help in locating the right researcher and obtaining his/her academic information from the correct homepage, and indexing bibliographic database more accurately and efficiently. In our paper, we think that the role of name disambiguation is that the application of Search Engines, Social Networks, Credit Evaluation, Conflict of Interest and so on. The Table 1 lists the five records with the authors' name and the paper title. It is difficulty for us to make sure the author "Wen Gao" is the same person. Beyond the problem of sharing the same name among different people, name abbreviations and other reference variations compound the challenge of name disambiguation. This paper is to solve this problem of sharing the same name among different people through the coauthors and paper title.

Table 1. An example of name disambiguation

1.Xin Liu,Hongxun Yao,**Wen Gao**
Title: An Adaptive Dandelion Model for Reconstructing Spherical Terrain-Like Visual Hull Surfaces
2.Baochang Zhang,**Wen Gao**,Shiguang Shan,Wei Wang
Title: Constraint Shape Model Using Edge Constraint and Gabor Wavelet Based Search
3.**Wen Gao**,Baocai Yin
Title: Sequence Decomposition Method for Computing a Grobner Basis and Its Application to Bivariate Spline
4.Deng Lei,**Wen Gao**,Ming-Zeng Hu,Zhenzhou Ji
Title: An Efficient VLSI Architecture for MC Interpolation in AVC Video Coding
5.**Wen Gao**,Xinyu Liu,Lei Wang,Takashi Nanya
Title: A Reconfigurable High Availability Infrastructure in Cluster for Grid

There are many existing approaches to solve the problem of name ambiguity in the former research. For example, Yang et al. [19] proposed two kinds of correlations between citations, namely, Topic Correlation and Web Correclation, to exploit relationships between records. However, they only considered the Web pages edited by humans to measure Web correlation without using other existent digital libraries. Tang et al. [18] formalized the problems in a unified framework and proposed a generalized probabilistic model to solve this problem. In the difinition section, they assigned six attributes to each paper. Unfortunately, those may not always be practical for all digital libraries because we can not get all information directly.

In the disambiguation features selection, previous works have employed coauthors, title of articles/publications, topics of articles, and years of publications that constitute basic citation data. Title of articles may epitomize research areas of their authors, thus under the assumption that namesakes do not heavily share their research areas, title similarity between articles may be employed in resolving homonymous author names. The same also applies to the case of title of books. In addition to the above record-internal features, some have utilized record-external features such as abstracts, self-records, and citations URLs. When the full text of articles is available, additional features such as e-mail addresses, affiliation, and keywords can be extracted and applied to the author name disambiguation, but it is not available in many digital libraries. In this study, we choose the coauthors and title of paper as the disambiguation features.

In this paper, to reduce errors in the process of hierarchical agglomerative clustering [17,21] on a list of citation records, we design the concept of ranking confidence to measure the confidence of the different similarity measurements. The ranking confidence can decides which similarity measures to use when merging clusters. We employ three similarity algorithm such as Jaccard Similarity [12,14], Cosine Similarity [12,13] and Euclidean Distance[1] to contrast the similarity between two clusters. In the process of clustering, we adapt a pair-wise grouping algorithm to group records into clusters, which repeatly merges the most similar pairs of clusters. In the experiments, we use PairPresicion, PairRecall and PairF1 score to evalute our method results and compare with other methods.

The rest of this paper is organized as follows. In Sect. 2, we discuss related work in name disambiguation. In Sect. 3, we decsribe details of our approach including String Similarity, Ranking Confidence and clustering procedure. In Sect. 4, we describe our experiments, evaluation methods and result analysis, and compare our approach with other methods. We also conclude and discuss this study in Sect. 5.

2   Related Work

This section will discuss recent works and prior works for name disambiguation. A great deal of research has focused on the name disambiguation problem in

---

[1] https://en.wikipedia.org/wiki/Euclidean_distance.

different types of data sets and adapted unsupervised learning approaches or semi-supervised learning methods or supervised learning measures and so on. If the name disambiguation results can be better, it is useful for evaluating faculty publications and calculating statistics of social network and anthors impacts.

There are many existing methods for name disambiguation problems. Kang et al. [8] explored the net effects of co-authorship on author clustering in bibliographic data. They proposed a web-assisted technique of acquiring implicit coauthors of the target author to be disambiguated and considered that the identity of an author can be determined by his/her coauthors. However, Korean was one of best suitable languages for this study and their work can not identity which pages are personal pages. Han et al. [7] proposed an unsupervised learning approach using K-way spectral clustering that disambiguates authors in citations. However, this clustering method may not work well when there are many ambiguous athors in the dataset and it is unsuitable for large-scale digital libraries since K is not known a priori for an ever increasing digital library.

Lei et al. [3] proposed new research for entity disambiguation with the focus of name disambiguation in digital libraries, they adapted the pairwise similarity and a novel Hierarchical Agglomerative Clustering approach with Adaptive Stopping Criterion. Zhu et al. [23] proposed an approach that can effectively identified and retrieved information from web pages and used the information to disambiguate authors. Initially, they also implement a web pages identification model by using a neural network classifier and traffic rank. However, the information of web pages are disordered and it is difficulty to extract effective information, especially when the information of author are lacking. Mann et al. [11] presented a set of algorithms for distinguishing personal names with multiple real referents in text, based on little or no supervision. The approach utilized an unsupervised clustering technique over a rich feature space of biographic facts.

Lin et al. [16] proposed a supervised method for exploiting all side information including co-author, organization, paper citation, title similarity, author's homepage, web constraint and user feedbacks. Although user feedbacks can increase more useful information, but when the amount of data is very large, the user feedbacks information are very difficult to collect and also expend much manpower and material resources in the process of collecting. Li et al. [9] presented a novel categorical set similarity measure named CSLR for two sets which both follow categorical distributions. It is applied in Author Name Disambiguation to measure the similarity between two venue sets or coauthor sets. However, there is only one kind of similarity which can not be persuasive for the different emphasis on the similarity.

Yang et al. [19] proposed two kinds of correlations between citations, namely, Topic Correlation and Web Correlation, to exploit relationships between citations, in order to identify whether two citations with the same author name refer to the same individual. They only extracted citations' topics from venue information and discover the topic-based relationships and which may not be typical. Yin et al. [20] developed a general object distinction methodology called DISTINCT, which combined two complementary measures for relational similarity:

set resemblance of neighbor tuples and random walk probability. Although the accuracy of this method is high but it relies heavily on the quality of the training data, which is difficult to obtain. On the other hand, the name of the Proceedings and Conferences also has a certain ambiguity, which may affect the calculation of the similarity.

In the attribute selection of data sets, many academic papers choose a variety of attributes. For example, Tang et al. [18] assigned six attributes to each paper, such as the title of the paper, the conference published of paper, the time published of paper, the abstract of paper, the authors of paper and the reference of paper. Han et al. [7] used three types of attributes to design features for name disambiguation: coauthor names, paper titles, and publication venue titles. Han et al. [5] preprocessed the datasets on author names, paper title words and journal title words as follows. Arif et al. [1] chose six attributes: the title of citation, authors of citation, e-mail of author, affiliations of publication and year of publication. In this paper, we only choose two attributes: coauthors and the title of paper for analysis.

3 Proposed Approach

3.1 String Similarity Metrics

In this section, we will use three methods such as Jaccard similarity, Cosine Similarity and Euclidean Distance to calculate the similarity of two paper title, then we adapt the confidence of ranking to determine which similarity values should be used. Our approach based on Hierarchical Agglomerative Clustering (HAC) that can effectively identify the disambiguation authors.

3.1.1 Jaccard Similarity

Among many possible token-based similarity measures, we use Jaccard similarity [12,14] to calculate the similarity of two paper title. Jaccard similarity coefficient is an index to measure the similarity of two sets. The smaller the value, the more similar between the two titles. We briefly describe the metric below. Using the terms of Table 2, Jaccard coefficient similarity function can be defined as follows:

$$Jaccard = \frac{|T_x \bigcap T_y|}{|T_x \bigcup T_y|} \tag{1}$$

Table 2. Terms in the set of paper title

Name	Description
x,y	The title of paper
T_x	All words of the title x (remove the stop words)

3.1.2 Cosine Similarity

In this approach, we use vector similarities to measure the similarity of the paper title. The cosine similarity [12,13] between two vectors (or two title of paper on the Vector Space) is a measure that calculates the cosine of the angle between them. This metric is a measurement of orientation and not magnitude, it can be seen as a comparison between title of paper on a normalized space because we're not taking into the consideration only the magnitude of each word count (TF-IDF) of each title of paper (we have removed the stop words), but the angle between the paper titles of records.

For example, suppose we have two paper title of citations, one is 'Why Does Unsupervised Pre-training Help Deep Learning', another is 'Deep Learning in Neural Networks: An Overview', all the words of them are why, does, Unsupervised Pre-training, help, Deep Learning, in, Neural Networks, an, overview, then we remove the stop words, so the all words of them are Unsupervised Pre-training, help, Deep Learning, Neural Networks, overview. Fially, we have a vector $V(title)_1 = [1, 1, 1, 0, 0]$ and another vector $V(title)_2 = [0, 0, 1, 1, 1]$. We use the simple cosine similarity, an angle between two vectors, defined as:

$$cos\theta = \frac{\overrightarrow{V(title)_1} \times \overrightarrow{V(title)_2}}{||V(title)_1|| \times ||V(title)_2||} \tag{2}$$

3.1.3 Euclidean Distance

The Euclidean Distance[2] between points x and y is the length of the line segment connecting them $(\overline{\mathbf{xy}})$. We first calculate the Euclidean distance, and then we can get the similarity between the two titles of paper. The farther the distance between them, the greater the difference among the titles. One of the most prominent instantiations of the function is as follows.

$$dist(X, Y) = \sqrt{\sum_{i=1}^{n}(x_i - y_i)^2} \tag{3}$$

3.2 Ranking Confidence

To minimize the risk of using only one similarity metric, we use the ranking of confidence to determine which metric should used for clustering. For example, there is one cluster R which includes three citation records r_1, r_2, r_3, if we want to know whether a citation record q can be merged into R, we should calculate the similarity between q and each record in R. Suppose that the similarity between r_1 and q is ranked lower than all other records using Jaccard Similarity, but it is ranked higher than other citation records using Cosine Similarity, we do not know the exact ranking should be in this case. Therefore, we adapt the confidence of ranking to determine which similarity values should be used.

---

[2] https://en.wikipedia.org/wiki/Euclidean_distance.

After calculating three similarity between two title of the paper respectively, we design an algorithm to calculate the ranking of confidence. As we know, the confidence of three similarity in rankings should be highly related to the distribution of similarity values [2]. The functions are defined as follows Eqs. (4), (5), (6):

$$d_{(r,q,f,k)} = |sim_f(r,q,k) - sim_f(r,q,t)|(k = 1,2,3\dots n), \tag{4}$$

$$C_{(r,q,f,k)} = \frac{1 - d_{(r,q,f,k)}}{\sum_{i=1}^{n} C_{(r,q,f,i)}}(k = 1,2,3\dots n), \tag{5}$$

$$C_{(r,q,k)final} = average(\sum_{i=1}^{n} C_{(r,q,f,k)}). \tag{6}$$

Where $C_{(r,q,f,k)}$ is the confidence of similarity method f in the ranking of cluster r. We model $C_{(r,q,f,k)}$ as the probability of cluster r is ranked at position k in the final ranking given that its similarity value is $sim_f(r,q,k)$, the $sim_f(r,q,k)$ is the similarity value of the method f. The $d_{(r,q,f,k)}$ tries to compute the difference between $sim_f(r,q,k)$ and $sim_f(r,q,t)$, and uses it to model the required confidence value, the $sim_f(r,q,t)$ indicates the top value of the $sim_f(r,q,k)$ (k = 1,2,3… n). In Eq. (6), $C_{(r,q,k)final}$ is the final value of confidence through calculating average.

The Fig. 1 shows the process of calculating the value of confidence. In this process, we want to calculate the confidence value between a citation record and the every citation record of cluster1, the $S(1,J)$ and $C(1,J)$ stands for the value of similarity using the Jaccard Simalarity and the value of confidence between the title of citation between the title1 of cluster1 respectively, the $S(1,C)$ and and $C(1,C)$ indicates the value of similarity using the Cosine Similarity and the value of confidence between the title of citation between the title1 of cluster1

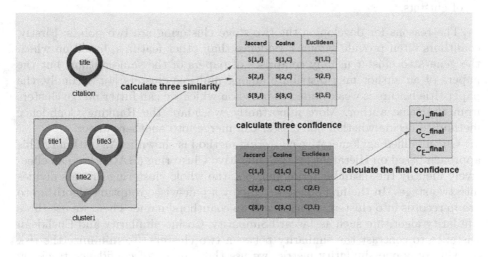

Fig. 1. The process of calculating the value of confidence

respectively, the $S(1, E)$ and $C(1, E)$ represents the value of similarity using the Euclidean Distance and the value of confidence between the title of citation between the title1 of cluster1 respectively. The value of confidence $C(1, J)$, $C(1, C)$ and $C(1, E)$ are calculated by Eqs. 4, 5. Finally, we get the final value C_{J_final}, C_{C_final} and C_{E_final} of confidence through Eq. 6.

For example, suppose that we hava a citation record Q, and a cluster R includes three citation records R_1, R_2, R_3, we first calculate the three similarity value by Jaccard similarity method between Q and R_1, R_2, R_3, the value is S_1, S_2, S_3, then we can get three corresponding confidence values C_1, C_2, C_3 by Eqs. (4), (5). Therefore, by Eq. (6), we can get the result C_{J_final} as the final value of Jaccard similarity to compare with other similarity value of cluster1. According to the confidence of ranking, suppose that the C_{J_final} value is the highest between C_{J_final}, C_{C_final} and C_{E_final}, then we use the Jaccard simalrity value as the final value of cluster1 to compare with other clusters' similarity value.

3.3 Clustering Procedure

In the paper, we employ Hierarchical Agglomerative Clustering (HAC) [17,21] as the basic framework. It starts with each paper being a cluster, we first cluster the citations on HAC, then we find the most similar (the used similarity measuers will be definded later) pairs of clusters, and merge them, until the maximal similarity falls below certain threshold. In this process, we choose two kinds of attributes to cluster: coauthors and paper title. The whole clustering process divides into two stages:

1. Merge based on the evidence from shared coauthors based on Hierarchical Agglomerative Clustering;
2. Merge based on the combined similarity defined on the title sets of each pair of clusters.

The reasons for developing the two-stage clustering are two points: Firstly, coauthors often provide stronger evidence than other features, based on which the generated cluster usually comprises of papers of the same author, but the papers of an author may distribute among multiple clusters [4]; Secondly, the paper title feature is weak evidence, based on which we can furter merge clusters from the same author. More importantly, we adapt the Ranking Confidence method to decide whether one cluster can merge into another cluster.

Our modified agglomerative clustering method is shown in Algorithm 1. This approach based on Hierarchical Agglomerative Clustering (HAC) that can effectively identify the disambiguation authors, the whole clustering process divides into two stages. In the first stage, we employ a pair-wise groupting algorithm to group records into clusters, which based on coauthors' name. Then, we use three similarity algorithm such as Jaccard Similarity, Cosine Similarity and Euclidean Distance to contrast the similarity between two clusters. To minimize the risk of using only one similarity metric, we use the ranking of confidence to determine which metric should used for clustering. When we get the final metric, we will check if two of three citation records are greater than the threshold we set for each similarity metric. If two of three citation records are greater than the threshold, we will merge the citation record into the cluster, otherwise we will stop the clustering process. This threshold is adjusted during the experiment.

Algorithm 1. Proposed Clustering Algorithm

---

1. Input

    $R_1, R_2, \ldots, R_m$: R_i represent the ith citation record,

    $A_1, A_2, \ldots, A_m$: A_i represent the coauthor set of the ith citation record,

    $T_1, T_2, \ldots, T_m$: T_i represent the title of the ith citation record,

    $C_1, C_2, \ldots, C_m$: C_i represent the ith cluster,

    $C\_A_1, C\_A_2, \ldots, C\_A_m$: C_A_i represent the coauthor set of the ith cluster,

    $C\_T_1, C\_T_2, \ldots, C\_T_m$: C_T_i represent the title set of the ith cluster,

    $alpha1, alpha2, alpha3$: $alphai$ represent the threshold of the ith similarity calculation method.

2. Intialize

    put R_i into the C_i,

    put A_i into the C_A_i,

    put T_i into the C_T_i.

3. Clustering Procedure

    (1)Repeat

        If $C\_A_i \cap C\_A_j \neq \phi$:

            $C_i = C_i \cup C_j$,

            $C\_A_i = C\_A_i \cup C\_A_j$,

            $C\_T_i = C\_T_i \cup C\_T_j$,

            $C_j = \phi$,

            $C\_A_j = \phi$,

            $C\_T_j = \phi$.

Until

    each two C_A have not same author.

    (2)Repeat

    Calculate the Jaccard, Cosine, Euclidean similarity between each citation record in C_i and each citation record in C_j,

        For R_ik in C_i:

            get R_jk from C_j by using the ranking of confidence

            cond1 : $Jaccard(R\_ik, R\_jk) > alpha1$

            cond2 : $Cosine(R\_ik, R\_jk) > alpha2$

            cond3 : $Euclidean(R\_ik, R\_jk) > alpha3$

            If at least two cond is true:

                $C_i = C_i \cup C_j$,

                $C\_T_i = C_Ti \cup C_Tj$,

                $C_j = \phi$,

                $C\_T_j = \phi$.

Until

    each two C_T can't merge.

4. Output

    Clusters $C_1, C_2, \ldots, C_k$, which is not empty.

---

4   Experiments

4.1   Data Sets

In our experiments, we perform evaluations on a dataset constructed by Tang et al. [18], which contain the citations collected from the DBLP Website. We downloaded this dataset from the Website[3]. Each citation record consists of three basic attributes: title, coauthors, and venue, but we only choose the two attributes of those three basic attributes: title of paper and coauthors to our experiments. In this paper, we collected 110 author citation records which includes 1723 individual authors and 8505 citation records. Some statistics of this data set are shown in Table 3. For example, there are 28 persons with the name "David Brown" and 40 persons named "Lei Chen".

Table 3. Evaluation dataset

Name	Num. authors	Num. records	Name	Num. authors	Num. records
Bin Li	60	181	Juan Carlos Lopez	1	36
Charles Smith	4	7	Kai Zhang	24	66
David Brown	28	61	Kuo Zhang	4	16
David C. Wilson	5	65	Lei Chen	40	196
David E. Goldberg	3	231	Lei Jin	8	20
Eric Martin	5	85	Lu Liu	17	58
F. Wang	17	19	Manuel Silva	4	74
Fan Wang	14	56	Michael Wagner	14	71
Fei Su	4	37	Qiang shen	3	70
Gang Luo	9	47	Robert Allen	11	24
Jianping Wang	5	37	S. Huang	14	16
Jie Tang	6	66	Sanjay Jain	5	217
Jie Yu	9	32	Shu lin	2	76
Jim Gray	6	192	Thomas D. Taylor	3	4
John F. McDonald	2	34	Thomas Hermann	9	44

4.2   Evaluation Results

As in [18,20], we use Pairwise Precision, Pairwise Recall, and Pairwise F1 score to evaluate the performance of our method and to compare with previous methods. The pairwise measures are adapted for evaluating name disambiguation by considering the number of pairs of papers assigned with the same label. For example, if author one and author two whose names are the same in the two papers, we think that those two papers are the same as the author one and is one pair with the same label. Specifically, any two papers that are annotated with the same label in the ground truth are called a correct pair, and any two papers are predicted with the same label. For two papers with the same label

---

[3] https://cn.aminer.org/data-sna.

predicted by an approach, but not have the same label in the ground truth, we call it a wrong pair. We note the counting is for pairs of papers with the same label (either predicted or labeled) only. Thus, we can define the measures as follows Eqs. 7, 8, 9:

$$Pair\,Precision = \frac{\#PairsCorrectlyPredictedToSameAuthor}{\#TotalPairsPredictedToSameAuthor} \tag{7}$$

$$Pair\,Recall = \frac{\#PairsCorrectlyPredictedToSameAuthor}{\#TotalPairsToSameAuthor} \tag{8}$$

$$Pair\,F1 = \frac{2 \times Pair\,Precision \times Pair\,Recall}{Pair\,Precision + Pair\,Recall} \tag{9}$$

In this paper, we compare several baseline methods, such as Hierarchical Agglomerative Clustering (HAC) [17,21], K-means [10], SACluster [22] to evaluate our approach. HAC only used the Jaccard Similarity to measure the similarity between the citations and based on a list of citations and utilized a search engine to help the disambiguation task. K-means is a process for partitioning an N-dimensional population into k sets on the basis of a sample. The process appears to give partitions which are reasonably efficient in the sense of within-class variance [10], In K-means algorithm, each citation is represented by a feature vector, with each coauthor name and each keyword of the paper title as a feature of the vector [6]. SACluster tries to partition the nodes in a graph into K clusters by using both structural and attributes information associated to each node [18].

For fair comparison, we try to input the same attribute features defined in our method in these methods. Our approach consider the baseline method Hierarchical Agglomerative Clustering (HAC) and Ranking Confidence to help the disambiguation task, with the feature coauthors and title of paper. We conducted disambiguation experiments for papers related to each of the author names in our data set. Table 4 shows the results of some examples in our data sets. It is clearly indicated that our approach outperforms the baseline methords for name disambiguation, (+13.15% over HAC, +46.62% over K-means, +20.18% over SACluster by average F1 score). On the other hand, our approach have much higher precision compared to other methods excepts HAC and higher recall than other three methods. As we can see, the recall and F1 of HAC with ranking confidence (our proposed method) higher than the HAC without ranking confidence. For example, the results of "Sanjay Jain" and "Qiang shen" demonstrate that our approach have much higher precision, recall and F1 than the method HAC. Obviously, the author "Charles Smith"'s result indicates that our approach much better than the K-means and the result of the author "Thomas D. Taylor" suggests that our approach have much better than the method K-means and SACluster.

Table 4. Results of name disambiguation (Percent)

Name	HAC			K-means			SACluster			Our approach		
	Prec	Rec	F1	Prec	Rec	F1	Prec	Rec	F1	Prec	Rec	F1
Bin Li	98.53	78.34	87.28	7.81	34.54	12.75	6.94	77.17	12.74	72.09	78.92	75.35
Charles Smith	100.00	100.00	100.00	16.67	16.67	16.67	62.50	83.33	71.43	100.00	100.00	100.00
David Brown	98.29	85.82	91.63	16.97	27.61	21.02	49.86	64.55	56.26	95.40	85.07	89.94
David C. Wilson	99.37	35.63	52.45	73.63	34.78	47.24	79.02	87.03	82.83	92.45	78.02	84.62
David E. Goldberg	99.84	63.16	77.37	98.34	22.69	36.88	97.98	79.33	87.67	99.86	91.45	95.47
Eric Martin	99.79	84.12	91.29	49.55	38.70	43.46	72.09	67.72	69.84	99.74	88.44	93.75
F. Wang	88.89	94.12	91.43	88.89	94.12	91.43	78.95	88.24	83.33	88.89	94.12	91.43
Fan Wang	98.72	76.07	85.93	27.17	38.03	31.69	72.65	53.11	61.36	98.66	96.39	97.51
Fei Su	98.97	80.62	88.85	82.90	44.94	58.29	88.54	80.34	84.24	98.97	80.62	88.85
Gang Luo	98.39	70.18	81.93	50.81	36.01	42.15	77.52	52.98	62.94	98.70	87.39	92.70
Jianping Wang	98.80	49.20	65.69	72.10	33.60	45.84	82.45	87.40	84.85	86.71	82.20	84.39
Jie Tang	99.79	83.76	91.08	71.94	26.72	38.96	79.33	93.13	85.68	99.76	96.62	98.17
Jie Yu	97.28	97.28	97.28	60.63	41.85	49.52	3.03	95.65	59.36	97.28	97.28	97.28
Jim Gray	99.06	31.12	47.36	93.35	34.55	50.43	93.77	85.45	89.42	98.39	72.07	83.20
John F. McDonald	99.57	88.09	93.48	90.75	29.68	44.73	93.76	88.09	90.84	99.57	88.09	93.48
Juan Carlos Lopez	98.80	39.37	56.30	100.00	28.57	44.44	100.00	89.21	94.30	98.94	73.97	84.65
Kai Zhang	93.24	83.64	88.18	25.19	20.00	22.30	12.99	64.85	21.64	93.24	83.64	88.18
Kuo Zhang	96.55	82.35	88.89	60.87	41.18	49.12	5.68	82.35	78.87	96.55	82.35	88.89
Lei Chen	99.45	47.92	64.67	40.08	33.63	36.57	40.53	99.64	57.62	78.38	78.24	78.31
Lei Jin	100.00	83.33	90.91	41.38	28.57	33.80	72.73	76.19	74.42	100.00	83.33	90.91
Lu Liu	94.94	74.26	83.33	21.67	34.65	26.67	49.82	69.80	58.14	95.57	74.75	83.89
Manuel Silva	99.69	62.41	76.77	90.61	41.98	57.38	90.11	44.53	59.60	99.68	90.91	95.09
Michael Wagner	94.63	25.09	39.66	35.48	31.32	33.27	35.88	65.12	46.27	85.54	63.17	72.67
Qiang shen	93.66	9.49	17.24	79.83	31.69	45.36	83.33	93.87	88.28	98.61	73.60	84.29
Robert Allen	98.68	51.37	67.57	42.98	33.56	37.69	70.10	46.58	55.97	98.02	67.81	80.16
S. Huang	93.33	100.00	96.55	73.33	78.57	75.86	81.25	92.86	86.67	93.33	100.00	96.55
Sanjay Jain	99.71	65.84	79.31	92.56	34.94	50.73	78.27	97.17	86.70	99.76	91.89	95.67
Shu lin	99.29	35.37	52.16	95.69	29.57	45.18	96.91	51.91	67.60	99.60	79.90	88.67
Thomas D. Taylor	100.00	100.00	100.00	50.10	49.31	42.15	50.00	66.67	57.14	100.00	100.00	100.00
Thomas Hermann	83.54	18.44	30.21	43.14	36.03	39.27	33.99	81.84	48.03	93.78	63.13	75.46
Average	97.36	66.55	75.83	58.14	59.81	36.94	42.36	76.87	68.80	95.25	84.11	88.98

5 Conclusion and Discussion

Name disambiguation in databases is a non-trivial task because different person can share the same name and one person can have many name variations. What's more, in most case only limited information is associated with each name in the database. This paper describes a clustering approach for name disambiguation in DBLP, which only use the coauthor and paper title information of each person. Firstly, we group records of the same name into different clusters according to coauthors. Then, merging two clusters if the similarity of their titles reach the threshold. To reduce chance of selecting similarity algorithm, we propose an algorithm called Ranking Confidence. In the experiments, we use PairPresicion, PairRecall and PairF1 score to evaluate our method and compare with other methods. Experiment results show that the approach efficiently differentiate authors with the same name and generate better results than baseline methods: HAC, K-means, SACluster when using only two attributes: coauthors and title of paper. If the name disambiguation results can be better, it is useful for evaluating faculty publications and calculating statistics of social network and anthors impacts. In the future, we will pay more attention to the merge method of different clusters.

Acknowledgement. This work was supported by the Natural Science Foundation of Guangdong Province, China (2015A030310509), the Public Research and Capacity Building in Guangdong Province, China (2016A030303055), the Major Science and Technology projects of Guangdong Province, China (2016B030305004, 2016B010109008, 2016B010124008) and the National Natural Science Foundation of China (61272067).

References

1. Arif, T., Ali, R., Asger, M.: Author name disambiguation using vector space model and hybrid similarity measures. In: International Conference on Contemporary Computing-IC, pp. 135–140 (2014)
2. Bishop, T.A., Dudewicz, E.J.: Complete ranking of reliability-related distributions. IEEE Trans. Reliab. **R–26**(5), 362–365 (1977)
3. Cen, L., Dragut, E.C., Si, L., Ouzzani, M.: Author disambiguation by hierarchical agglomerative clustering with adaptive stopping criterion. In: International ACM SIGIR Conference on Research and Development in Information Retrieval (2013)
4. Cota, R.G., Ferreira, A.A., Nascimento, C., Goncalves, M.A., Laender, A.H.F.: An unsupervised heuristic-based hierarchical method for name disambiguation in bibliographic citations. J. Am. Soc. Inf. Sci. Technol. **61**(9), 1853–1870 (2010)
5. Han, H., Giles, L., Zha, H., Li, C., Tsioutsiouliklis, K.: Two supervised learning approaches for name disambiguation in author citations. In: Proceedings of the Joint ACM/IEEE Conference on Digital Libraries, pp. 296–305 (2004)
6. Han, H., Zha, H., Giles, C.L.: A model-based k-means algorithm for name disambiguation. In: International Semantic Web Conference (2003)
7. Han, H., Zha, H., Giles, C.L.: Name disambiguation spectral in author citations using a k-way clustering method. In: Proceedings of the ACM/IEEE Joint Conference on Digital Libraries, JCDL, Denver, CO, USA, 7–11 June, pp. 334–343 (2005)
8. Kang, I.S., Na, S.H., Lee, S., Jung, H., Kim, P., Sung, W.K., Lee, J.H.: On co-authorship for author disambiguation. Inf. Process. Manag. **45**(1), 84–97 (2009)
9. Li, S., Cong, G., Miao, C.: Author name disambiguation using a new categorical distribution similarity. In: Flach, P.A., Bie, T., Cristianini, N. (eds.) ECML PKDD 2012. LNCS (LNAI), vol. 7523, pp. 569–584. Springer, Heidelberg (2012). doi:10.1007/978-3-642-33460-3_42
10. Macqueen, J.: Some methods for classification and analysis of multivariate observations. In: Proceedings of the Berkeley Symposium on Mathematical Statistics and Probability, pp. 281–297 (1967)
11. Mann, G.S., Yarowsky, D.: Unsupervised personal name disambiguation, pp. 33–40 (2004)
12. Nadimi, M.H., Mosakhani, M.: A more accurate clustering method by using co-author social networks for author name disambiguation. J. Comput. Secur. **1**, 307–317 (2015)

13. On, B.W.: Social network analysis on name disambiguation and more. In: International Conference on Convergence and Hybrid Information Technology, pp. 1081–1088 (2008)

14. On, B.W., Lee, I.: Meta similarity. Appl. Intell. **35**(3), 359–374 (2011)

15. Pasula, H., Marthi, B., Milch, B., Russell, S., Shpitser, I.: Identity uncertainty and citation matching. In: NIPS, pp. 1425–1432 (2003)

16. Quan, L., Bo, W., Yuan, D.U., Wang, X., Yuhua, L.I.: Disambiguating authors by pairwise classification. Tsinghua Sci. Technol. **15**(6), 668–677 (2010)

17. Tan, Y.F., Kan, M.Y., Lee, D.: Search engine driven author disambiguation. In: Proceedings of the ACM/IEEE Joint Conference on Digital Libraries, JCDL, Chapel Hill, NC, USA, 11–15, June, pp. 314–315 (2006)

18. Tang, J., Fong, A.C.M., Wang, B., Zhang, J.: A unified probabilistic framework for name disambiguation in digital library. IEEE Trans. Knowl. Data Eng. **24**(6), 975–987 (2011)

19. Yang, K.-H., Peng, H.-T., Jiang, J.-Y., Lee, H.-M., Ho, J.-M.: Author name disambiguation for citations using topic and web correlation. In: Christensen-Dalsgaard, B., Castelli, D., Ammitzbøll Jurik, B., Lippincott, J. (eds.) ECDL 2008. LNCS, vol. 5173, pp. 185–196. Springer, Heidelberg (2008). doi:10.1007/978-3-540-87599-4_19

20. Yin, X., Han, J., Yu, P.S.: Object distinction: distinguishing objects with identical names. In: International Conference on Data Engineering, ICDE, The Marmara Hotel, Istanbul, Turkey, April, pp. 1242–1246 (2007)

21. Zepeda-Mendoza, M.L., Resendis-Antonio, O.: Hierarchical agglomerative clustering. Encycl. Syst. Biol. **43**(1), 886–887 (2013)

22. Zhou, Y., Cheng, H., Yu, J.X.: Graph clustering based on structural/attribute similarities. Proc. VLDB Endow. **2**(1), 718–729 (2009)

23. Zhu, J., Fung, G., Wang, L.: Efficient name disambiguation in digital libraries. In: Wang, H., Li, S., Oyama, S., Hu, X., Qian, T. (eds.) WAIM 2011. LNCS, vol. 6897, pp. 430–441. Springer, Heidelberg (2011). doi:10.1007/978-3-642-23535-1_37

24. Zhu, J., Cheong Fung, G.P., Zhou, X.: Anddy: a system for author name disambiguation in digital library. In: Kitagawa, H., Ishikawa, Y., Li, Q., Watanabe, C. (eds.) DASFAA 2010. LNCS, vol. 5982, pp. 444–447. Springer, Heidelberg (2010). doi:10.1007/978-3-642-12098-5_46

25. Zhu, J., Zhou, X., Fung, G.P.C.: A term-based driven clustering approach for name disambiguation. In: Li, Q., Feng, L., Pei, J., Wang, S.X., Zhou, X., Zhu, Q.-M. (eds.) APWeb/WAIM -2009. LNCS, vol. 5446, pp. 320–331. Springer, Heidelberg (2009). doi:10.1007/978-3-642-00672-2_29

BDQM

Capture Missing Values with Inference on Knowledge Base

Zhixin Qi, Hongzhi Wang$^{(\boxtimes)}$, Fanshan Meng, Jianzhong Li, and Hong Gao

Department of Computer Science and Technology, Harbin Institute of Technology,
Harbin, China
zhixin.qi@foxmail.com,fanshan.mfs@gmail.com,
{wangzh,lijzh,honggao}@hit.edu.cn

Abstract. Data imputation is a basic step for data cleaning. Traditional data imputation approaches are lack of accuracy in the absence of knowledge. Involving knowledge base in imputation could overcome this shortcoming. A challenge is that the missing value could be hardly found directly in the knowledge bases (KBs). To use knowledge base sufficiently for imputation, we present FOKES, an inference algorithm on knowledge bases. The inference not only makes full use of true facts in KBs, but also utilizes types to ensure the accuracy of captured missing values. Extensive experiments show that our proposed algorithm can capture missing values efficiently and effectively.

Keywords: Knowledge base · Missing values · Inference · Imputation · Data quality

1 Introduction

Missing values have become an inevitable problem that cannot be overlooked in the research field as well as in the industrial one. The expansion of data size makes data quality even harder to guarantee, of which missing data is a serious issue. Missing values are caused by many factors such as wrong measurements, limitation of data collection, human fault [7], which make it more difficult to prevent.

Thus, missing value imputation is in demand. During imputation, the type of the missing value should be considered. Here, type means the semantic type of a value. For instance, the type of *Washington* is *city* while the type of *U.S.A* is *country*. With the consideration of type, how can we capture missing values accurately? This problem brings following challenges.

1. When we gain a captured value, if we fill it in the table without any consideration about its type, it will possibly be a wrong value. For example, we want to capture which country *Alice* was born in. Without the consideration of type, we could take *Washington* as the missing value from KBs. However, the type of the missing value is *country* while the type of *Washington* is *city*. Therefore, the first challenge is how to judge whether the type of a candidate value matches corresponding attribute.

© Springer International Publishing AG 2017
Z. Bao et al. (Eds.): DASFAA 2017 Workshops, LNCS 10179, pp. 185–194, 2017.
DOI: 10.1007/978-3-319-55705-2_14

2. During imputation, when our knowledge is not enough, missing values can hardly be imputed accurately. For example, in the process of imputing which country *Bob* was born in, we obtain that *Bob* was born in *New York*, and *New York* is located in *U.S.A.* Without potential knowledge mining, we can only fill *New York* as the missing value. Hence, the second challenge is how we can mine potential knowledge for imputation with available conditions.

These two challenges are related to the accuracy of missing value imputation and should not be overlooked. Many missing value imputation algorithms [8,10] have been proposed. Even though they could solve the problem on many data sets, there exists some drawbacks in some cases.

Motivated by these, we attempt to fill missing values based on not only extra knowledge but also inferential knowledge from KBs to capture the missing values with complex semantics, which are difficult to be provided by web, KBs or human directly. To achieve this goal, we propose a novel framework for imputation based on inference on knowledge base.

2   Preliminaries

Knowledge Base is a structural knowledge storage system. We consider KBs as RDF-based data, whose schema is defined using Resource Description Framework Schema (RDFS) [5]. A resource is a unique identifier for a real-world entity. Two different resources are represented differently using Uniform Resource Identifiers (URIs) in a knowledge base.

Knowledge Base Pattern is a pattern described with a triple (S, P, O), where S, P, O are the subject, predicate and object, respectively. Usually, we view a knowledge base as a directed graph G. G is made up of a great deal of directed edges, containing starting points and ending points. Each edge is considered as a knowledge base pattern, where each starting point means a subject, ending point means an object, and relationship on the edge means a predicate. Consider a knowledge base $\mathcal{K}$. $\mathcal{K}$ has following two basic patterns.

For simplicity, we define several functions about SPO pattern used in this paper.

Definition 1. *Subject s and object o are initialized as two resources in KBs. Given two values x_i, y_i, the function setPairs(s, o, x_i, y_i) sets $s = x_i, o = y_i$, $i \in \{1, 2, \ldots n\}$.*

Definition 2. *Given a SPO pattern (s, p, o), the function getPredicate(s, p, o) returns predicate p in (s, p, o).*

Definition 3. *Given a SPO pattern (s, p, o), the function getObject(s, p, o) returns object o in (s, p, o).*

3   Framework

In this section, we introduce the framework of our approach. We first detect missing values in the input tables. After, we aim at capturing missing values. Beforehand missing value imputation, we should acquire knowledge base patterns of missing values. Thus, we add a pattern mining module so as to mine relationship patterns and type patterns for further imputation and inference. The framework is sketched in Fig. 1. It contains three components, (a) missing values detection, (b) pattern mining and (c) FOKES.

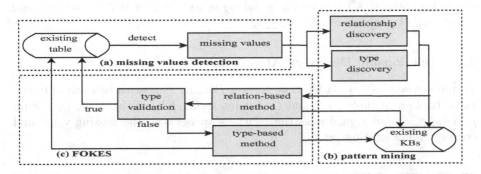

Fig. 1. The sketch of the framework: (a) missing values detection; (b) pattern mining; (c) FOKES.

Before introducing these components, we define the notation v_{ij} used through the paper. v_{ij} is the value in the ith attribute and the jth tuple in a table $\mathcal{T}_{m \times n}$, where $i = 1, 2, \ldots n$, $j = 1, 2, \ldots m$.

Missing Values Detection. We first detect which values are missing from existing dirty tables. Consider a dataset $D(t_1, t_2, \cdots, t_m)$, where t_i is a tuple in D. For each tuple $t(a_1, a_2, \cdots, a_n)$, where a_j is an attribute in t_i, if $a_j = null$, we view a_j as a missing value.

Pattern Mining. Aiming to mine patterns with known values, we first extract correct tuples from missing values detection. Then, we make hash sampling for these tuples to gain a small amount of typical data for knowledge base pattern discovery. Beforehand pattern mining, we create a q-gram index as an interface to access KBs. In this way, we can use index-based retrieval to obtain effective patterns with extracted samples. We will discuss this step in detail in Sect. 4.

FOKES. In the last step, we aim to capture missing values with mined patterns via KBs. Note that in the paper, we take *candidate value* as a captured value which is not validated by type matching mechanism. We first search candidate values for missing attributes using *relation-based method* with patterns obtained in pattern mining. Then, *type validation* process judges whether the type of the candidate value matches the type of corresponding attribute. If not, *type* *validation* will return a relationship between these two types. According to the relationship, we use *type-based method* to make further inference in order to capture accurate missing values. We will discuss this step in detail in Sect. 5.

4 Pattern Mining

First, we define an operator *access* used throughout this paper. Note that in this paper, $ed(s_1, s_2)$ is edit distance between two strings s_1 and s_2.

Definition 4. *Given a resource s_0, a knowledge base $\mathcal{K}$ and a threshold ϵ, the operator $access(s_0, \mathcal{K})$ retrieves in $\mathcal{K}$ taking s_0 as a subject in SPO pattern, and returns all (s, p, o) where $ed(s, s_0) \leq \epsilon$.*

4.1 Relationship Discovery Algorithm

To mine relationships via KBs, we form a *Pattern Pair* with two values in the same tuple in a table as S and O. By retrieving KBs with (s, o), we obtain p in (s, p, o) and regard it as the relationship between the missing value and corresponding existing value.

Algorithm 1. Relationship discovery

Input:
    a dirty table $\mathcal{T}_{m \times n}$, a knowledge base $\mathcal{K}$, a threshold ϵ, group_id g_i.
Output:
    relationships $\mathcal{P}$
1: Let $\mathcal{P}$ be relationship set, initialized empty
2: $a_k \leftarrow$ attribute_id of $\mathcal{T}_{m \times n}$, $t_j \leftarrow$ tuple_id of $\mathcal{T}_{m \times n}$
3: $v_{kj} \leftarrow \mathcal{T}_{m \times n[a_k][t_j]}$, $v_{ij} \leftarrow \mathcal{T}_{m \times n[g_i][t_j]}$
4: **for** each g_i **do**
5:     **for** each $a_k (k \neq i)$ **do**
6:         Let S_p be existing pair set, P_{ki} be candidate relationships between a_k and g_i, both initialized empty
7:         **if** $(v_{kj} \neq null) \wedge (v_{ij} \neq null)$ **then**
8:             add (v_{kj}, v_{ij}) in S_p
9:         $H \leftarrow hash(S_p)$
10:        **for** each $(v_{kj}, v_{ij}) \in H$ **do**
11:            $setPairs(s, o, v_{kj}, v_{ij})$
12:            **if** $ed(o, getObject(access(s, \mathcal{K}))) \leqslant \epsilon$ **then**
13:                add $getPredicate(access(s, \mathcal{K}))$ in P_{ki}
14:        add $max(P_{ki})$ in $\mathcal{P}$
15:        break
16: **return** $\mathcal{P}$

The pseudo code is shown in Algorithm 1. Given a table $\mathcal{T}_{m \times n}$, a knowledge base $\mathcal{K}$, a threshold ϵ and group_id g_i, the algorithm produces a relationship set

$\mathcal{P}$. First, we initialize an empty relationship set $\mathcal{P}$ to store relations of *Pattern Pairs* (Line 1). We continue to use attribute_id a_k and tuple_id t_j to locate values in $\mathcal{T}_{m \times n}$ (Line 2–3). For each a_k and g_i, existing *Pattern Pairs* (v_{kj}, v_{ij}) are added in pair set S_p (Line 4–8). Then, we make hash sampling for S_p to obtain a sample pair set H (Line 9). For each pair (v_{kj}, v_{ij}) in H, we retrieve it in $\mathcal{K}$ via *access* and add the corresponding predicate in P_{ki} (Line 10–13). Finally, we figure out the predicate with the maximum frequency in P_{ki} and add it in $\mathcal{P}$ (Line 14).

4.2 Type Discovery Algorithm

Similar as relationship, the type of a resource also matches SPO pattern in KBs. We take a known value in the same attribute with the missing value as S and <rdfs:type> as P in a SPO pattern. Then, we obtain the value of O in SPO as the type of the missing value. Thus, our basic idea for the section is to mine SPO patterns of missing attributes and take the objects of the patterns as types.

The pseudo code is shown in Algorithm 2. Given a table $\mathcal{T}_{m \times n}$, a knowledge base $\mathcal{K}$ and group_id g_i as the input, the algorithm produces types of g_i. We first initialize an empty type set $\mathcal{E}$ to store each type of g_i (Line 1). We still use tuple_id t_j to locate values in $\mathcal{T}_{m \times n}$ (Line 2–3). For each g_i, existing v_{ij}s are added in value set S_t (Line 4–7). Then, we make hash sampling for S_t to obtain a sample value set H (Line 8). For each value v_{ij} in H, we retrieve it in $\mathcal{K}$ via *access* and add the corresponding type in E_i (Line 9–11). Finally, we figure out the type with the maximum frequency in E_i and add it in $\mathcal{E}$ (Line 12).

Algorithm 2. Type discovery

Input:
     a dirty table $\mathcal{T}_{m \times n}$, a knowledge base $\mathcal{K}$, group_id g_i.
Output:
     types $\mathcal{E}$
1: Let $\mathcal{E}$ be type set, initialized empty
2: $t_j \leftarrow$ tuple_id of $\mathcal{T}_{m \times n}$
3: $v_{ij} \leftarrow \mathcal{T}_{m \times n[g_i][t_j]}$
4: **for** each g_i **do**
5:    Let S_t be existing value set, E_i be candidate types of g_i, both initialized empty
6:    **if** $v_{ij} \neq null$ **then**
7:      add v_{ij} in S_t
8:    $H \leftarrow hash(S_t)$
9:    **for** each $v_{ij} \in H$ **do**
10:      **if** $getPredicate(access(v_{ij}, \mathcal{K}))$=rdfs:type **then**
11:        add $getObject(access(v_{ij}, \mathcal{K}))$ in E_i
12:    add $max(E_i)$ in $\mathcal{E}$
13: **return** $\mathcal{E}$

---

5   FOKES (Inference Algorithm on Knowledge Bases)

Relation-Based Method. We start to capture missing values using discovered relation and known values. As each edge represents a relation of two nodes in KBs, such relation-based approach is feasible to capture missing values. For instance, in Fig. 2, if we want to know *"Where was John born in?"*, we can easily capture the value by setting *John* and *wasBornIn* as the subject and predicate, respectively, and retrieving them in KBs. Thus, we obtain the result that *New York* is where *John wasBornIn*. Similarly, when we retrieve the *capital* of *U.S.A* taking *U.S.A* as the subject and *hasCapital* as the predicate, we get *Washington* as a result.

Type Validation. To make captured missing values more accurate, we propose a process of type validation. If the captured value matches the type of the attribute, it will be more certain to be correct. For type validation, we first define a function *Compare* to judge the relation between two types.

Definition 5. *Given two types t_1, t_2 and a knowledge base $\mathcal{K}$ with type taxonomy in a hierarchical structure, $Compare(t_1, t_2, \mathcal{K})$ judges the relation of t_1 and t_2 according to $\mathcal{K}$. It has five possible returned results denoting different relations between t_1 and t_2. (1) SAMETYPE: t_1 is the same type as t_2 in $\mathcal{K}$. (2) SAMECLASS: t_1 and t_2 are in the same class but not the same type in $\mathcal{K}$. (3) SUBCLASS: t_1 is a subclass of t_2 in $\mathcal{K}$. (4) SUPERCLASS: t_1 is the superclass of t_2 in $\mathcal{K}$. (5) NULL: there is not a relation between t_1 and t_2 in $\mathcal{K}$.*

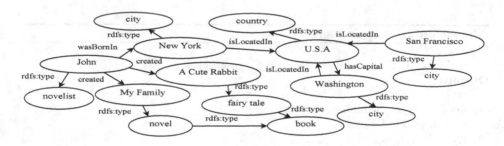

Fig. 2. The sketch of KBs

For instance, in Fig. 2, if we want to know the hierarchical relation between *city* and *country*, we can compute $Compare(city, country, \mathcal{K})$. It will return SUBCLASS, which means that *city* is a subclass of *country*. Similarly, if we compute $Compare(country, city, \mathcal{K})$, SUPERCLASS will be returned, which means that *country* is the superclass of *city*. Also, we can compute $Compare(novel, fairytale, \mathcal{K})$ to gain the hierarchical relation between *novel* and *fairy tale*. It will return SAMECLASS, which means that they are in the same class. Undoubtedly, if we compute $Compare(fairytale, fairytale, \mathcal{K})$, it will return SAMETYPE to show that they are of the same type.

Type-Based Method. Type-based imputation aims to capture the missing values according to their type patterns. First, we define a function to obtain the type of a resource in KBs.

Definition 6. *Given a resource x and a knowledge base $\mathcal{K}$, searchType(x, $\mathcal{K}$) returns getObject(access(x, $\mathcal{K}$)) when getPredicate(access(x, $\mathcal{K}$)) = type.*

In the process of type-based method, we perform different operations according to the results of $Compare(t_1, t_2, \mathcal{K})$ in type validation. First, if it returns SUBCLASS, which means that t_1 is a subclass of t_2, we will proceed type-based method centering on the captured value in relation-based step. If a proper value is obtained, we take it as the missing value. Intuitively, in Fig. 2, we obtain *New York* as a result of relation-based imputation, but the type of missing value is *country* while the type of *New York* is *city*. Then, we use type-based method taking *New York* as the subject and *country* as the type. In this way, we can get accurate value *U.S.A* for imputation.

Second, if it returns SAMECLASS, which means that t_1 and t_2 are in the same class but not the same type, we will impute missing values with type-based method centering on the known value in relation-based step. As illustrated in Fig. 2, we get *My Family* as the result of relation-based method. However, the type of the missing value is *fairy tale* while the type of *My Family* is *novel*. Both of these two types belong to class *Book*. Then, we execute type-based method regarding *John* as subject again and *fairy tale* as the type. In this way, an accurate missing value *A Cute Rabbit* is obtained.

Third, if it returns SUPERCLASS, which means that t_1 is the superclass of t_2, we cannot capture more accurate missing value by inference. The reason is that a superclass cannot infer a subclass. For instance, we want to impute which city *John* was in. When we capture *U.S.A* as a result. We can judge that its type is *country* but the type of missing value is *city*. If we impute with type-based method centering on *U.S.A*. We can obtain many cities as results and cannot judge which is the correct value. Thus, we cannot use SUPERCLASS to infer SUBCLASS.

FOKES Algorithm. The pseudo code is shown in Algorithm 3. Given a dirty table $\mathcal{T}_{m \times n}$, a knowledge base $\mathcal{K}$, group_id g_i, missing tuple set G_i, relationships $\mathcal{P}$ from Algorithm 2 and types $\mathcal{E}$ from Algorithm 3, the algorithm produces a clean table $\mathcal{T}'_{m \times n}$.

For each g_i, we take P_{ki} as the relationship P between a_k and g_i, and E_i as the type of g_i (Line 1–2). We first execute relation-based method and obtain x as the result (Line 3–5). Then, we gain t as the type of x (Line 6). If $Compare(t, E, \mathcal{K})$ returns SAMETYPE, we will fill x in $\mathcal{T}_{m \times n}$ as v_{ij} (Line 7–8). If it returns SUBCLASS, we will execute type-based method centering on x, and then obtain y as the value of v_{ij} (Line 9–12). If it returns SAMECLASS, we will execute type-based method centering on v_{kj}, and then fill z as the value of v_{ij} in $\mathcal{T}_{m \times n}$ (Line 13–16). If it returns others, we will fill *null* in $\mathcal{T}_{m \times n}$ (Line 17–18).

Algorithm 3. Inference on knowledge base
Input:
　　a dirty table $\mathcal{T}_{m \times n}$, a knowledge base $\mathcal{K}$, group_id g_i, missing tuple set G_i, relationships $\mathcal{P}$, types $\mathcal{E}$.
Output:
　　a clean table $\mathcal{T}'_{m \times n}$.
```
 1: for each gi do
 2:     P = Pki ∈ P, E = Ei ∈ E
 3:     for each tj(tj∈Gi) do
 4:         if getPredicate(access(vkj, K)) = P then
 5:             x ← getObject(access(vkj, K))
 6:             t ← searchType(x, K)
 7:             if Compare(t, E, K) = SAMETYPE then
 8:                 vij ← x
 9:             else if Compare(t, E, K) = SUBCLASS then
10:                 searchType(getObject(access(x, K)), K)=E
11:                 y ← getObject(access(x, K))
12:                 vij ← y
13:             else if Compare(t, E, K) = SAMECLASS then
14:                 searchType(getObject(access(vkj, K)), K)=E
15:                 z ← getObject(access(vkj, K))
16:                 vij ← z
17:             else if then
18:                 vij ← null
19: return T'm×n
```

6　Experiments

Knowledge Bases. We used **Yago** [6] and **Wikidata** [1] as knowledge bases. Wikidata has been converted by Freebase [4] lately. We conduct experiments on these KBs since there are abundant and various knowledge in them.

Datasets. We used three datasets as the test dataset: (*i*) **University** has 800 tuples about universities around the world and their countries [2]. (*ii*) **Movie** contains 22086 tuples from the Web [3]. (*iii*) **WikiTables** contains 28 tables from Wikipedia pages. The average number of tuples is 32.

Setup. All experiments were conducted on a PC with an Intel i7 CPU@2.4 Ghz, 8 GB of memory, and a 1 TB hard disk. Algorithms were implemented in JAVA.

Metrics. We use standard F-measure to evaluate effectiveness of our algorithm and use running time to evaluate efficiency of algorithm.

For the comparison with state-of-the-art methods, we choose an approach with KBs and crowdsourcing [5] (i.e. KATARA), and a Bayesian Network algorithm [9]. These two algorithms are universally acknowledged in missing values imputation. We compare these three methods on various datasets, including University, Movie and WikiTables, in terms of accuracy. Experimental results are shown in Fig. 3.

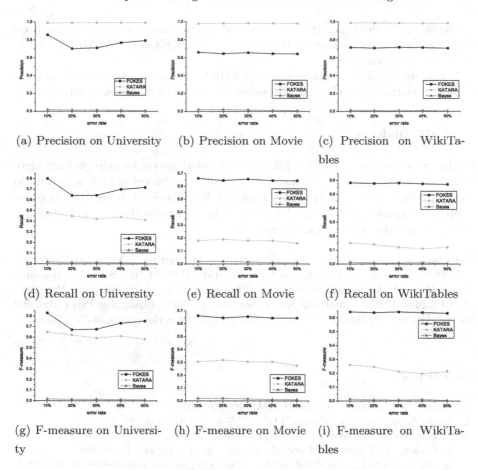

(a) Precision on University   (b) Precision on Movie   (c) Precision on WikiTables

(d) Recall on University   (e) Recall on Movie   (f) Recall on WikiTables

(g) F-measure on Universi-   (h) F-measure on Movie   (i) F-measure on WikiTables
ty

Fig. 3. Results of ours method v.s. state-of-the-art methods

As illustrated in Fig. 3(a)–(c), we observe that KATARA achieves higher precision than FOKES and Bayes. High precision of KATARA is due to its use of both KBs and experts. Both existing coverage in the KBs and answers from experts help ensure the correctness of captured values. But inference process of FOKES leads to the large number of captured values since it can infer missing values when KBs cannot provide enough knowledge.

However, from Fig. 3(d)–(f), we can observe that recall of FOKES is considerably higher than KATARA and Bayes. This is because KATARA depends much on the coverage of KBs while FOKES contains inference process, which reduces the dependence on KBs. When direct knowledge in KBs is insufficient, FOKES can infer missing values with known knowledge. This leads to the large number of corrected captured values and high recall.

Then, as depicted in Fig. 3(g)–(i), we observe that F-measure of FOKES outperforms KATARA and Bayes in all datasets. We also find that F-measure of Bayes has a declining trend as error rate rises since it depends much on the number of known values while accuracy of FOKES and KATARA is nearly unaffected by error rate.

Thus, we can draw a conclusion that FOKES is suitable for various datasets and outperforms KATARA and Bayesian Network algorithm on accuracy.

7   Conclusions

In this paper, we propose FOKES, the first inference on knowledge base algorithm to capture missing values. The algorithm achieves a high accuracy in missing value imputation due to its type matching mechanism and its process of inference. Experimental results demonstrate both the effectiveness and efficiency of our proposed data imputation method.

Acknowledgement. This paper was partially supported by NSFC grant U1509216, 61472099, National Sci-Tech Support Plan 2015BAH10F01, the Scientific Research Foundation for the Returned Overseas Chinese Scholars of Heilongjiang Provience LC2016026 and MOE - Microsoft Key Laboratory of Natural Language Processing and Speech, Harbin Institute of Technology. Hongzhi Wang is the corresponding author of this paper.

References

1. https://www.wikidata.org/wiki/Wikidata:Main_Page
2. http://rankings.betteredu.net/THE/2015-2016/top-800.html
3. http://files.grouplens.org/datasets/movielens
4. Bollacker, K., Evans, C., Paritosh, P., Sturge, T., Taylor, J.: Freebase: a collaboratively created graph database for structuring human knowledge. In: SIGMOD (2008)
5. Chu, X., Morcos, J., Ilyas, I.F., Ouzzani, M., Papotti, P., Tang, N., Ye, Y.: KATARA: a data cleaning system powered by knowledge bases and crowdsourcing. In: SIGMOD (2015)
6. Hoffart, J., Suchanek, F.M., Berberich, K., Weikum, G.: YAGO2: a spatially and temporally enhanced knowledge base from wikipedia. Artif. Intell. **194**, 28–61 (2013)
7. Hua, M., Pei, J.: DiMaC: a system for cleaning disguised missing data. In: SIGMOD (2008)
8. Lakshminarayan, K., Harp, S.A., Goldman, R.P., Samad, T.: Imputation of missing data using machine learning techniques. In: KDD (1996)
9. Mayfield, C., Neville, J., Prabhakar, S.: ERACER: a database approach for statistical inference and data cleaning. In: SIGMOD (2010)
10. Yang, K., Li, J., Wang, C.: Missing values estimation in microarray data with partial least squares regression. In: Alexandrov, V.N., Albada, G.D., Sloot, P.M.A., Dongarra, J. (eds.) ICCS 2006. LNCS, vol. 3992, pp. 662–669. Springer, Heidelberg (2006). doi:10.1007/11758525_90

Weakly-Supervised Named Entity Extraction Using Word Representations

Kejun Deng[1], Dongsheng Wang[1], and Junfei Liu[2(✉)]

[1] School of Electronics Engineering and Computer Science,
Peking University, Beijing, China
{kejund,wangdsh}@pku.edu.cn
[2] National Engineering Research Center for Software Engineering,
Peking University, Beijing, China
liujunfei@pku.edu.cn

Abstract. Named entity extraction is a key subtask of Information Extraction (IE), and also an important component for many Natural Language Processing (NLP) and Information Retrieval (IR) tasks. This paper proposes a weakly-supervised named entity extraction method by learning word representations on web-scale corpus. The highlights of our method include: (1) Word representations could be trained on either web documents or query logs; (2) Finding correct named entities is guided by a small set of seed entities, without any need for domain knowledge or human labor, allowing for the acquisition of named entities of any domain. Extensive experiments have been conducted to verify the effectiveness and efficiency of our method, comparing with the state-of-art approaches.

Keywords: Named entity · Word representations · Weakly-supervised extraction · Web documents · Query logs

1 Introduction

A named entity is an atomic element in text, which represents the name of an object of a certain class, such as an organization or a person. Named entity extraction is a key subtask of Information Extraction (IE), and also an important component for many Natural Language Processing (NLP) and Information Retrieval (IR) tasks [1].

The supervised methods, extracting named entities from texts into predefined categories [2–4], have achieved a better performance, but these methods depend on manual annotated corpus. The unsupervised methods don't need domain knowledge or human labor [1,5,6], but have a poor performance and cannot identify the class of each extracted entity. To solve these problems, weakly-supervised methods are referenced to extract named entity. The extraction is guided by a small set of seed named entities, without any need for domain-specific knowledge or handcrafted extraction patterns, allowing for the acquisition of named entities pertaining to various classes of interest [7].

© Springer International Publishing AG 2017
Z. Bao et al. (Eds.): DASFAA 2017 Workshops, LNCS 10179, pp. 195–203, 2017.
DOI: 10.1007/978-3-319-55705-2_15

Some weakly-supervised named entity extraction methods have already been proposed [1,7–9]. Which usually consist of three processes: (1) inducing templates from seed named entities based on syntactic or contextual features in the corpus; (2) generating candidate entities from the corpus using induced templates; (3) selecting the candidate entities with high confidence as output result. For every new input seeds set, the whole above processes need to be repeated. Most of the current methods follow the above processes, and the difference among these studies lies in using different types of data resources. The mainstream work is to extract entities from web documents [8,9]. Another body of research focuses on extracting entities from user search queries [1,7]. As reported in both [6,10], entity extraction from query log outperforms systems based on web documents.

In this paper, we propose a weakly-supervised named entity extraction method, which could extract named entities from both web documents and query logs, and does not need to build templates for seed entities. Our method uses word2vec [11] to learn distributed representations of words into a vector space on large un-annotated corpus. The word representations computed by neural networks could be used to calculate the similarity between two words because the learned vectors explicitly encode many linguistic regularities and patterns. Word representations have been shown to successfully improve various NLP tasks [12], such as part-of-speech tagging, chunking, named entity recognition, and semantic role labeling.

To the best of our knowledge, we are the first to using word representations for weakly-supervised named entity extraction from the web. In summary, the main contributions of this paper includes:

1. We introduce word representations into the task of named entity extraction. Our weakly-supervised method could output a large number of semantic relevant entities for any seed instances of any domain by scanning the corpus once, which get rid of template generating work.
2. We demonstrate that: (1) our method can extract named entities from either web documents and query logs based on a small set of seed entities with high precision and less time; (2) our method can also extract fine-grained entities from the same domain, for word representations capture a large number of precise syntactic and semantic word relationships. In both setting, our method outperforms the state-of-art approached of weakly-supervised entity extraction.

The rest of this paper is organized as follows. In Sect. 2, we introduce the related work. In Sect. 3, we propose our method. We show our experimental settings in Sect. 4 and evaluation in Sect. 5. We conclude this paper in Sect. 6.

2   Related Work

2.1   Entity Extrction Methods

Traditional entity extraction is to identify and classify entities from domain-specific texts into predefined categories [2–4,8]. Recently, entity extraction methods have been paying much attention on discovering named entities from massive web corpus. Because supervised methods heavily rely on human annotated corpus, which is incapable of web-scale corpus, extensive methods have been proposed to extracts entities from different data with weak supervision. A lot of work tries to extract entities from web documents [13,14]. For example, in [8], a weakly supervised system is presented to create keyword queries for a given class and extract named entities from search result document pages with predefined extracting rules. Some work focus on extracting entities from search queries. In [7], a method is proposed to extract candidate entities from search queries by context patterns learned from queries based on the seeds. Besides, unsupervised methods have been proposed to extract entities from web documents or query logs without human intervention and domain knowledge [1,15,16]. These methods usually use sophisticated NLP techniques or language rules to determine the boundaries of entities.

2.2 Word Representation Model

Word representations have a successful history of use in semantic computation and information retrieval based on the long-standing linguistic hypothesis that words occurring in similar contexts tend to have similar meanings. Different word representation models have been shown to successfully improve various NLP tasks by grouping similar words [12]. Recently, Mikolov et al. [11,17] introduced CBOW and Skip-gram models for learning high quality word representations of words from large amounts of unstructured text data, which make the training extremely efficient: an optimized single-machine implementation could train on more than 100 billion words in one day. These studies provide a reasonable basis for our work.

3 Proposed Method

Our method learns word representations by training on large un-annotated corpus for weakly-supervised named entity extraction. The extractor takes a small set of seed entities and web corpus as input. Different from other weakly-supervised methods, the input web corpus could be either web documents or query logs. The extractor output a subset of entities that are most possibly domain-relevant to the seeds. This process involves two main stages: learning word representations on the input web corpus (Step 1); and finding domain named entities based on the input seed entities and word representations (Step 2).

3.1 Learning Word Representations

In this paper, we used word2vec to learn distributed representation of words. Word2vec is a famous tool based on deep learning and released in 2013. This tool implements two models architectures, continuous bag-of-words (CBOW)

Input   Projection   Output

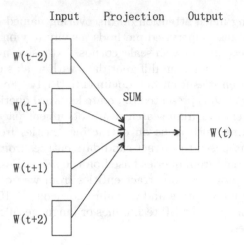

Fig. 1. The CBOW model architecture, which predicts the current word based on the context

and continuous Skip-gram model, to learn the word representations. The CBOW model predicts the current word based on the context, and the Skip-gram model predicts surrounding words given the current word [11,17]. For this paper, we used the CBOW model as Fig. 1, a neural network model that avoids multiple hidden layers for extremely fast and efficient training.

3.2 Finding Domain Named Entites

After we obtained the word representations, we can calculate the similarity scores among the word vectors to find the most semantic relevant entities for the seed instances. The similarity scores are computed based on a similarity function, named Cosine Similarity, which is used frequently in text processing. Given two word vectors A and B, the cosine similarity, $cos(\theta)$, is represented using a dot product and magnitude as:

$$Similarity = \cos(\theta) = \frac{A \bullet B}{||A|| \bullet ||B||} = \frac{\sum_{i=1}^{n} A_i B_i}{\sqrt{\sum_{i=1}^{n} A_i^2} \sqrt{\sum_{i=1}^{n} B_i^2}}$$

Where A_i and B_i are components of word vector A and B respectively.

4   Experimental Setting

4.1   Data

The goal of the experiment is twofold: (1) evaluate the ability of our weakly-supervised extraction method by learning word representations on both web documents and query logs; (2) verify if our method is applicable for fine-grained entities extraction. We use web documents and query logs data provided by Sogou Lab [18], who devotes to providing the most authoritative data for information processing in Chinese.

1. *Web documents*: We use news data collected from a number of news websites during June to July 2012 among 20 channels, such as sports, social, entertainment and so on. The original corpus is 2.1G, which provide URL and content information. We extract the context information from news file. Because the data is in Chinese, we use word segmentation tool to pre-process. The pre-processed news corpus is 3.31G.
2. *Query logs*: We use the search engine query log database from Sogou search engine. There are around 25 million queries in Chinese submitted by Web users in June 2008. We also use word segmentation tool to pre-process in our method.

4.2 Compared Method

We compare our method with the method proposed in [7], which is the state-of-art weakly-supervised domain entity extraction method based on web search logs, noted as WSLOG. First, WSLOG generates query template learned from queries based on the seeds. In detail, each query that contains a seed entity generates a query template composed of the prefix and the postfix around the matched seed instance of the query. Then, WSLOG uses these templates to extract more entities from queries. Each candidate entity is represented into a vector with a set of templates which extract it. Finally, the entities are ranked according to the distance between their representations.

4.3 Parameters

We use word2vec to learn words representations on corpus, which has several parameters to be set. As noted above, we choose the CBOW training model. We set dimension of vectors to be 200, and size of training window as 5.

5 Evaluation

In this section, we compare the experiment results of our method using both web documents and query logs with the result of WSLOG method using query logs.

5.1 Domain Entities Extraction Precison

We conducted experiments on 6 target domains that are the same as examined in [7], including City, Country, Food, Movie, Person, and University. To better test the ability of the extraction method to discover useful name entities, each target domain is specified through only five seed instances.

We conducted our method using both Web documents and query logs introduced in Sect. 4. We re-implemented WSLOG and conducted experiments using the same query logs data.

For each domain, we chose the top ranked 500 extracted entities from each method and pooled them together. We judged the quality of these pooled entities of each domain by human annotation. Table 1 shows the performance of our method and WSLOG, and our method train on both news documents and query logs.

Table 1. Comparison on domain entities extraction.

Domain	Extraction precision					
	WSLOG		Ours-logs		Ours-news	
	@250	@500	@250	@500	@250	@500
City	0.91	0.88	0.98	0.94	0.92	0.87
Country	0.28	0.17	0.52	0.30	0.69	0.41
Food	0.63	0.48	0.99	0.96	0.98	0.94
Movie	0.77	0.76	0.99	0.98	1.00	1.00
Person	0.75	0.69	0.97	0.90	0.94	0.91
University	0.64	0.58	0.74	0.57	0.76	0.58
Average-domain	0.66	0.59	0.87	0.78	0.88	0.79

As expected, in every domain, our method outperforms WSLOG in precision. For WSLOG and Ours-Logs, using the same input data consisting of around 25 million query logs, our system has a significant increase in precision. It shows that our word representation based method is better than WSLOG using templates in capturing the syntactic and semantic similarity between entities.

For ours-Logs and ours-News, using the same method on different corpus, their performance is equivalent in this experiment, although the latter using a large input corpus. This is because search queries usually take named entities as keywords.

In general, the extraction precision in Country domain is worse than other domains, for the amount of named entities in Country domain is limit and small.

5.2   Fine-Grained Entities Extraction Precision

In order to measure the extraction performance of our method at different granularity, we designed a sub-domain entity extraction experiment. We divided Person domain into five sub-domains, including Politician, Scientist, Athlete, Entertainer, and Character. We also used five seed instances for each sub-domain, and conducted our method and WSLOG using the same data in the above experiment. For each sub-domain, we chose the top 100 extracted entities, and judged the result by human annotation. To clarify, we defined Scientist to experts from a broad of areas, and Character to famous people in a film, book or play. Table 2 shows the results.

Table 2. Comparison on sub-domain entities extraction.

Sub-domain	Fine-grained extraction precision		
	WSLOG	Ours-logs	Ours-news
Politician	0.28	0.75	0.85
Scientist	0.36	0.43	0.68
Athlete	0.12	0.54	0.86
Entertainer	0.94	0.91	0.99
Character	0.07	0.48	0.81
Average-sub domain	0.35	0.62	0.84

Can be seen from the result, our method has better performance than WSLOG in fine-grained entities extraction, especially training on web documents, for web documents contains more entities and semantic features. The result of WSLOG in sub-domain entities extraction is obvious worse than the result in domain extraction. For example, the most extracted entities based on character seeds are related books or films instead of other characters. It proves that word representations contains more accurate semantic relations between entities than templates.

5.3 Running Time

As mentioned above, for each input set of seed instances, the WSLOG system repeats the three processes: (1) templates inducing; (2) candidate entities generating; (3) entities ranking. The most time-consuming part is step 2. The more templates are induced with input seed instances, the more time is spent. In our experiment on entity extraction with around 25 million query logs, the average time of candidate entities generating is about 5 h of the six domains. The longest time is more than 15 h for City domain, and the shortest time is 1.5 h for Movie.

As to our system, we train corpus only once. For each input seeds set, we load the vector model and return the output entities by comparing their word vectors with seed instances vectors. The training time for words in 25 million query logs is 1292s. Apparently, the method based word representation is more suitable for entities extraction from web-scale corpus than the template-based approach. The main reason is that word2vec using some speedup techniques for the neural network training, such as Huffman tree.

6   Conclusion and Future Work

In this paper we introduced word representation in named entities extraction task. Our weakly-supervised method can extract entities with a small set of seed entities of any domain by learning word representations on both web documents and query logs. Experimental results show that our method could extract entities in different granularity with high precision and less time.

In the future, we plan to improve our method from weakly-supervised to unsupervised. Unsupervised entity extraction can identify and classify entities from web corpus without seed instances. The main challenge of unsupervised method is to precisely determine the boundaries of entities [18]. As to our method, we should learning entity representation instead of word representation from corpus.

References

1. Song, W., Zhao, S., Zhang, C., Wu, H., Wang, H., Liu, L., Wang, H.: Exploiting collective hidden structures in webpage titles for open domain entity extraction. In: Proceedings of the 24th International Conference on World Wide Web, WWW 2015, Florence, Italy, 18–22 May 2015, pp. 1014–1024 (2015)
2. Bikel, D.M., Miller, S., Schwartz, R.M., Weischedel, R.M.: Nymble: a high-performance learning name-finder. In: ANLP, pp. 194–201 (1997)
3. Chieu, H.L., Ng, H.T.: Named entity recognition: a maximum entropy approach using global information. In: 19th International Conference on Computational Linguistics, COLING 2002, Howard International House and Academia Sinica, Taipei, Taiwan, 24 August–1 September 2002 (2002)
4. McCallum, A., Li, W.: Early results for named entity recognition with conditional random fields, feature induction and web-enhanced lexicons. In: Proceedings of the Seventh Conference on Natural Language Learning, CoNLL 2003, Held in Cooperation with HLT-NAACL 2003, Edmonton, Canada, 31 May–1 June, pp. 188–191 (2003)
5. Downey, D., Broadhead, M., Etzioni, O.: Locating complex named entities in web text. In: Proceedings of the 20th International Joint Conference on Artificial Intelligence, IJCAI 2007, Hyderabad, India, 6–12 January 2007, pp. 2733–2739 (2007)
6. Jain, A., Pennacchiotti, M.: Open entity extraction from web search query logs. In: Proceedings of the Conference on 23rd International Conference on Computational Linguistics, COLING 2010, Beijing, China, 23–27 August 2010, pp. 510–518 (2010)
7. Pasca, M.: Weakly-supervised discovery of named entities using web search queries. In: Proceedings of the Sixteenth ACM Conference on Information and Knowledge Management, CIKM 2007, Lisbon, Portugal, 6–10 November 2007, pp. 683–690 (2007)
8. Collins, M., Singer, Y.: Unsupervised models for named entity classification. In: Proceedings of the Joint SIGDAT Conference on Empirical Methods in Natural Language Processing and Very Large Corpora, pp. 100–110. Citeseer (1999)
9. Etzioni, O., Cafarella, M.J., Downey, D., Popescu, A.-M., Shaked, T., Soderland, S., Weld, D.S., Yates, A.: Unsupervised named-entity extraction from the web: an experimental study. Artif. Intell **165**(1), 91–134 (2005)

10. Parameswaran, A.G., Garcia-Molina, H., Rajaraman, A.: Towards the web of concepts: extracting concepts from large datasets. PVLDB **3**(1), 566–577 (2010)
11. Mikolov, T., Sutskever, I., Chen, K., Corrado, G.S., Dean, J.: Distributed representations of words and phrases and their compositionality. In: Advances in Neural Information Processing Systems 26: 27th Annual Conference on Neural Information Processing Systems 2013. Proceedings of a Meeting Held Lake Tahoe, Nevada, United States, 5–8 December 2013, pp. 3111–3119 (2013)
12. Collobert, R., Weston, J., Bottou, L., Karlen, M., Kavukcuoglu, K., Kuksa, P.P.: Natural language processing (almost) from scratch. J. Mach. Learn. Res. **12**, 2493–2537 (2011)
13. Dalvi, B.B., Cohen, W.W., Callan, J.: Websets: extracting sets of entities from the web using unsupervised information extraction. CoRR, abs/1307.0261 (2013)
14. Talukdar, P.P., Reisinger, J., Pasca, M., Ravichandran, D., Bhagat, R., Pereira, F.C.N.: Weakly-supervised acquisition of labeled class instances using graph random walks. In: 2008 Proceedings of the Conference on Empirical Methods in Natural Language Processing, EMNLP 2008, Honolulu, Hawaii, USA, 25–27 October 2008 A Meeting of SIGDAT, a Special Interest Group of the ACL, pp. 582–590 (2008)
15. Etzioni, O., Fader, A., Christensen, J., Soderland, S., Mausam, M.: Open information extraction: the second generation. In: Proceedings of the 22nd International Joint Conference on Artificial Intelligence, IJCAI 2011, Barcelona, Catalonia, Spain, 16–22 July 2011, pp. 3–10 (2011)
16. Banko, M., Cafarella, M.J., Soderland, S., Broadhead, M., Etzioni, O.: Open information extraction from the web. In: IJCAI, vol. 7, pp. 2670–2676 (2007)
17. Mikolov, T., Chen, K., Corrado, G., Dean, J.: Efficient estimation of word representations in vector space. CoRR, abs/1301.3781 (2013)
18. Sougou Labs. http://www.sogou.com/labs/

RDF Data Assessment Based on Metrics and Improved PageRank Algorithm

Kai Wei[✉], Pingfang Tian, Jinguang Gu, and Li Huang

College of Computer Science and Technology,
Wuhan University of Science and Technology, Wuhan, China
lydiasjbd@gmail.com, 24958320@qq.com,
{simon,huangli82}@wust.edu.cn

Abstract. With the development of the Internet, lots of data appears on the Internet. But these data can't be efficiently used owing to the lack of validity and believability, so data trust assessment has become a hot topic in the current research in the field of web. Considering the close relationship between the data credibility and its provenance, this paper proposes its own quantization rules with the existing trust evaluation model. And because of the similarities between web pages and RDF data, the improved PageRank algorithm is put forwarded in order to filter invalid data set. At last, the DataHub dataset is used to carry out a comprehensive experiment. The experimental results which carried out with DataHub set show that the proposed quantization rule and the improved PageRank algorithm can greatly improve the sorting result of the data set and reduce the effect of invalid data set on the sorting result.

Keywords: RDF data · Provenance · Assessment · Trust · PageRank

1 Introduction

With the rapid development of Web technology, the information exchanging and sharing between people becoming more smooth and convenient. Nevertheless, the increasing information in the Web has made people seem overwhelmed in the mass information processing. So people began to think about the traditional Web innovation, in order to allow humans and machines understand the content of the Web more easily, Semantic Web concept is proposed.

The semantic Web was first proposed by Tim Berbers-Lee to address interoperability issues between different applications, enterprises, and communities. Figure 1 shows the initial hierarchical structure of the Semantic Web model [1].

As the top layer of the whole Semantic Web hierarchy model, the research of the trust layer is of great significance. Although the existing Semantic Web technology has made great progress, but the RDF data trust [2] is still a difficult problem.

As a basic data model framework for describing web resources in the Semantic Web, RDF (Resource Description Framework) is widely used. In order to make the semantic Web play a greater role, Lee proposed the idea of Linked Data (Linked Data) [3], under his initiative, Linking Open Data (LOD) project is gradually developed. The goal of design LOD is to promote the self-development of Semantic Web by publishing RDF

Z. Bao et al. (Eds.): DASFAA 2017 Workshops, LNCS 10179, pp. 204–212, 2017.
DOI: 10.1007/978-3-319-55705-2_16

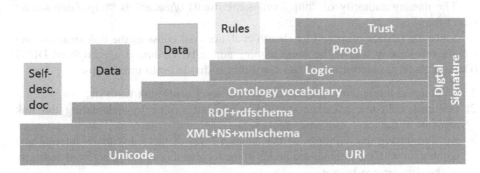

Fig. 1. Semantic Web hierarchy model

data sets on the web and creating links between the data sets which can lead to the publication of the whole community link data [4]. However, due to the openness of the semantic Web, which allows anyone to publish any content in any environment, resulting in the semantic Web RDF data quality and credibility of the uneven. At present, there is no effective method to evaluate the reliability of RDF data in Semantic Web. Therefore, the reliability of RDF data in Semantic Web has become the focus of people's research. Existing trusted evaluation of RDF data is generally carried out from two aspects [5]: one is the content-based trust for RDF data content, and the other is metadata-based trust.

Considering the content-based Trust evaluation is too subjective, the metadata-based evaluation is more convincing, so this paper starts from the following two points:

- Based on the existing trust evaluation model, the quantization standard is put forward according to the DataHub dataset;
- Optimize the PageRank algorithm to filter the invalid data set.

2 Related Work

Referencing the PageRank algorithm, Harth et al. proposed the concept of Naming Authority, which corresponded URI authoritative data sources, and then calculated the authoritative values for each authoritative data element by PageRank algorithm. Finally, according to the data source the URI uses the relationship to compute the authoritative value for each URI. As naming authority definition is abstract, to make it simple, you can take it as an ability to define a specific structure of the data source URI. A naming authority can have multiple forms, such as provenance information, documents, hosts, people, organizations, or other entities. In order to express convenience, the following methods are usually used to get the naming authority of the RDF data:

If the URI contains #, the front part of the URI is used as the authoritative name of the URI, for example:

The naming authority of "http://www.example.org/~allen/test.rdf#allenpage" is "http://www.example.org/~allen/test.rdf";

If the URI does not contain #, then the entire URI is defined as its naming authority, for example:

The naming authority of "http://xmlns.com/foaf/1.0/maker" is "http://xmlns.com/foaf/1.0/maker";

Since the scheme proposed by Hearth et al. does not consider the link structure and weight relationship within the same naming authority, Delbru et al. proposed DING (Dataset Ranking) algorithm. DING algorithm is divided into three steps:

1. Calculate authoritative values at the dataset level using link analysis;
2. For each data set, calculate the object authority value by calculating the link analysis of the local semantic object set;
3. By transferring the authority of the data set level of the value to its internal and combining with the internal semantics of the local authority itself, a global authority of the data set can be got.

In recent years, more and more researches have been done on the Provenance Model in the web environment. These models consider adding Provenance information from the Web data creation, and propose a set of ontology vocabulary to describe Provenance. Hartig and Zhao [6] extended the Provenance Model, which covers information of data creation and access. The Hartig model uses the Provenance diagram to describe the provenance information for a specific data entry, where the node is the provenance element that describes the fragment Provenance information about the data, such as the true creator of the data set. The Provenance diagram in this model define different types of Provenance elements and the possible associations between these types, and defines three different types of Provenance elements: Actors, Executions, and Artifacts. Based on the provenance model, this paper develops the quantitative criteria, and the Sect. 3 will give the concrete explanation.

On the other hand, consider the existing PageRank algorithm may exist in the low-quality data sets, and because of the commonness between data set ranking and page ranking, a new method, QPR, is used to dynamically evaluate the quality of each data set after learning from the experience of Xiaofe et al. [7] in removing the page spam links. The PR value of each data set is fairly distributed according to the quality of the data set. See Sect. 4 for details.

The experimental results in Sect. 5 show that the proposed QPR algorithm can greatly improve the ranking of query results, and can effectively reduce the spam data set on the query results.

3 Quantification of Evaluation Factors

As is known to all, the source information can be obtained by analyzing the metadata carried by the data. Although there are many models of credibility evaluation, the effective quantization means is lacking, so this paper proposes a quantization scheme by combining DataHub dataset.

DataHub [8] is a powerful data management platform that provides a range of services for creating, searching, managing, and updating data, as well as retaining source-related information about the data. In this paper, the evaluation factor quantization scheme is mainly developed from Zaveri et al. [9] and other data quality assessment factors, covering all the validation information in the process of data set generation:

Reputation, reputation is a user's evaluation of resources. It is primarily related to the data publisher (person or organization) and not to the characteristics of the data set. Credibility is usually a fraction. Assuming that the reputation value is a value from 0 (low) to 1 (high), according to Amrapali Zaveri, we can use the values of page ranks as a quantification standard by calculating the original PageRank for each dataset and then normalized the PR values as the value of the reputation for each data set.

Believability, believability is defined as the degree to which the information is determined to be correct and true. In this paper, we use the method of content trust [10] to extract n questions from the data set, and then ask the participating surveyors to give the answers to the three choices of correctness, uncertainty, error, and statistics. Suppose the number of correct answer is i, the uncertainty answer is j, the wrong answer is k, finally the believability of the data set is:

$$V(b) = \frac{i * 1 + j * 0.5 + k * 0}{n} \tag{1}$$

Objectivity is defined as the degree of impartiality of information. This information and the source, author, and maintainers are inseparable. So consider the sources, authors, and maintainers of the statistical dataset, examine their names, have more objectivity with org, gov suffixes, and have more objectivity with datasets with government tags.

Verifiability, the verifiability of the dataset, the completeness (absence or not) of the source, the maintainer, the author, and the license can be a degree.

4 Improved PageRank Algorithm

The existing PageRank algorithm is the quality response to the data set in the sorting algorithm. In the iterative process, the weights of the datasets are evenly distributed to the datasets it refers to, without taking into account the differences of the datasets. And with the rapid development of the Semantic Web, there are a large number of data sets in the network, some data sets may be garbage data, the difference of the reliability of the data sets become more obvious. This makes it even more unreasonable to distribute the weights evenly during the iterative process of PageRank, and the current algorithm can't effectively filter the invalid datasets. In addition the data set of the PageRank algorithm derived PR does not absolutely measure the reliability of the data set. As shown in Fig. 2, each circle represents a data set, the size of the circle represents the size of the PR value of the data set. Data set A and B have the same PR value, A is referenced by a small number of data sets with very high PR values, and B is referenced by a large number of data sets with very low PR values. Obviously in this case, A should be assigned a higher PR value than B's.

In order to solve these problems, this paper proposes a page quality based PageRank algorithm (QPR) based on the reliability of data set, and to make it clear, this paper brings the concept of relative reliability. By dynamically utilizing the value of the page PR after each iteration combined with RDF data link structure, the PR of each data set can be calculated.

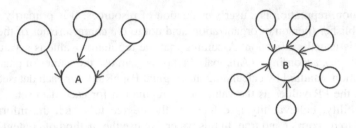

Fig. 2. Dataset link diagram

For convenience, this paper define the following variables. First, Q(p) is defined as the relative reliability, which is used as a parameter to measure the relative reliability during the iterative process. In order to reduce the effect of invalid data set on weight assignment, according to the concept of invalid dataset, the data sets with larger PR value are given higher quality, and the data sets with the largest value Set the quality of the relationship between the concept of invalid coefficient is proposed, this paper assumes that the invalid coefficient is calculated as

$$IP(p) = \frac{MPR(p)}{\left(\sum_{q \in B(p)} PR(q) \right)} \qquad (2)$$

$$MPR(p) = \max_{q \in B(p)} PR(q) \qquad (3)$$

Where B(p) represents the set of inbound chains of the data set p, as shown in Eq. (2). MPR(p) represents the maximum data set PR value in the inbound set of p.

And then according to the quality of the data set itself and the relative coefficient of failure to obtain the relative quality, the formula is

$$Q(p) = PR(p) * IP(p) \qquad (4)$$

The relative confidence value Q(p) is taken into account in the process of iteration. The data sets with the same PR value are considered as the better PR value and on next iteration it will get a higher PR value.

Using the traditional PageRank algorithm, PR value of the formula is:

$$PR(p) = d \sum_{q \in B(p)} \frac{PR(q)}{OutDeg(q)} + \frac{(1-d)}{N} \qquad (5)$$

PR(p) denotes the PR value of the data set p, N is the total number of linked datasets, and OutDeg (q) denotes the out-degree of the data set q. D is the damping factor, usually 0.85.

According to the relative quality of the data set, combined with PageRank algorithm to obtain the dynamic estimation of PR value, the formula is:

$$PR(p) = d \sum_{q \in B(p)} PR(q) \Pr(q \to p) + \frac{(1-d)}{N} \qquad (6)$$

$$\Pr(q \to p) = Q(p)/ \sum_{r \in F(q)} Q(r) \qquad (7)$$

Where Pr(q -> p) represents the PR weight of the data set q assigned to the data set q. F(q) denotes the set of outgoing chains of the data set q, that is

$$F(q) = \{p \mid (q,p) \in E\} \qquad (8)$$

The iteration can be described as follows:

1. Calculate the Q(p) value for each data set P;
2. For each link(q, p) calculate Pr(q -> p);
3. Calculate PR(p) value for each data set P;
4. Go to the next iteration.

Figure 3 is an example based on the improved PageRank algorithm. For example, suppose dataset A is linked by dataset C and the other three datasets. Data set B is linked by dataset C and another dataset. Then according to the conventional PageRank algorithm, the data set C will be assigned to A in the next round of iteration, and the PR value of B will be calculated according to the following algorithm:

$$Q(A) = \frac{1}{1+1} * 5 = 2.5$$

$$Q(B) = \frac{1}{1+1} * 5 = 2.5$$

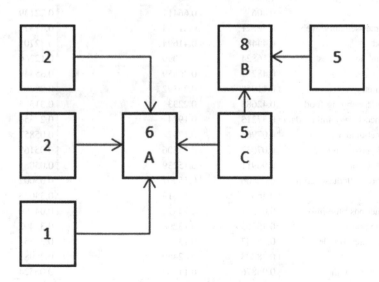

Fig. 3. QPR algorithm example

And according to the QPR algorithm, the result will be:

$$Q(A) = \frac{6}{(2+2+1+5)/5} = 3;$$

$$Q(B) = \frac{8}{(5+5)/5} = 4;$$

$$\Pr(C \rightarrow A) = 3/(3+4) = 3/7;$$

$$\Pr(C \rightarrow B) = 4/(3+4) = 4/7;$$

Therefore, with this algorithm, the PR value of data set C assigned to A is $5 \times 3/7$, and the PR value assigned to B is $5 \times 4/7$. From the results, it can be seen that the data set B is linked by two higher-quality data sets, the PR values are 5, and the dataset A is linked to the data set quality of 2, 2, 1, 5, respectively, in the real, Data set A is likely to be linked with other data sets, so the data set A should be lower than the data set B according to the idea of removing the invalid data set, so the algorithm is in accordance with expected outcome.

5   Result of Experiment

In this paper, the initial data credibility is obtained by using the evaluation model. The initial credibility is mainly inspected in several aspects: credibility, objectivity, credibility, and inspection. Then the data link graph constructed by the data set is used

Table 1. Partial dataset ranking of original PageRank algorithm

	Source of dataset	Original trustness	Trustness of improved algorithm	Initial value of seed
1	dbpedia	1.000	1.00	0.73504
2	w3c	0.406	0.66417	0.22139
3	geonames-semantic-web	0.56221	0.61154	0.41325
4	eea-rod	0.44489	0.41694	0.32701
5	personal-homepages	0.16824	0.37099	0.12366
6	lexvo	0.48227	0.26859	0.35449
7	europa-eu	0.11968	0.26392	0.08797
8	semantic-web-dog-food	0.42685	0.2233	0.31375
9	ordnance-survey-linked-data	0.37213	0.1924	0.27354
10	statusnet-quitter-se	0.07968	0.1757	0.05857
11	bio2rdf-taxonomy	0.07032	0.15506	0.05169
12	bio2rdf-taxon	0.06911	0.15239	0.0508
13	statusnet-skilledtests-com	0.06806	0.15007	0.05002
14	eunis	0.463	0.14919	0.34033
15	statusnet-postblue-info	0.06517	0.1437	0.04790
16	hellenic-police	0.45565	0.13298	0.33492
17	hellenic-fire-brigade	0.45507	0.1317	0.3345
18	lcsh	0.38208	0.12459	0.28085
19	lod2-project-wiki	0.08876	0.1188	0.06524
20	statusnet-fragdev-com	0.05132	0.11316	0.03772

Table 2. Partial dataset ranking of improved PageRank algorithm

	Source of dataset	Original trustness	Trustness of improved algorithm	Initial value of seed
1	dbpedia	1.000	1.00	0.73504
2	w3c	0.406	0.66417	0.22139
3	geonames-semantic-web	0.56221	0.61154	0.41325
4	eea-rod	0.44489	0.41694	0.32701
5	personal-homepages	0.16824	0.37099	0.12366
6	lexvo	0.48227	0.26859	0.35449
7	europa-eu	0.11968	0.26392	0.08797
8	semantic-web-dog-food	0.42685	0.2233	0.31375
9	ordnance-survey-linked-data	0.37213	0.1924	0.27354
10	statusnet-quitter-se	0.07968	0.1757	0.05857
11	bio2rdf-taxonomy	0.07032	0.15506	0.05169
12	bio2rdf-taxon	0.06911	0.15239	0.0508
13	statusnet-skilledtests-com	0.06806	0.15007	0.05002
14	eunis	0.463	0.14919	0.34033
15	statusnet-postblue-info	0.06517	0.1437	0.04790
16	hellenic-police	0.45565	0.13298	0.33492
17	hellenic-fire-brigade	0.45507	0.1317	0.3345
18	lcsh	0.38208	0.12459	0.28085
19	lod2-project-wiki	0.08876	0.1188	0.06524
20	statusnet-fragdev-com	0.05132	0.11316	0.03772

to obtain the final data credibility by the improved PageRank algorithm. In this paper, taking DataHub set as an example, the relevant experimental results are as follows (due to article space constraints, the top 20 data sets are selected), see Tables 1 and 2:

The initial values of the seeds obtained from the conventional features in the dataset are listed, the PR value obtained by PageRank and the improved PageRank algorithm.

It can be seen that dbpedia, geonames-semantic-web and other data sets have a high degree of credibility ranking, these data sets have common characteristics, links and linked data sets, maintainers with high authority, as well as richer source information. By comparing the two tables of data, some data sets such as statusnet in improved algorithm result ranking in the top, because this type of data sets are same type, and has same type of links. As with data sets muninn-world-war-i, dbpedia-nl, although the data sets between each link-intensive, but the lack of authoritative data set links, maintainers are individuals, the lack of authority. And thus the overall credibility is low, it can be seen improved PageRank algorithm to a large extent filter these invalid data sets. It can also be concluded from Table 2 that the comprehensive confidence of the data set can't be completely determined by the initial seed value. Considering the relative weights between the data sets and the breadth of the data sets themselves, the data sets are ranked Key factor.

6   Conclusion

Former evaluation model usually lack of effective evaluation criteria. Based on the deep analysis of DataHub dataset, this paper puts forward its own quantitative evaluation scheme. At the same time, we note that the traditional trustworthiness PageRank algorithm distributes the PR value evenly in the iterative process, ignoring the difference of data set reliability. The PageRank algorithm based on data quality is proposed, and the reliability of the data set is analyzed dynamically by using the updated PR value and the link structure between the data sets.

At present, the data set evaluation criterion has not been able to fully utilize the information disclosed in the data set, and more evaluation criteria should be considered in the future. At the same time, the relative reliability parameters of the improved PageRank algorithm can be further improved and remove more invalid data.

References

1. Gogulakrishnan, R., Thirumalaivasan, K., Nithiya, S., et al.: An investigation on semantic Web. Int. J. **2**(3) (2013)
2. Tomaszuk, D., Pąk, K., Rybiński, H.: Trust in RDF graphs. In: Morzy, T., Härder, T., Wrembel, R. (eds.) Advances in Databases and Information Systems, pp. 273–283. Springer, Heidelberg (2013)
3. Chiarcos, C., Mccrae, J., Osenova, P., et al.: Introduction and overview. In: 3rd Workshop on Linked Data in Linguistics: Multilingual Knowledge Resources and Natural Language Processing (2014)
4. Auer, S., Lehmann, J., Ngonga Ngomo, A.-C., Zaveri, A.: Introduction to linked data and its lifecycle on the web. In: Rudolph, S., Gottlob, G., Horrocks, I., Harmelen, F. (eds.) Reasoning Web 2013. LNCS, vol. 8067, pp. 1–90. Springer, Heidelberg (2013). doi:10. 1007/978-3-642-39784-4_1
5. Jacobi, I., Kagal, L., Khandelwal, A.: Rule-based trust assessment on the semantic web. In: Bassiliades, N., Governatori, G., Paschke, A. (eds.) RuleML 2011. LNCS, vol. 6826, pp. 227–241. Springer, Heidelberg (2011). doi:10.1007/978-3-642-22546-8_18
6. Hartig, O., Zhao, J.: Publishing and consuming provenance metadata on the web of linked data. In: McGuinness, D.L., Michaelis, J.R., Moreau, L. (eds.) IPAW 2010. LNCS, vol. 6378, pp. 78–90. Springer, Heidelberg (2010). doi:10.1007/978-3-642-17819-1_10
7. Xiaofei, C., Yitong, W., Xiaojun, F.: An improvement of PageRank algorithm based on page quality. J. Comput. Res. Dev. **46**(suppl.), 381–387 (2009)
8. Schmachtenberg, M., Bizer, C., Paulheim, H.: Adoption of the linked data best practices in different topical domains. In: Mika, P., Tudorache, T., Bernstein, A., Welty, C., Knoblock, C., Vrandečić, D., Groth, P., Noy, N., Janowicz, K., Goble, C. (eds.) ISWC 2014. LNCS, vol. 8796, pp. 245–260. Springer, Heidelberg (2014). doi:10.1007/978-3-319-11964-9_16
9. Zaveri, A., Rula, A., Maurino, A., et al.: Quality assessment for linked data: a survey. Semant. Web **7**(1), 63–93 (2015)
10. Pattanaphanchai, J.: DC proposal: evaluating trustworthiness of web content using semantic web technologies. In: Aroyo, L., Welty, C., Alani, H., Taylor, J., Bernstein, A., Kagal, L., Noy, N., Blomqvist, E. (eds.) ISWC 2011. LNCS, vol. 7032, pp. 325–332. Springer, Heidelberg (2011). doi:10.1007/978-3-642-25093-4_25

Efficient Web-Based Data Imputation
with Graph Model

Yiwen Tang, Hongzhi Wang$^{(\boxtimes)}$, Shiwei Zhang, Huijun Zhang, and Ruoxi Shi

Harbin Institute of Technology, Harbin, China
{wangzh,shiruoxi}@hit.edu.cn, isabeltang147@gmail.com, ylxdzsw@gmail.com,
zhjsss12@hotmail.com

Abstract. A challenge for data imputation is the lack of knowledge. In this paper, we attempt to address this challenge by involving extra knowledge from web. To achieve high-performance web-based imputation, we use the dependency, i.e. FDs and CFDs, to impute as many as possible values automatically and fill in the other missing values with the minimal access of web, whose cost is relatively large. To make sufficient use of dependencies, we model the dependency set on the data as a graph and perform automatical imputation and keywords generation for web-based imputation based on such graph model. With the generated keywords, we design two algorithms to extract values for imputation from the search results. Extensive experimental results based on real-world data collections show that the proposed approach could impute missing values efficiently and effectively compared to existing approach.

1 Introduction

According to recent statistics, the quality of database degenerates over time and causes loss or even disasters [1]. Data quality issues are to be solved to ensure the usability of data. For data quality issues, data incompleteness is one of the most pervasive data quality problems to handle [2]. Due to its importance, data imputation has been widely studied [3,4]. However, big data era brings new challenges for data imputation.

Firstly, for big data, knowledge is often insufficient for imputation, especially for the case with many missing values. Thus, extra knowledge is often required for big data imputation.

Second, even with sufficient knowledge, the accuracy of imputed value could not be ensured due to inconsistency and outdated data. For example, missing value could be filled according to existing knowledge base. With inconsistent or outdated data in corresponding item the knowledge base, the correct value could hardly be imputed.

Last but not least, efficiency is a non-negligible issue for big data imputation. Big data may contain many missing values to imputation. Timely imputation requires efficient imputation algorithms.

Facing to these challenges, some approaches have been proposed. Web is often adopted as supplementary knowledge since web contains a large number of data

© Springer International Publishing AG 2017
Z. Bao et al. (Eds.): DASFAA 2017 Workshops, LNCS 10179, pp. 213–226, 2017.
DOI: 10.1007/978-3-319-55705-2_17

sources. An example is WebPut [8]. However, WebPut requires many times of web search, whose cost is too large to meet the need of high efficiency.

Recently, crowdsourcing is very popular which organizations use contributions from Internet users to obtain needed services or ideas. However there has limitation on this method. Crowdsourcing allows anyone to participate, allowing for many unqualified participants and resulting in large quantities of unusable contributions which may reduce the accuracy.

Other approaches utilize initial data, in which approximate values are selected as the imputing values, according to the distribution characteristics or the constrains between attributes [3,10,11]. However, these methods are only suitable for numerical attributes but fail to impute category attributes.

In order to solve such problems, we introduce an optimized web-based data cleaning method. For both efficiency and effectiveness issues, we attempt to select proper missing values to be imputed according to the web based on the dependency between values, i.e. functional dependency (FD) or conditional dependency (CFD), such that other missing values could be imputed accurately based on the imputed values. With the consideration of complex dependency relationships among missing values, we model such relationships as a directed graph, which is called statistical dependency graph (SDG). In a SDG, we introduce three kinds of nodes, attribute nodes, condition nodes and logic relation nodes to represent attributes, condition from CFD and the relationship among attributes, respectively. Since the value of an attribute may be implied from other values according to the relationship, it is necessary for us to add a logic node between attribute nodes.

Furthermore, to achieve high filling ratio according to web, we leverage the capabilities of web search engines towards the goal of completing missing attribute values based on the keyword group obtained from SDG. We input the keyword group in the search engine to get the text dependency from Internet. Text dependency is the relation between attribute and text or attribute and attribute, which is the pattern for data cleaning. For some data set, it is difficult to find pattern. In this case, we just use keyword group values for searching.

The contributions of this paper are summarized as follows.

- We propose an optimized web-based big data imputation approach based on the dependency among missing values to increase the efficiency without the loss of effectiveness. As we know, this is the first work to combine the dependency with web-based imputation.
- To increase the filling ratio, we use multiple search engines to cover as many data sources as possible. With the consideration of the variety in representation, we develop pattern discovery and keyword-group-based search algorithms to extract proper information for imputation from the search results.
- We conduct extensive experiments to test the performance of proposed approaches. Experimental results show that our approach could impute large data sets efficiently and effectively for various data types.

The remaining parts of this paper are organized as follows. Section 2 introduce background and overview the approach. We define graph model in Sect. 3. The imputation approach based on web search is proposed in Sect. 4. Experimental results and analysis are presented in Sect. 5. Section 6 draws the conclusions.

2   Overview

In this section, we introduce background and overview our approach.

2.1   Introduce to Functional Dependency

A functional dependency is a constraint that describes the relationship between attributes in a relation. A functional dependency $FD : X \rightarrow Y$ means that the values of Y are determined by the values of X. Conditional functional dependency (CFD) [9] is proposed as a novel extension of FDs. FD holds on all the tuples in the relation, while CFD is an FD which holds on the subset of tuples satisfying a certain condition. Compared with FDs, CFDs incorporate the bindings of semantically related values which can effectively capture the consistency of data. Thus, CFD represents the dependency relationship among attributes more subtly. As a result, We can use FDs or CFDs to find the missing data based on known values according to their dependencies.

Table 1. NBA team example

ID	Team	Start-End	Arena	Location	Capacity	Coach
t_1	Golden state warriors	1964–1966	CivicAuditorium	SanFrancsicoCA	7500	
t_2	Golden state warriors	1964–1966	USFMemorialGym		6000	A.Hannum
t_3	Oklahoma city thunder	2007–2014		OklahomaCityOK	18203	
t_4		1966–1967	CivicAuditorium			
t_5	Atlanta Hawks	1949–1951	WheatonFieldHouse			Arnold Jacob

We use an example to illustrate FD and CFD.

Example 1. Consider Table 1, which specifies the NBA team VS arena in terms of the Team, Start-End, Arena, Location and Capacity. A set of CFDs and FDs on this data set is as follows.

$f_1 : [Arena] \rightarrow [Location, Capacity]$

$f_2 : [Start - End, Arena] \rightarrow [Team, Location, Capacity]$

$f_3 : [Start - End, Team] \rightarrow [Arena]$

$f_4 : [Arcna] \rightarrow [Team], 80\%$

$f_5 : [Capacity] \rightarrow [Location], 70\%$

$f_6 : [Coach = A.Hannum, Start - End] \rightarrow [Team]$

f_1 is an FD and shows that Arena determines the value of *Location* and *Capacity*. In contrast, f_6 holds on the subset of tuples that satisfies the constraint "*Coach* = *A.Hannum*", rather than on the entire table. Compared with FDs, f_6 cannot be considered an FD since f_6 includes a constraint with data values in its specification. In a word, we can utilize the values of some attribute to determine missing values for imputation according to FDs, and CFDs can compensate FDs for achieving potential relations conditionally to find missing value in a domain.

2.2   Overview of Our Approach

For efficient and effective imputation, we develop the web-based imputation framework in Fig. 1. The basic flow of the framework is to construct some keyword groups according to the attribute names, values and dependencies, retrieve results from search engines with keyword groups and extract values for imputation from search results.

To reduce the number of values to be imputed according to web, we adopt dependencies. With the consideration of multiple dependencies on the data set, we design a graph model for dependencies, *Statistical Dependency Graph* (SDG), which will be defined in Sect. 3.

As shown in Fig. 1, the first step constructs the SDG according to the dependencies among data. Through the statistics information, confidence, the statistics of the data set representing the possibility that a parent node determines a child node, is introduced as the weight of the SDG. Such confidences are used to generate the optimal keyword group for the further step. The details of SDG construction is described in Sect. 3.1. We generate keyword groups by the largest confident single sink graph discovery algorithm to the target vertex on SDG (Sect. 3.2), where a single sink graph is a special subgraph of SDG with the sink as the attribute node corresponding to a missing value and the sources as the attribute with existing values in the original data.

The selected optimal keyword group is submitted to web search engines and the results are used for imputation. To obtain proper keywords for the search engine, we obtain the pattern though keywords group value from clean tuples. Otherwise, it means that the missing value could hardly be extracted with some fixed pattern. Thus, we submit the keywords group with the attribute name corresponding to the missing value to the engine and extract the value for imputation with dictionary.

We use an example to illustrate the flow.

Example 2. We attempt to imputation missing values in the NBA data set in Table 1. According to our model, we firstly construct the SDG based on *FDs* (f_1, f_2, f_3, f_4, f_5) and CFD (f_6) shown in Fig. 2. In this example, we assume that all of the weights of f_1, f_2, f_3, f_6 are 100%, and the weights of f_4, f_5 are 80% and 70%, respectively.

Then we obtain the keyword group from the single sink graph according to SDG. In the SDG, the single sink graph with maximum confidence for attribute

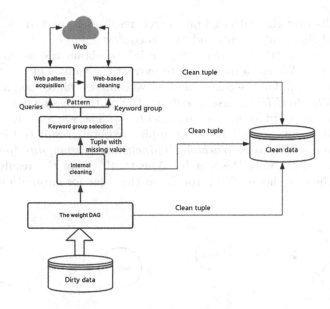

Fig. 1. The imputation framework

Arena, Location, Capacity, Team is shown in Fig. 3(a), (b), (c) and (d), respectively. The confidence of the four graphs are all 100% based on Fig. 2. Since the vertex with short path to the sink have high confidence to imply the value corresponding to the sink, we obtain the single sink graph through breadth first search (BFS). The weight of a graph is the product of confidences of all edges in the graph. From the single sink graph, we generate four keyword groups $\{Start - End, Team, Arena\}\{Arena, Location\}\{Arena, Capacity\}\{A.Hannum, Start - End, Team\}$, respectively.

For example, if we want to find the missing value of *Location* in t_5, We firstly submit keyword group $\{CivicAuditorium, SanFrancsicoCA\}$ to the search

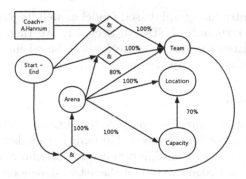

Fig. 2. SDG based on Table 1

engine and discover the pattern from search result. The pattern is "$A_1 in A_2$", where A_1 and A_2 are attributes and "in" is context.

Next, we search "$Wheaton FieldHouse$ in" to obtain the missing value of *Location* on web. We use a dictionary to extract the value which is the closest with keyword group. On the other hand, if we cannot achieve the pattern, then we search $\{Wheaton FieldHouse, Location\}$ to find the missing value according to the optimal keyword group $\{Arena, Location\}$. The way of value extraction is based on dictionary as well. In the example, we obtain result *Get information about WheatonFieldHouse in Wheaton, IL, including location, directions, reviews and photos ...* from web. We match "WheatonIL" from the result based on dictionary. Then we choose "WheatonIL" as the value for imputation.

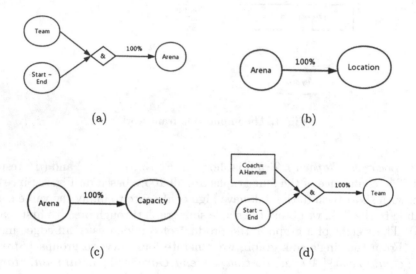

(a)   (b)

(c)   (d)

Fig. 3. Single sink graph

3   Graph-Based Model

In this section, we introduce graph-based model as well its usage in imputation. We first define the structure of SDG in Sect. 3.1. We then generate keyword groups to be submitted to search engine for web-based imputation. Keyword group selection approach is proposed in Sect. 3.2.

3.1   SDG Definition

The goal of SDG is to capture the complex (conditional) dependency relationships among attributes for further internal imputation and keyword group generation. Since a dependency relationship between two attributes could be naturally modelled as an directed edge, we model the set of dependencies as a weighted directed graph. The SDG based for f_1 to f_6 is shown in Fig. 2.

With such considerations, we define SDG as follows.

Definition 1. *A Statistics Dependency Graph (SDG) is a weighted directed graph $G = (V_a, V_l, V_c, E, W)$ corresponding to a CFD set S, where V_a is the set of attribute nodes, V_l is the set of logic nodes, V_c is the set of condition nodes, E is the set of directed edges, and W is the weight function of edges. W is the confidence which is the ratio of number of tuples which satisfied the DFs and the total number of tuples. For each CFD $f \in S$, $f : \{C, a_1, a_2, \ldots a_n\} \rightarrow \{d_1, d_2, \ldots d_m, confidence\}$. $a_1, a_2, \ldots a_n, d_1, d_2, \ldots d_m$ are attributes which are represented as attribute nodes $(v_a \in V_a)$. C represents the condition, denoted as the condition node $v_c \in V_c$. When multiple attributes determine other attributes, we introduce a logic node $v_l \in V_l$ with the parents as $a_1, a_2, \ldots a_n$ and the children as $d_1, d_2, \ldots d_m$. Confidence is represented as the weight on the graph of the graph.*

The structure of SDG based on Example 1 is shown in Fig. 2. According to f_1–f_6, there are five attributes in FDs and CFD. Additionally, in f_2, f_3 and f_6, multiple attributes determine other attributes. For example, when both attribute $Start - End$ and $Arena$ are known, we can determine attribute $Team$, $Location$ and $Capacity$ according to f_2. We can see that attribute $Start - End$ and $Arena$ has logical relationship. Also f_6 has a condition to restrain the dependency. $[Coach = A.Hannum, Start - End] \rightarrow [Team]$, which consists of a pattern tuple $(A.Hannum, \_, \_)$. The condition is that the value of an attribute $Coach$ is $A.Hannum$. As a result, we construct a SDG with five attribute nodes (V_a), three logic nodes (V_l) and a condition node (V_c). We can obtain the weight of FDs and CFD according to the statistics of the data sets. In Example 1, the confidence of f_1–f_6 are 100%, 100%, 100%, 80%, 70% and 100%, respectively according to the weights of SDG.

From this example, we could discover that SDG could effectively represent the depending relationship among attributes according to FDs and CFDs with their confidences.

3.2   Keyword Group Selection

Since we attempt to use search engine for web-based imputation. To make sufficient and efficient use of search engine, proper keywords generation is crucial. In this section, we define single sink graph and discuss the generation approaches for proper keyword groups. The definition is as follows.

Definition 2. *A Single sink graph is a subgraph of SDG, $G* \subset G$. $G*$ has a sink and a set of source. We define the attribute whose value v_t needs to be imputed as the sink node t. The attributes corresponding to the existing values are the sources while each other node from a source s to t corresponds to an attributes with missing values.*

We find that one node may have many previous nodes in SDG. It means that many dependencies correspond to one attribute. If we randomly choose a single sink graph for data imputation, the accuracy is affected because the source node may not have the closest relation with the sink node. As a result, using such keyword to search may get the wrong value in higher probability. Therefore, it requires to choose a proper dependency for imputation. As discussed in Sect. 3.1, a SDG denotes the relation among attributes and also reflects the dependency of the attributes. We can find the most confident dependency through SDG. In the optimal keyword group, the confidence as the production of all edges should be maximized. To this end, we develop an efficient algorithm for finding the optimal keyword group by breadth first search to select the single sink graph with maximum production of confidences in SDG.

In this algorithm, the confidence means how much the probability for a node to determine its child. Therefore, a higher confidence means higher dependency. For efficiency and accuracy, we should choose the single sink graph with highest confidence from SDG. Considering that the value of parent node may miss as well, we should continue to find the previous node of the parent node until we find the node with existing data as the source node. During BFS, we find that the dependency will be weak with continuous tracking back to the previous node. As a result, we define the production of confidences of all edges as the weight of the single sink graph. We start at an attribute with missing values. Then we find all its parent nodes and record the corresponding confidences. If the values of some parent nodes are missing, we continue to find the previous node and record the production of the confidence. We introduce a threshold k to denote the credibility of the keyword group. If the cost is less than k, the keyword group should not be submitted to the search engine. Finally, we select the single sink graph with maximum weight and obtain the optimal keyword group which contains all nodes in the graph.

Algorithm 1. Optimal Keyword Group Selection

Input:
  missing value, SDG, threshold K
Output:
  $G*$
1: Initialize $CandidateRule = \varphi$
2: **for** each $rule \in RuleList$ **do**
3:   **for** each $condition \in rule.ConditionList$ **do**
4:     search for the attribute in $condition$
5:     **if** meet $condition$ **then**
6:       **for** each $Dependentproperties \in rule.DependentPropertiesList$ **do**
7:         $c = rule_c \times \sqcap c (c \in condition\ of\ Dependent\ properties)$
8:         add $rule$ to $CandidateRule$
9:     **else**
10:        add $rule$ to $CandidateRule$, $c = 0$
11: $G* = maxrule_c$
12: **if** $G*_c < K$then **then**
13:    result NA

Algorithm 1 shows the optimal keyword group selection algorithm. For each *rule* in *RuleList*, If meet *condition* in correspondence *ConditionList*, we calculate the production of *confidence* and add *rule* to the *CandidateRule* (Line 2–8). Otherwise, we set confidence as 0 (Line 9, 10). After that, we obtain the optimal $G*$ with the maximum confidence (Line 11). If the confidence is less than threshold K, the optimal $G*$ is abandoned (Line 12, 13).

We use an example to illustrate this algorithm.

As is shown in Fig. 2, the SDG is based on Table 1. In t_5, the value of attribute *Location* is missing. According to Fig. 2, two attributes *Capacity* and *Arena* determine the attribute *Location*. Since the value of *Capacity* is missing as well, we need to find its parent which is attribute *Arena*. The first single sink graph is $\{Arena \rightarrow Capacity \rightarrow Location\}$, with weight $W = 100\% \times 70\%$. It is obvious that we can determine the value of *Capacity* according to attribute *Arena*, but the probability is only 70% to determine value of *Location* according to attribute *Capacity*. In a word, we may have probability 30% to impute the wrong value. The second single sink graph is $\{Arena \rightarrow Location\}$. Compared with the first graph, the confidence is 100% which outnumbers the first graph confidence. For accuracy, we define threshold $K = 80\%$ to constrain the graph. In this situation, the first graph will be abandoned.

Keyword group selection is important for pattern mining and web searching. We can effectively find the missing value according to the keyword group based on the single sink graph with maximum confidence.

Time and Space Complexity. In this algorithm, we scan all missing values. In this part, the time and space complexity is $O(n)$. For missing value, we obtain a single sink graph from SDG by BFD. As a result, the time complexity is $O(|V|+|E|)$ and the space complexity is $O(B)$, where B is the maximum branch coefficient.

4   Data Imputation Based on Web

To find proper information, we submit generated keyword groups to search engine and extract values for imputation from returned results. Most information on the web is text, and we require to extract proper value from text for imputation. For such extraction, we define the pattern in Sect. 4.1. For the special case without significant pattern, we also give the solution to achieve high filling ratio.

4.1   Text Dependency

Intuitively, the values in the text with special context corresponding to some entities in a tuple. For instance, in the Table 1, when we search keywords *CivicAuditorium* and *SanFrancsicoCA*, we will achieve the results that Civic Auditorium is a multi-purpose venue in San Francisco which reflect the relation between attribute *Arena* and *Location*. Thus, we attempt to use such context with variables representing values in the tuple as the pattern. With such pattern, we could extract the value for imputation according to existing values in the tuple. For example, we want to find the missing value in $t_5[Location]$ based on f_1. We firstly search $CivicAuditorium\ SanFrancsicoCA$ and mine pattern from the results. The obtained pattern is "$[Arena]$ in $[Loacation]$". After that, we search "$WheatonFieldHouse\ in$" and obtain result "$Get\ information\ about\ WheatonFieldHouse\ in\ Wheaton,\ IL,\ including\ location,\ directions,\ reviews\ and\ photos \dots$". From this phrase, we know that $Wheaton$ is in the position of $[Loacation]$. We match it with dictionary and find the imputed value "$WheatonIL$".

Motivated by this, we define text dependency as the pattern as follows.

Definition 3. *A pattern is in form of* $P = X_1A_1X_2A_2 \dots X_nA_kX_{k+1}$, *where* X_i *represents a phrase and* A_i *is an attribute in the table* T *to impute. For a tuple* $t \in T$, *we denote* "$X_1t[A_1]X_2t[A_2]\dots X_nt[A_k]X_{k+1}$" *as* $P(t)$. *For a text set* $\mathbb{S}$, *a text dependency* P *and a table* T, $S_{P,T} = \{S | S \in \mathbb{S} \wedge \exists t \in T,\ P(t) \in S\}$. *For a pattern* P, *if* $|S(P,T)|/|\mathbb{S}| \geq Q$, *where* Q *is a threshold,* P *is called a text dependency on* $\mathbb{S}$.

For example, a text dependency could be "$WheatonFieldHousein$" as is mentioned above. With such patter, $WheatonIL$ could be extracted from the text in the return result according to the value $CivicAuditorium$ and $SanFrancsicoCA$. Such value could be imputed to the value of attribute $Location$ in t_5 in Table 1.

4.2 Keyword-Group-Based Search

In practice, the pattern could hardly be obtained. To handle such case, we develop keyword-group-based search algorithm.

In practice, we may find that we cannot extract the pattern for some data set. In a word, data cleaning based on pattern may not suitable for all data sets. To this end, we develop an efficient method to solve this problem.

In this method, we firstly obtain the optimal keyword group and search it on web. Since the attributes from the single sink graph with the largest confidence has the closest relation with the attribute of missing value, we input source node values from the graph and the attribute of the missing value (sink node) to the search engine. We can get the value from the return result.

The pseudo code is shown in Algorithm 2. We first construct the keyword group which is "$sourcenodevalue + sinknodefromG*$" (Line 1). Then we construct URL and request for website (Line 2). After that, we extract words from the results with the dictionary (Line 3). Finally, we compare the average distance from keyword and select the minimum for data imputation (Line 4–6).

For example, according to the Table 1, we want to find the value of attribute $Team$ in t_4. We generate the optimal keyword group which is $\{Start - End, Arena, Team\}$. Therefore, we use $\{1966$–$1967,\ CivicAuditorium,\ Team\}$ as the keyword to search on web. After searching on web, we extract words in the return result by dictionary. We select the words with closest distance with the keyword in dictionary. Note that the dictionary is built by extracting all possible values in the column from web and existing values in the table.

Algorithm 2. Keyword-group-based Search

Input:
  G^*, D, source node value
Output:
  missing value w
  1: Initialize $k = source\ node\ value + sink\ node\ from\ G^*$
  2: $T = \mathrm{getPage}(K)$
  3: extract words(T,D)
  4: avgDistance(w, K, T)
  5: select min(avgDistance)
  6: **return** missing value w

---

5 Experimental Study

To test the performance of the proposed approach, we conduct extensive experiments in this section.

We implement all proposed approaches with Python and use baidu search engine. We run our experiments on a PC with an Intel Core i5-3470 CPU and 8 GB memory, running 64 bit Ubuntu 12.04. We compare state-of-art web-based data cleaning method [8].

To test the proposed methods comprehensively, we use following data sets.

The data sets above are complete relational tables. To test the performance of imputation, we omit values at random from the data sets and keep key attribute value in each tuple according to the FDs. We ensure that the previous node of missing value is imputed at least. Each proposed result is the average of 5 evaluations, that is, for each missing value percentage (1, 5, 20, 30, 40, 50 and 60%) refer to other experiments in web-based cleaning. In each evaluation, we remove data at random positions which means that we generate 5 incomplete tables. We then impute these incomplete tables using our model and evaluate the performance (Table 2).

Table 2. Experimental results on accuracy

(a) The Accuracy Of Disney Dataset

Missing ratio	1	2	3	4	5	average
5		1	1	0.75	1	1	0.95
10		1	1	1	1	0.83	0.97
20		0.91	0.89	0.92	0.89	1	0.92
30		0.87	0.93	0.95	1	0.92	0.93
40		1	0.88	0.95	1	0.78	0.92
50		1	0.82	0.85	0.89	0.92	0.90
60		0.93	0.91	0.89	0.8	0.96	0.90

(b) The Accuracy Of University Dataset

Missing ratio	1	2	3	4	5	average
5		1	1	0.75	1	1	0.95
10		1	1	1	1	0.83	0.97
20		0.91	0.89	0.92	0.89	1	0.92
30		0.87	0.93	0.95	1	0.92	0.93
40		1	0.88	0.95	1	0.78	0.92
50		1	0.82	0.85	0.89	0.92	0.90
60		0.93	0.91	0.89	0.8	0.96	0.90

1. Multilingual Disney Cartoon Table (Disney) [8]: This table contains names of 51 classical Disney cartoons in 8 different languages collected from Wikipedia.

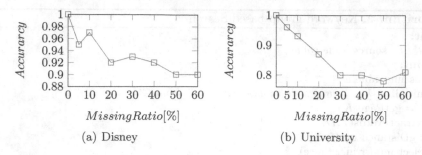

Fig. 4. The accuracy of web-based data imputation with graph model

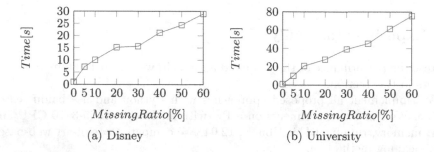

Fig. 5. The time cost of web-based data imputation with graph model

2. University Principle Information Table (Principle): This table contains 100 Chinese university and information of university includes address, city and principal which are collected from Wikipedia.

In this experiment, we first evaluate the accuracy of our model over two data sets. We propose tables of two datasets which contains specific results of 5 evaluations. As is shown in Fig. 4(a) and (b), the accuracy is pretty high and because of the random removal of value, there is no obvious relation between missing ratio and accuracy. That is good for cleaning big data in real world because our model ensure high accuracy of data sets regardless of the size of the data. The run time on these two data sets are shown in Fig. 5(a) and (b). We observe that the time cost is nearly linear with the missing ratio. As mentioned above, we introduce threshold k to restrain the data imputation for accuracy. Since there has 100% dependency relation, the threshold k does not affect the result in the two data sets according to our experiment.

Finally, We compare the performance of our model and Greedy Iterative as is shown in Fig. 6. We find that the accuracy of GreedyIter is better than One-Pass and Iterative, so we only propose the comparison with Greedy Iterative. It is clear that our approach reaches better accuracy at any missing ratio. The accuracy of Greedy Iterative drops with the increase of missing ratio. By comparison, our approach is not affected by the missing ratio.

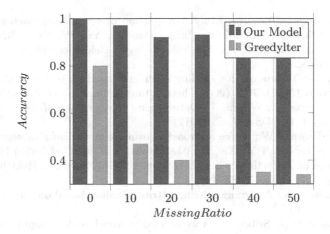

Fig. 6. Comparing the accuracy of our model and greedy iterative

6   Conclusion

We present the web-based model for processing dirty data effectively and efficiently. We have developed SDG to express FDs and CFDs. Our model has two parts for data cleaning. First we process dirty data by internal cleaning which can reduce waste time on web searching. We find that sometimes we can use known value to find missing value according to the sematic relations between the attributes. Then we can find the optimal keyword group from SDG which help us to find the missing value with high accuracy on web. The experiments show that our model can impute missing values more accurately for a relational datasets. In our future work, we will extend our model to deal with multiple kinds of datasets.

Acknowledgement. This paper was partially supported by NSFC grant U1509216, 61472099, National Sci-Tech Support Plan 2015BAH10F01, the Scientific Research Foundation for the Returned Overseas Chinese Scholars of Heilongjiang Provience LC2016026 and MOE-Microsoft Key Laboratory of Natural Language Processing and Speech, Harbin Institute of Technology. Hongzhi Wang is the corresponding author of this paper. We thanks Dr. Zhixu Li for providing the data set of Disney.

References

1. Eckerson, W.W.: Data quality and the bottom line: achieving business success through a commitment to high quality data. Technical report, The Data Warehousing Institute (2002)
2. Loshin, D.: The data quality business case: projecting return on investment. Informatica (2008)

3. Grzymala-Busse, J.W., Hu, M.: A comparison of several approaches to missing attribute values in data mining. In: Ziarko, W., Yao, Y. (eds.) RSCTC 2000. LNCS (LNAI), vol. 2005, pp. 378–385. Springer, Heidelberg (2001). doi:10.1007/3-540-45554-X_46

4. Magnani, M.: Techniques for dealing with missing data in knowledge discovery tasks. Obtido 15(01), 2007 (2004). http://magnanim.web.cs.unibo.it/index.html

5. Fan, W., Geerts, F., Wilsen, J.: Determining the currency of data. ACM Trans. Database Syst. (TODS) 37, 25 (2012)

6. Grzymala-Busse, J.W.: Three approaches to missing attribute values: a rough set perspective. In: Lin, T.Y., Xie, Y., Wasilewska, A., Liau, C.-J. (eds.) Data Mining: Foundations and Practice. SCI, vol. 118, pp. 139–152. Springer, Heidelberg (2008). doi:10.1007/978-3-540-78488-3_8

7. Li, J., Cercone, N.: Assigning missing attribute values based on rough sets theory. In: ICGC (2006)

8. Li, Z., Sharaf, M.A., Sitbon, L.: A web-based approach to data imputation. WWWJ 17, 873–897 (2014)

9. Bohannon, P., Fan, W., Geerts, F.: Conditional functional dependencies for data cleaning. In: ICDE (2007)

10. Ramoni, M., Sebastiani, P.: Robust learning with missing data. Mach. Learn. 45(2), 147–170 (2001)

11. Zhu, X.F., Zhang, S.C., Jin, Z., Zhang, Z.L., Xu, Z.M.: Missing value estimation for mixed-attribute data sets. IEEE Trans. Knowl. Data Eng. 23(1), 110–121 (2011)

12. Shi, H., Wang, Z., Webb, G.I., Huang, H.: A new restricted bayesian network classifier. In: Whang, K.-Y., Jeon, J., Shim, K., Srivastava, J. (eds.) PAKDD 2003. LNCS (LNAI), vol. 2637, pp. 265–270. Springer, Heidelberg (2003). doi:10.1007/3-540-36175-8_26

A New Schema Design Method for Multi-tenant Database

Yaoqiang Xu[1] and Jiacai Ni[2(✉)]

[1] East China Grid, Shanghai, China
xuyq@ec.sgcc.com.cn
[2] Department of Computer Science, Tsinghua University, Beijing, China
njc10@mails.thinghua.edu.cn

Abstract. Existing multi-tenant database systems either emphasize on high performance and scalability at the expense of limited customization or provide enough customization at the cost of low performance and scalability. It calls for new efficient methods to address these limitations. In this paper, we propose a *customized* database schema design framework which supports schema customization for different tenants without sacrificing performance and scalability. We propose a customized schema integration method to help tenants better design their customized schema. To effectively integrate the customized schemas, we devise the interactive-based recommendation technique, hierarchical agglomerative clustering algorithm and multi-tenancy integration algorithm based on the schema and instance information. We propose the graph partition method to reorganize the integrated tables and develop optimization techniques from both the space and the workload perspectives. Besides our customized method can adapt to any schemas and query workloads. Further, our method can be easily applied to existing databases with minor revisions. Experimental results show that our method achieves better performance and higher scalability with schema customization property than the state-of-the-art methods.

1 Introduction

Since the multi-tenant database system can amortize the cost of hardware, software and professional services to a large number of tenants, it significantly reduces per-tenant cost by increasing the scale and thus has attracted more and more attention from both industrial and academic communities.

As the service provided by the multi-tenant system is usually topic related, the system will predefine some tables which are called *predefined tables*. In the predefined tables, the system also provides some attributes which are called *predefined attributes*. However, it is impossible for the service provider to cover all the tables or attributes required by different tenants. In order to design the schema which fully satisfies tenants' requirements, the tenants require to configure their customized tables or add some new customized attributes in the predefined tables. The tables newly configured by the tenants are called *customized tables*, and the

Z. Bao et al. (Eds.): DASFAA 2017 Workshops, LNCS 10179, pp. 227–240, 2017.
DOI: 10.1007/978-3-319-55705-2_18

attributes configured by the tenants both in the predefined tables and customized tables are called *customized attributes*. Providing a high-degree customization to satisfy each tenant's needs is one of the important features in the multi-tenant database system in order to be competitive in the software service, and thus this feature must be well supported. The well-known leaders in multi-tenant database systems, e.g., Salesforce, Oracle, SAP, IBM, EMC and Microsoft, start to provide such features to boost their competitiveness. Aulbach et al. [2] emphasized that a high-level customization can be more attractive for the tenants. However, one big challenge for the multi-tenant database is to not only support the tenant customization but also to devise a high-quality database schema, achieving excellent performance, low space requirements and high scalability.

To our best knowledge, state-of-the-art approaches on the multi-tenant schema design methods can be broadly divided into three categories [7,15]. The first one is Independent Tables Shared Instances (ITSI). It maintains a physical schema for each customized schema. Then the number of tables grows in proportion to the number of tenants. The scalability relies on the table limitation that one database server can handle. The modern database system allocates the buffer pool pages for each table. Then buffer contention often occurs among different tables and thus this method has the scalability problem [1]. Admittedly, for in-memory database system, there is no buffer pool problem. But in the real multi-tenant application we still often use the common disk-based database. The second method is Shared Tables Shared Instances (STSI). In this method different customized tenants share only one table to store the data. This method has poor performance and consumes more space since it will involve large numbers of NULLs. Although modern database system can manage large numbers of NULL values, the performance is still far from satisfactory, the space can also be wasted to some extent, and the index cannot be efficiently utilized. The third method is to simplify the service system and do not allow tenants to precisely configure their private schema in order to achieve high performance, low space and excellent scalability. Thus existing multi-tenant database systems either emphasize on high performance and scalability at the expense of limited customization or provide enough customization at the cost of low performance and scalability.

To address these limitations, in this paper we propose a *customized* framework to enable efficient schema customization. We not only allow the tenants to fully configure the customized schemas but also recommend the appropriate customized tables or attributes to the tenants. Meanwhile we use the tenant configuration information as well as the schema information and instance information to integrate the customized schemas in order to enhance the scalability of the system. We propose the graph partition method to reorganize the integrated tables and do optimization from both the space and workload perspectives. To summarize we make the following contributions. (1) We propose a *customized* framework to design the customized database schema. Our framework has the following advantages. First, our method provides tenants with opportunities to fully configure their customization which satisfies their requirements closely and help to precisely design their private schema. Second, our framework can adapt to any schemas and query workloads and make full use of the space. Third, our method can be easily applied to existing databases (e.g., MySQL) with minor revisions. (2) We discuss how to accurately integrate the customized schemas and devise the recommendation techniques, hierarchical agglomerative clustering algorithm and multi-tenancy integration algorithm to integrate the schemas based on the schema and instance information. (3) We propose the graph partition method to reorganize the integrated tables, and optimize the schema design from both the space and workload perspectives. (4) Experimental results show that our method with full schema customization property achieves high performance and good scalability with low space requirements.

2 Our Customized Framework

2.1 Problem Formulation

In the multi-tenant system, each tenant outsources a database with a set of tables $\{T_1, T_2, \cdots, T_n\}$. Each table is called a *source table*. The source tables include both the predefined tables and the customized tables.

To provide each tenant with fast responses, the multi-tenant database requires to achieve high performance. To serve large numbers of tenants, the multi-tenant database needs to have good scalability and low space requirement. To achieve these goals, one big challenge of multi-tenant databases is to devise high-quality schemas. In other words the service provider needs to redesign the database schema based on the source tables to efficiently manage the data. The redesigned tables in the multi-tenant databases are called *physical tables*. In the paper, we focus on how to design effective physical tables to enable schema customization with high performance and scalability while keeping the customization property.

2.2 Basic Idea

The schema design for the multi-tenant database system can be divided into three categories. The first category is how to organize the predefined attributes in the predefined tables. The second is how to manage the customized attributes in the predefined tables. The third is the design for the customized tables. Next we give a brief discussion for each aspect.

(1) The predefined attributes in the predefined tables are often the compulsory part for different tenants. Thus we can just organize these attributes from different tenants in one table, the table is dense and the performance can be guaranteed. In addition, the scalability is excellent. So the design for the predefined attributes is straightforward and effective.
(2) For the customized attributes in the predefined tables, if we put them into the predefined tables, the predefined tables can be extremely wide or even cannot hold all the customized attributes. Because nearly all the tenants will configure the customized attributes and thus the number of customized attributes for each predefined tables can be incredibly large. Because each

customized attribute is usually shared by a small number of tenants. Then the table is very sparse and contains large numbers of NULLs. On the one hand, it dramatically decreases the performance. On the other hand, it wastes lots of space. On the contrary, if we maintain one extension table for the customized attributes in the predefined tables, the number of NULLs is reduced. However, the number of extension tables usually equals to the number of tenants. The scalability of the system becomes poor and will further negatively affect the performance. Existing studies [7] can do some optimization for these attributes, but they make the assumption that tenants have limited option and cannot fully configure customized schema.

(3) For the customized tables, similar to the customized attributes in the predefined tables, if the system maintains one table for each customized table, the number of tables grow in proportion to the number of tenants. If tenants configure plenty of customized tables, the maintenance cost is extremely large. To our best knowledge, there are few studies in this field.

To sum up, the first category is easy to solve. The second and the third categories are both the customized schemas in the system and pose great challenges to the service provider. In this paper we focus on how to design high-quality physical tables for the customized tables and attributes. Considering large numbers of customized tables and attributes, designing physical tables can follow two steps.

First, we integrate customized schemas of different tenants. The traditional schema integration can significantly avoid large numbers of different heterogeneous representations of different users and make it generic in different application domains. However, compared with traditional schema integration, the schema integration in the multi-tenant system is different:

- The result of the traditional schema integration is usually to build a global view, while the integrated schema in the multi-tenant system reflect the physical data organization.
- Traditionally, the schemas coming from different organizations may follow different data type standards to define their schemas. However, in the multi-tenant system, different tenants will follow the same data type rule to build their own schemas. Thus the system does have the possibility to conveniently integrate the data from different tenants with the same data type constraint.
- The traditional purpose of the schema integration only focuses on the accuracy. However, the schema integration in the multi-tenant database system not only focuses on accuracy but also needs to consider the performance and scalability of the system as well as space requirements.

Second, after integrating the schemas from different tenants, we notice that in one integrated table, the integrated attributes can have different sharing degrees by different numbers of tenants. In addition, different attributes may occur in different kinds of workloads. Thus, based on the integrated schema, we should do further optimization from both space and performance perspectives, and achieve the final physical schema.

• Account	Account_Detail
• Customer	Account_Priority
• Contact	
• Product	Account_Source
• Activity	Account_Stage
• Case	Account_Status
• Lead	Account_History
• Asset	
• Approval	Account_Share
• Campaign	Account_Access_Level
• Contract	Account_Territory_Assignment
• Partner	Account_Team_Member
• Opportunity	Account_Owner_Sharing

Fig. 1. Customized tables recommendation list

2.3 Our Customized Framework

In this section, we will briefly introduce our Customized framework. We first introduce how to integrate the customized schemas in Sect. 2.3 and then discuss how to optimize the integrated schema in Sect. 2.3.

Schema Integration. The schema integration in our customized framework can be divided into two categories. Next we discuss each of them respectively in general.

Interactive-Based Schema Integration. Consider a new tenant who starts to configure customized tables and attributes. This is the compulsory part the tenant will participate into the system. In our framework, we use the former configuration information from other tenants and the online algorithm to recommend the customized tables and attributes to the current tenant. The recommendation techniques can help tenants better design their schema, meanwhile, the selection of the recommendation tables or attributes just do the integration work with no interruption. Figure 1 depicts how the system recommends the customized tables to the tenants, where the left is the recommended customized tables list, while the right is the recommended customized attribute list. For instance, when the tenant wants to configure customized tables, the system recommends a list of customized tables configured by former tenants. When the current tenant touches the customized table Account, the system displays the customized attributes in the Account table in the right list. If one of the customized table matches the tenant's needs, the tenant can select it. Otherwise, the tenant can add a new customized table. Similarly, when the tenant wants to add new customized attributes in predefined tables, the system will maintain a list of customized attributes. In Sect. 3, we clarify how to rank and recommend the customized tables and attributes to the tenants in detail.

Table 1. Similar customized tables in one cluster

Company table				
Attribute	Type	Unique	Nullable	Len
C_NO	ID	Yes	No	18
C_Op_Info	String	No	Yes	60
C_Phone	Phone	No	Yes	20
Organization table				
Attribute	Type	Unique	Nullable	Len
O_ID	ID	Yes	No	18
O_Info	String	Yes	No	120
O_Email	Email	No	Yes	255

Cluster-Based Schema Integration. Considering that multiple tenants have customized many tables and the integration degree in the interactive-based method is limited, we need to further integrate these tables to increase the scalability. In multi-tenant databases, the integrated schema reflects the physical data storage to some extent. For the customized tables it is good for us to first cluster similar ones together and then integrate the schemas in each cluster. The clustering step results in a significant reduction of the search space and thus improve efficiency. In addition, integrating the attributes in each cluster reduces the memory requirements compared to integrating the full schemas directly. Furthermore, when recommending customized tables to the tenants, we can use one representative table in each cluster to ensure the recommendation and integration quality. In the multi-tenant system an important feature provides the possibility for the idea, i.e., the accurately defined data type information. Because the data type in the typical multi-tenant application can precisely represent the information of each attribute especially for the string attributes. For customized tables, we firstly build the *table feature vector* for each of them based on the data type information, and in Sect. 3 we give the formal definition. According to the vectors we group them into an appropriate number of similar clusters. Then in each cluster we integrate the schemas as an integrated table using both constraint and instance information of the customized attributes. For the customized attributes in the predefined tables, we do not cluster them, instead we directly do integration. Therefore, compared with the chunk tables in [1] and the method in Salesforce [20], the attribute integration algorithm in our customized method is obviously more accurate because those methods only consider the data type information and the data type categorization is not accurate.

Table 1 gives a running example. For ease of presentation, the table just lists a part of the information. After the first step clustering, the customized tables Company and Organization are grouped into the same cluster. Then the attributes C_NO and O_ID not only have the same data type, but also have the same constraint information, so we can finally integrate them as one integrated attribute. Although C_Op_Info and O_Info have the same data type, the remain-

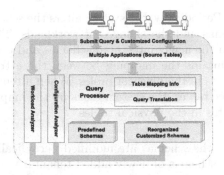

Fig. 2. Architecture of customized

ing constraint information are completely different, then we still put them into different columns in the integrated table. C_Phone and O_Email obviously have different data types, and we can directly put them into two separate columns. In our running example, the data type, uniqueness, nullable, data length are all constraint information. In the real application, we also adopt the cardinality information which can represent the distinct value percent in each field. Further the cardinality information can be easily obtained just by calling an aggregate function. In the interactive-based schema integration, it novelly utilizes the configuration information of the tenants to integrate the schemas. Thus to sum up, our proposed schema integration in the multi-tenant system includes the schema-based method, instance-based method, and the manual method. The combined integration methods can efficiently and precisely integrate the customized schemas which considers both the accuracy of integration and the scalability of the system.

Optimize the Integrated Schema. After integration, the current schema cannot be adopted as the final schema. We should do further optimizations from both the space and performance perspectives at the same time.

Space Perspective. In each integrated table, the number of columns can be very large and each integrated attribute can be shared by a small number of tenants. Thus the integrated table containing large numbers of NULLs is still sparse which wastes space and has poor performance. We can put those integrated attributes shared by similar tenants together.

Workload Perspective. Our customized method should have the ability to adapt to various workloads. Thus in customized we should make reorganized schemas workload aware. For example, some integrated attributes frequently occur in the same submitted workload, if we put them into separate tables the cost can be increased. Specifically for the workload with many costly operations, the increased cost can tremendously degrade the performance.

Query Processing. Periodically, the system alters the schema gradually based on newly added schemas and the changes of workloads. The recommendation list is altered based on the newly integrated schemas and the tenant customization. The system maintains the mapping between the attribute in the source tables and the attributes in the physical tables. When a tenant submits a query based on its own source tables, the system looks up the mapping and return the corresponding attributes in the physical tables.

Data Migration. As the schemas are modified periodically, the modification happen mainly on the schemas of newly added tenants. We modify the schema from the attribute level and most of attributes are shared by a part of the tenants. Thus the attribute number and the corresponding tuple size to be modified are both pretty small. Admittedly, when migrating the attributes, it will interrupt the service to some extent, however, the migration does not happen during the busy hours, the cost is still acceptable. And if some of the tables need to support $365 \times 24 \times 7$, we can duplicate both the source and target tables to ensure the normal operation. In the real application, each piece of data may have several number of copies at different sites. To sum up, the periodical minor modification is practical and adaptive in the real multi-tenant database system.

Figure 2 depicts the brief architecture of the system. Besides the table mapping between the source tables and physical tables the system also maintains the workload and configuration analyzer to collect the corresponding information so as to efficiently help to reorganize the customized schemas.

3 Experimental Study

In this section, we conduct extensive experiments to discuss how to integrate the customized schemas, reorganize integrated tables and make comparisons with the state-of-the-art methods.

3.1 Benchmark

It is very important to use standard benchmarks to compare the performance of different approaches. TPC-H and TPC-C are two well-known benchmarks, however, they are not well suited to evaluating the multi-tenant database since they can not be configured by different tenants. Currently, there is no ideal benchmark supporting the multi-tenancy feature especially for the full customization of different tenants and few efforts have been made in this aspect either in the academic or industrial communities. Kiefer et al. [9] proposed an approach to benchmark the multi-tenant database system and applies the same schema for different tenants. It does not consider the differences of different tenants and is not practical in the real applications. Salesforce is a well-known company which well supports the customization of different tenants. Therefore, in our paper we design our full customized multi-tenant benchmark *FC-MultiTenancy* based on the typical tables and the data type information in the Salesforce. The *FC-MultiTenancy* consists of 3 components.

Schema Generator. We use schema generator to produce the schema for each tenant. The method to manage customized attributes in the predefined tables can directly use the techniques of managing customized tables, thus for simplicity in our experimental studies we only generate a number of customized tables. When generating the customized tables we define the following parameters. The number of total tenants is T_{num}. The average number of customized tables for each tenant is Tab_{num}. The average number of attributes for each customized table satisfies the normal distribution $N(\mu, \sigma^2)$, where μ represents the average number of attributes and σ is standard variance. Based on the common schemas in the Salesforce tables, we also synthetically design customized schemas of different tenants to simulate the full customization.

Workload Generator. In order to observe the performance of our proposed method under different workloads, we use the workload generator to generate different operations, e.g, simple operations on a single table or complicated operations on more tables involving grouping, sorting or aggregation. In addition, we can also specify the ratios of different operations.

Tenanter. The last module in our benchmark is Tenanter. It is conceptually equivalent to the Driver in the TPC-H benchmark. Each Tenanter submits queries to the multi-tenant database and reports the execution time of those queries. We run Tenanter and multi-tenant database in a̧rclient/serverą́s configuration. We place the Tenanter and the database system in different machines interconnected by a network.

3.2 Experimental Settings

We generate private database schemas for tenants by running schema generator. For each tenant, we generate 3 sets of schemas by setting μ as 10, 15, 20 respectively and fixing $\sigma = 2$. We finally generate 3 groups of schemas for 1000, 2000 and 4000 tenants. Each table contains 2000 tuples.

The Tenanters and the database system are run on two machines with the same configurations. Our algorithms were implemented with C++. Each machine is running 64 bit 12.04 LTS version with an Intel(R) Xeon(R) X5670 2.93 GHz processor and 8 GB memory. We use MySQL database and InnoDB engine. To compare different methods in a fair way, for each experimental we repeat 5 times and obtain the average.

3.3 Table Clustering

In this section, we discuss how to cluster the customized tables. Figure 3 shows the number of clusters under different values of τ_{tab}. In our experiments, Tab_{num} is set 5 and μ is set 15. We vary the number of tenants T_{num} from 1000 to 4000. When determining the number of clusters, our customized method adopts the

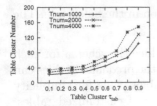
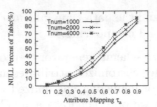
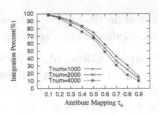

Fig. 3. Table clustering quality

Fig. 4. Integrated table
quality

Fig. 5. Integration degree

"Elbow" point method. We can see that when the number of tenants is 1000 and 2000, there is a sharp increase of the cluster number when τ_{tab} is altered from 0.8 to 0.9. Thus $\tau_{tab} = 0.8$ is the "Elbow" point. Thus in our experiment when we use the clustering-based method to integrate the schema, we set $\tau_{tab} = 0.8$. Similarly, when the number of tenants is 4000, the "Elbow" point is 0.7.

3.4   Attribute Integration Analysis

In this section, we discuss the quality of the integrated tables. After determining the number of clusters, we can integrate the customized schemas in each cluster. Figure 4 shows the average NULL percent of each integrated table. We can see that as τ_a increases, the integrated tables become more sparse containing many NULLs. As the number of tenants increases, the integrated table becomes more sparse.

Before we reorganize integrated tables, we need to make it clear that what integrated tables can be reorganized. Then Fig. 5 shows the corresponding integration percent $\mathcal{P}$ under different τ_a. Here $\mathcal{P}$ is used to reflect the integration degrees, it is computed as the ratio of the number of integrated attributes after integration and the total number of attributes of all the customized tables in this cluster. If $\mathcal{P}$ is too low, it means we do not make full use of the similarity of different customized attributes. Besides, we can directly build a wide table whose width is just total number of attributes from different customized tables in this cluster. While if $\mathcal{P}$ is too high, it means the integration accuracy is low, then for each integrated attribute the data is not similar and performance is poor. Thus, before we start to reorganize the integrated tables, we choose the integrated tables with appropriate $\mathcal{P}$. In our experiment, we choose those relatively high corresponding τ_a values when $\mathcal{P}$ is over 50%, then in our experiments for τ_a we choose 0.4, 0.5, 0.6.

3.5   Performance Comparison with Existing Methods

In this section, we compare the performance of our customized framework with two of the state-of-the-art methods, STSI and ITSI under different settings. Figures 6, 7 and 8 show the experimental results.

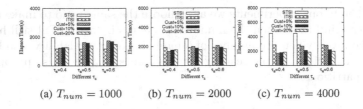

Fig. 6. Performance comparisons I (Different integration accuracy)

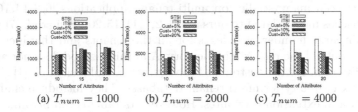

Fig. 7. Performance comparisons II (Different μ)

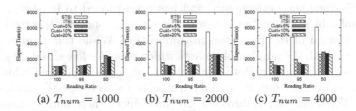

Fig. 8. Performance comparisons III (Different reading ratio)

First, we compare the performance of STSI, ITSI and our customized method under different τ_a. The workload contains 8000 operation, the reading operation ratio is 80% while the write operation is 20% which well represents the typical workload ratio in the real application. Here μ is set 15. We compare the performance when τ_a is 0.4, 0.5 and 0.6 respectively. We observe that as τ_a becomes larger more partitions have better performance. Because the corresponding integration degree is low and the integrated tables becomes wider. In addition, we find that STSI always perform poor under different cases. As the number of tenants become larger the scalability becomes the bottleneck for ITSI. Second, we compare the performance under different μ. The workload ratio is the same as that in the former experiments. Here τ_a is set 0.6. We find that when μ becomes larger, we need to sacrifice the scalability to some extent and use more partitions when deciding to reorganize the tables. Third, we compare the performance under different workloads. Here μ is set 15 and τ_a is set 0.6. Based on the analysis in the YCSB benchmark [3], we vary the ratio of reading operations. In the reading only workload, the reading ratio is 100%, in the reading heavy workload, the reading ratio is 95% and in the update heavy workload, the reading ratio is 50%. Then in our experiments we also choose the same ratio to make comparison. We can find that when the number of tenants is small, ITSI still has some priority, however, when the number of tenants is larger, scalability also becomes the problem. Besides, as the reading ratio becomes smaller, it is appropriate to use more partitions.

4   Related Work

There have been many studies on multi-tenant applications [6,8,14,16–18,21]. The approaches most related to ours are [1,2,7,11,15,20]. Hui et al. [7] proposes to maintain large number of indexes for all the tenants and this increases the chances to contend the resource in memory. And this method does not allow tenants to fully configure the customized configuration. Aulbach et al. [1] focuses on providing extensibility for multi-tenant databases. They introduce chunk tables to store data in extension attributes. However, its table is too big and plenty of costly self-joining operations happen on it. In addition, the attribute mapping is not accurate because the data types totally contain three categories which will obviously degrade the performance and has been proved to be slower than the conventional tables. The extension table layout mentioned in [1] is well suited to the customization of the multi-tenant application. While the number of tables grows linearly with the number of tenants which provides low scalability. Aulbach et al. [2] uses the object-oriented thoughts to design the schema which can be extended and evolved but its main purpose is to serve for the main-memory database. Weissman and Bobrowski [20] introduces their multi-tenancy architecture, when organizing the data from different tenants they do not consider the differences of the data from different tenants and the indexes techniques cannot be directly utilized, thus, their performance cannot reach the high level. Ni et al. [15] proposes the ADAPT schema design method. However, ADAPT needs to define a high threshold to first extract some common attributes and this cannot efficiently manage customized schemas.

Besides schema design, the multi-tenant application involves some other important issues. Elmore et al. [5] discusses how to accurately detect the changes of tenant behavior. Lang et al. [10] focuses on how to deploy resources for different tenants with various performance requirements. Data migration problem is discussed in [4,19]. Replacement and load balancing strategies are discussed in [12,13].

5   Conclusion and Future Work

In this paper, we propose the customized schema design framework for multi-tenant database system. We recommend customized tables and attributes to the tenant when they begin to configure the database service. Facing large numbers of customized schemas, we propose the interactive-based and clustering-based techniques to integrate the schemas with the tenant configuration information, schema and instance information. Based on the integrated schema, we further optimize the schema from the space and workload perspectives. Experimental results show that our framework with the fully customized property achieves high performance and good scalability with low space requirements across all kinds of workloads.

References

1. Aulbach, S., Grust, T., Jacobs, D., Kemper, A., Rittinger, J.: Multi-tenant databases for software as a service: schema-mapping techniques. In: SIGMOD Conference, pp. 1195–1206 (2008)
2. Aulbach, S., Seibold, M., Jacobs, D., Kemper, A.: Extensibility and data sharing in evolving multi-tenant databases. In: ICDE, pp. 99–110 (2011)
3. Cooper, B.F., Silberstein, A., Tam, E., Ramakrishnan, R., Sears, R.: Benchmarking cloud serving systems with YCSB. In: SoCC, pp. 143–154 (2010)
4. Elmore, A.J., Das, S., Agrawal, D., Abbadi, A.E.: Zephyr: live migration in shared nothing databases for elastic cloud platforms. In: SIGMOD Conference, pp. 301–312 (2011)
5. Elmore, A.J., Das, S., Pucher, A., Agrawal, D., El Abbadi, A., Yan, X.: Characterizing tenant behavior for placement and crisis mitigation in multitenant DBMSS. In: SIGMOD Conference, pp. 517–528 (2013)
6. Färber, F., Mathis, C., Culp, D.D., Kleis, W., Schaffner, J.: An in-memory database system for multi-tenant applications. In: BTW, pp. 650–666 (2011)
7. Hui, M., Jiang, D., Li, G., Zhou, Y.: Supporting database applications as a service. In: ICDE, pp. 832–843 (2009)
8. Jacobs, D., Aulbach, S.: Ruminations on multi-tenant databases. In: BTW, pp. 514–521 (2007)
9. Kiefer, T., Schlegel, B., Lehner, W.: MulTe: a multi-tenancy database benchmark framework. In: Nambiar, R., Poess, M. (eds.) TPCTC 2012. LNCS, vol. 7755, pp. 92–107. Springer, Heidelberg (2013). doi:10.1007/978-3-642-36727-4_7
10. Lang, W., Shankar, S., Patel, J.M., Kalhan, A.: Towards multi-tenant performance SLOs. In: ICDE, pp. 702–713 (2012)
11. Li, G., Ooi, B.C., Feng, J., Wang, J., Zhou, L.: EASE: an effective 3-in-1 keyword search method for unstructured, semi-structured and structured data. In: SIGMOD Conference, pp. 903–914 (2008)
12. Liu, Z., Hacigümüs, H., Moon, H.J., Chi, Y., Hsiung, W.-P.: PMAX: tenant placement in multitenant databases for profit maximization. In: EDBT, pp. 442–453 (2013)
13. Moon, H.J., Hacigümüs, H., Chi, Y., Hsiung, W.-P.: SWAT: a lightweight load balancing method for multitenant databases. In: EDBT, pp. 65–76 (2013)
14. Narasayya, V.R., Das, S., Syamala, M., Chandramouli, B., Chaudhuri, S.: SQLVM: performance isolation in multi-tenant relational database-as-a-service. In: CIDR (2013)
15. Ni, J., Li, G., Zhang, J., Li, L., Feng, J.: Adapt: adaptive database schema design for multi-tenant applications. In: CIKM, pp. 2199–2203 (2012)
16. Ooi, B.C., Yu, B., Li, G.: One table stores all: enabling painless free-and-easy data publishing and sharing. In: CIDR, pp. 142–153 (2007)
17. Schaffner, J., Eckart, B., Schwarz, C., Brunnert, J., Jacobs, D., Zeier, A., Plattner, H.: Simulating multi-tenant OLAP database clusters. In: BTW, pp. 410–429 (2011)

18. Schaffner, J., Januschowski, T., Kercher, M., Kraska, T., Plattner, H., Franklin, M.J., Jacobs, D.: RTP: robust tenant placement for elastic in-memory database clusters. In: SIGMOD Conference, pp. 773–784 (2013)
19. Schiller, O., Cipriani, N., Mitschang, B.: ProRea: live database migration for multi-tenant RDBMS with snapshot isolation. In: EDBT, pp. 53–64 (2013)
20. Weissman, C.D., Bobrowski, S.: The design of the force.com multitenant internet application development platform. In: SIGMOD Conference, pp. 889–896 (2009)
21. Zhang, F., Chen, J., Chen, H., Zang, B.: CloudVisor: retrofitting protection of virtual machines in multi-tenant cloud with nested virtualization. In: SOSP, pp. 203–216 (2011)

SeCoP

Reader's Choice
A Recommendation Platform

Sayar Kumar Dey[1] and Günter Fahrnberger[2(✉)]

[1] KIIT University, Bhubaneswar, Odisha, India
sayarkumardey@gmail.com
[2] University of Hagen, Hagen, North Rhine-Westphalia, Germany
guenter.fahrnberger@studium.fernuni-hagen.de

Abstract. The majority of book sellers usually abstains from offering books that are of too little interest to potential customers. Thence, such sellers might face profit losses, because the product popularity can vary from place to place. In order to avoid these losses, this disquisition introduces Reader's Choice – a system that recommends sellers to offer books based on the interest of people in different locations. Generally, most residents in a proximity share similar interests. In accordance with the search trends, Reader's Choice can learn and output the vogue of books in various regions. Thereby, the searches and purchases help Reader's Choice to determine where books are frequently sought respectively bought. Accordingly, Reader's Choice can suggest products in regions where they were more often searched and merchandised. Basically, Reader's Choice analyzes trends in datasets to draw insights. It employs Hadoop for the storage and analysis of search results and deals. A prudent performance scrutiny has testified Reader's Choice for the best functionality and the second-best information retrieval metrics among competitive book recommendation systems.

Keywords: Book · Recommendation engine · Recommendation platform · Recommendation system · Recommender engine · Recommender platform · Recommender system

1 Introduction

"A world where everyone creates content gets confusing pretty quickly without a good search engine [23]."

This quotation deserves closer attention in respect of books. There exist many sites as well as recommendation systems that provide search functionality for books. These sites use different types of recommendation techniques and can display results for any particular search phrase. Often enough, one may face a situation where they require a reference book on some topic, and they do not have a specific name for the book to search. Therefore, a good book recommendation platform is most welcome, which can suggest relevant books to users as per their interests, profession, or requirements.

© Springer International Publishing AG 2017
Z. Bao et al. (Eds.): DASFAA 2017 Workshops, LNCS 10179, pp. 243–255, 2017.
DOI: 10.1007/978-3-319-55705-2_19

Profession-based search results proffer the major advantage for users of obtaining information with more relevance to their professional life. A professional may only be interested in books that are more related to their profession. Similarly, a student may only require books with pertinence to their subject. Until now, nobody has invented a recommendation platform for books that can show customized results to a person based on their profession.

Very few book recommendation sites provide a personalized experience to the users. Even those with a customized search facility for individuals do not have the functionality of grouping users as per their geographical location to mine these data for sellers who seek current regional books trends.

The basic requirement of any recommendation system are meaningful data. The more reasonable data a system has available, the better the results it can draw out of them. The number of usable fields (that a recommender system can take into account for data mining) represents another requirement. The more data attributes a recommendation system takes into account, the more customized results it can derive. This seemingly simple process may cause an immense CPU-load during operations on huge datasets. Even frugal search queries might result in extremely long waiting times. To avoid such latencies, designers let their recommendation systems take less fields into consideration when executing a certain query. This leaves the lack of a system that always provides search results at user-satisfying accuracy and speed.

A book recommendation system, which

- groups users as per their geographical location for sellers,
- shows customized results to a person based on their profession, and
- provides search results at user-satisfying accuracy and speed,

does currently not exist.

Hence, this scholarly piece introduces Reader's Choice to mitigate this deficit. Reader's Choice denominates a highly customizable and efficient recommendation platform for books that evenhandedly targets both readers and sellers of books. For this purpose, Reader's Choice uses hybrid recommendation. It takes advantage of inputted user keywords to show them choices based on their personality or profession. Every conducted book search of users becomes stored and later used for suggesting them books based on their past searches. There also exists an option to show them books that may be related to their current search. In many cases users may not like the results that are displayed. In that case there is an option to rate the shown results. The ratings are collected and stored for enhancing the experience of all users.

The analyzed ratings are used to help a seller by displaying the popular searches by the users in a particular location. A seller can apply many filters on the result attributes. The filter includes sorting books as per the number of times they are searched by the users, ratings, and the popularity of genre. Every result shows the books that are likely to be popular in a region. This helps sellers to get an idea of the books that are popular in their targeted regions and, thus, to accordingly manage their stocks. This will help them in maximizing their profits, as they will be able to plan their business according to the demand.

At one sight, Reader's Choice coruscates with the following merits:

- Reader's Choice works on a Hive database [19], which helps in efficient processing of very large datasets.
- It comes along with the three main algorithms *Login, User Search,* and *Seller Search.*
- A comparative performance analysis evaluated the recommendation ability of Reader's Choice.

On that account, this paper is organized as follows. Section 2 explains the related work about recommendation systems for books. Section 3 describes the design of Reader's Choice. Section 4 explicates the performance analysis of the proposed system. Conclusions and recommendations for future work follow in Sect. 5.

2 Related Work

Recommender/recommendation systems/platforms/engines represent a subclass of information filtering systems that seek to predict the rating or preference that a user would have for an item. Recommender systems have become extremely common in recent years and utilized in a variety of areas, such as movies, music, news, books, research articles, search queries, social tags, and products in general [17]. There also exist recommender systems for experts, collaborators, jokes, restaurants, financial services, life insurances, romantic partners (online dating), and Twitter pages.

Recommender systems turn out to be a useful alternative to search algorithms, because they help users discovering items that they would not have found by their own. Interestingly enough, implementations of recommender systems often retrofit search engines with indexes of non-traditional data.

Recommender systems typically produce lists of recommendations in one of two ways, either through collaborative or through content-based filtering.

A collaborative filtering approach builds a model from a user's past behavior as well as from similar decisions made by other users [1]. On this account, a model incorporates previously purchased items or ratings given to them. Such a model can predict items (or ratings for them) with probable interest for a user.

A content-based filtering approach utilizes a series of discrete characteristics of an item in order to recommend additional items with similar properties [13]. Personality-based approaches as a special case of content-based filtering derive product and service preferences from a user's personality.

Planners often amalgamate both approach ilks to hybrid recommender systems [4,15].

The comparison of the two popular music recommender systems Last.fm [8] and Pandora Radio [10] demonstrates the differences between collaborative and content-based filtering.

- Last.fm creates a virtual station of recommended songs for a user by observing what bands and individual tracks they have listened to on a regular basis.

Further, it opposes the suggestions to the listening behavior of other users. Last.fm will play tracks that do not appear in a user's library and are often listened by other users with similar interests. As this approach leverages the behavior of users, it typifies an example of a collaborative filtering technique.
– Pandora uses the properties of a song or artist in order to spawn a virtual station that plays music with similar properties. For example, the Music Genome Project provides approximately 400 attributes per song. Pandora exploits user feedback to refine a station's playlist, to deemphasize certain attributes if a user *dislikes* a particular song, and to emphasize particular attributes if a user *likes* a song. This approach stands for an example of content-based filtering.

Each type of system has its own strengths and weaknesses. Last.fm requires a large amount of information about a user in order to afford them accurate recommendations. This entails a cold start problem that collaborative filtering systems commonly suffer from. On the contrary, Pandora needs very little information to get started, but it is far more limited in scope. For instance, it can only make recommendations that are similar to the original seed.

Apart from the above-mentioned classification, the appropriate scientific literature also contains the following five trend analysis systems for books.

2.1   Shelfari

Shelfari denotes a former social network built around books [5]. It was used to browse reviews and get suggested additions to reading lists. The site was also used to add one's own library contents and reviews into the mix. Anyone could pick a book they liked and see what additional titles Shelfari recommended. It often met with more success by browsing through the extensive lists, tag clouds, and collections of popular titles for additional reading. Amazon, the acquirer of Shelfari, decommissioned it in June 2016 and redirected all its links to Goodreads (see Subsect. 2.4).

2.2   LibraryThing

LibraryThing signifies a massively powerful online book-cataloging tool [9]. Even nonmembers can browse the extensive lists and suggestions. To take advantage of the massive catalog of millions of books and the tastes of their owners, one should really sign up and enter in some of their favorites to build a personal library. Uncannily good suggestions of LibraryThing stem from its extensive user base of over a million book lovers. Such a large pool most likely incloses somebody with similar tastes and a recommendation list to share.

2.3   Amazon

Amazon offers a list tool with what people can assemble lists, e.g. *Best Summer Reads* [11]. Undoubtedly, most people get their recommendations from Amazon by suggestions like *Customers Who Bought This* .... Buying a book on container gardening, Amazon will suggest other interesting books that people have purchased about container and backyard gardening. Amazon comes along with another recommendation feature that introduces one to totally new topics because of other shoppers who bought books in pairings, such as backyard gardening and composting or composting and chicken tending.

2.4 Goodreads

Goodreads names a book lover's site with great features, like interviews with authors, book trivia, book swap events, and more [14]. Most notably, Goodreads compiles lists with the input of thousands and thousands of users. Recommended historical fiction, best space operas, best young adult novels of 2010, books that were better than their movie counterparts, and other interesting lists abound on the site. By way of example, one obtains the opinion of thousands about the top humorous non-fiction travel books rather than merely of one.

2.5 tvtag

tvtag (formerly GetGlue) requires a registration to get any recommendations [3]. It does not support casual browsing or previewing of the database. After enrollment one can become a part of the large tvtag network with people who review and rank everything from books to music media with a simple like/dislike-toggle. tvtag compares their (dis)likes to other users and returns an entire battery of suggestions that they can again *like* or *dislike* to further refine the process. One of the rather novel features of tvtag is the *Achievement Unlocked! sticker* system. Therein one can earn badges for accomplishments in the GetGlue network, like being a first adopter that recommends a lot of new material or ranking a lot of media.

Albeit the aforesaid recommender systems certainly achieved marvelous merits, nary one of them dedicates itself to aid sellers in maximizing their profit.

3 Design

Reader's Choice tries to find out the popular trends of books by logging data from the searches of prospective customers. Thereby, it achieves the twofold purpose of providing them with a highly customizable book search engine and additionally helping sellers to achieve maximum profit. On account of this, the latter can plan their business strategies as per demand in the market. Beyond that, they acquire a fair idea of the quantity for each book that they should expect to sell.

Despite the current prototypical status of Reader's Choice, it can be already ascertained that this novel book recommender system will become a (cloud) web service in live operation. Hence, this treatise focuses on the database design (see Subsect. 3.1) and procedural design (see Subsect. 3.2) rather than on a comprehensive architectural design. In accordance with Fig. 1, Reader's Choice consists of two main functional units: a common GUI (Graphical User Interface) and a database. Reader's Choice requires an enormous amount of computation power when dealing with huge datasets, just as any recommendation system does. To overcome this issue, a Hive database based on Hadoop takes over the necessary computations [21].

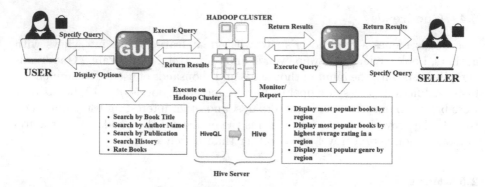

Fig. 1. Design of Reader's Choice

3.1 Hive Database Design

The Hive database of Reader's Choice comprises of four different file types. The subsequent paragraphs specify the use case for each file type.

User File. Each user is indexed in the user file with a unique user identity number. The enrollment process must ensure that each user obligatorily declares a secure password, their genre of main interest, and their physical address and, thereby, their geographical region.

Book File. A user can search in the book file by keying in either a book title, an author name, or a publishing company. Due to the support of auto completion, the partial input of one book attribute suffices. A user can rate any books that they have previously read. Finally, a user also has the option to rate the search results and to suggest improvements.

Query File. Reader's Choice logs all queries in diverse query files. There exists a dedicated query file for each user and every region.

Seller File. Each seller is indexed in the seller file with their unique seller identity number. The registration process coerces each seller to specify a safe password. On the contrary to users, sellers do not search for specific books. Rather they are interested in regions where they can make business. Owing to the query files, they can request the number of queries in all geographical regions in which they are interested. Moreover, they have the option to get the following counts per region: mostly searched books, highest rated books, and most often searched genres.

3.2 Procedural Design

This subsection specifies the simplified routines of Reader's Choice in textual and formal notation. For the sake of lucidity, Table 1 summarizes all introduced notations at a glance.

Table 1. Notations and their explanations

Notation	Explanation
Σ	Alphabet
$B \subsetneq \mathbb{N} \times \Sigma^* \times \Sigma^* \times \Sigma^* \times \Sigma^*$	Book file with the attributes ISBN, Title, Author, Publisher, and Genre
$G \subsetneq \Sigma^*$	Genre set
$I \subsetneq \mathbb{N} \mid (\forall i \in I)(\exists b \in B) b_I = i$	ISBN set
$Q_u \subsetneq \mathbb{N} \times \mathbb{N} \mid u \in U \wedge (\forall q_u \in Q_u)(\exists b \in B) b_I = q_{u_I}$	Query file of user u with the attributes ISBN and Rating
$Q_Z \subsetneq \mathbb{N} \times \mathbb{N} \mid (\forall q_Z \in Q_Z)(\exists b \in B) b_I = q_{Z_I}$	Query file of region Z with the attributes ISBN and Rating
$S \subsetneq \mathbb{N} \times \Sigma^* \times \mathbb{N}$	Seller file with the attributes Identity Number, Password, and Region Number
$U \subsetneq \mathbb{N} \times \Sigma^* \times \mathbb{N} \times \Sigma^*$	User file with the attributes Identity Number, Password, Region Number, and Main Interest
$b \in B$	Book with the attributes b_I (ISBN), b_T (Title), b_A (Author), b_P (Publisher), and b_G (Genre)
$g \in G$	Genre
$i \in I$	ISBN
$l \subset \mathbb{N}$	Entered identity number
$o \in \{$'A', 'D', 'G', 'H', 'P', 'R', 'T'$\}$	Selected menu option ('A' = Author search, 'D' = Demand popularity, 'G' = Genre popularity, 'H' = Search History, 'P' = Publisher search, 'R' = Book rating/Rating popularity, 'T' = Title search)
$p \in \Sigma^*$	Entered password
$q_u \in Q_u$	Element of user query file Q_u with the attributes q_{u_I} (ISBN) and q_{u_R} (Rating)
$q_z \in Q_Z$	Element of region query file Q_Z with the attributes q_{Z_I} (ISBN) and q_{Z_R} (Rating)
$s \in S$	Seller with the attributes s_I (Identity Number), s_P (Password), and s_Z (Region Number)
$t \in \{$'s', 'u'$\}$	Entered user type ('s' = seller, 'u' = user)
$u \in U$	User with the attributes u_I (Identity Number), u_P (Password), u_Z (Region Number), and u_G (Main Interest)
$w \in \Sigma^* \cup \mathbb{N}$	Entered search string

Login Routine. The GUI requests users and sellers to supply their login credentials (user or seller identity number $l \in \mathbb{N}$ plus password $p \in \Sigma^*$). Beyond that, they must declare the login type $t \in \{\text{'s'}, \text{'u'}\}$ (user or seller) by dint of a drop-down menu before Algorithm 1 proceeds.

Algorithm 1. Login Routine

Require: $l \neq$ " and $p \neq$ " and $t \in \{\text{'s'}, \text{'u'}\}$
1: **if** $t = \text{'u'}$ **and** $(\exists u \in U)l = u_I$ **and** $p = u_P$ **then** {User login}
2:     $u \in U | l = u_I; Z = u_Z$; Algorithm 2
3: **else if** $t = \text{'s'}$ **and** $(\exists s \in S)l = s_I$ **and** $p = s_P$ **then** {Seller login}
4:     $s \in S | l = s_I$; Algorithm 3
5: **else**
6:     **return** Login failed
7: **end if**

User Search Routine. According to Algorithm 2, a successfully authenticated user can choose from five different menu options: *Title Search, Author Search, Publisher Search, Search History,* and *Book Rating.* Reader's Choice stores the made decision in the variable $o \in \{\text{'A'}, \text{'H'}, \text{'P'}, \text{'R'}, \text{'T'}\}$.

Algorithm 2. User Search Routine

1: **loop**
Require: $o \in \{\text{'A'}, \text{'H'}, \text{'P'}, \text{'R'}, \text{'T'}\}$ **and** $w \neq$ " **and** $R \in \mathbb{N}$
2:   **if** $o = \text{'T'}$ **then** {Title search}
3:       $Q \leftarrow \{b_I | (\exists \alpha \in \Sigma^*)(\exists \beta \in \Sigma^*)(\exists b \in B)\alpha w \beta = b_T \wedge b_G = u_G\}$
4:       $Q_u \leftarrow Q_u \cup \{q, 0 | q \in Q\}; Q_Z \leftarrow Q_Z \cup \{q, 0 | q \in Q\}$
5:       **return** $\{b \in B | b_I \in Q\}$
6:   **else if** $o = \text{'A'}$ **then** {Author search}
7:       $Q \leftarrow \{b_I | (\exists \alpha \in \Sigma^*)(\exists \beta \in \Sigma^*)(\exists b \in B)\alpha w \beta = b_A \wedge b_G = u_G\}$
8:       $Q_u \leftarrow Q_u \cup \{q, 0 | q \in Q\}; Q_Z \leftarrow Q_Z \cup \{q, 0 | q \in Q\}$
9:       **return** $\{b \in B | b_I \in Q\}$
10:   **else if** $o = \text{'P'}$ **then** {Publisher search}
11:       $Q \leftarrow \{b_I | (\exists \alpha \in \Sigma^*)(\exists \beta \in \Sigma^*)(\exists b \in B)\alpha w \beta = b_P \wedge b_G = u_G\}$
12:       $Q_u \leftarrow Q_u \cup \{q, 0 | q \in Q\}; Q_Z \leftarrow Q_Z \cup \{q, 0 | q \in Q\}$
13:       **return** $\{b \in B | b_I \in Q\}$
14:   **else if** $o = \text{'H'}$ **then** {Search history}
15:       **return** $\{b \in B | b_I \in Q_u\}$
16:   **else if** $o = \text{'R'}$ **then** {Book rating}
17:       $Q \leftarrow \{b_I, R | R > 0 \wedge (\exists b \in B)b_I = w\}$
18:       $Q_u \leftarrow Q_u \cup Q; Q_Z \leftarrow Q_Z \cup Q$
19:   **else**
20:       **return** Invalid option
21:   **end if**
22: **end loop**

In case of A(uthor Search), P(ublisher Search), or T(itle Search), the GUI lets the user key a search string $w \in \Sigma^*$ in a text-field. Reader's Choice appends the same search results to two different query files, on the one hand, to the user's personal query file $Q_u \subsetneq \mathbb{N} \times \mathbb{N}$ and, on the other hand, to the query file $Q_Z \subsetneq \mathbb{N} \times \mathbb{N}$ that is dedicated to the user's region $u_Z \in \mathbb{N}$. Results whose queried attribute fully matches the inputted search string appear before those with partial matches in the sort sequence. What is more, merely books related to the user's main interest occur in the output $Q \subsetneq \mathbb{N}$. This functionality helps providing personalized search results.

The option *Search History* as a bookmark feature aggregates and displays all previous search results of the user.

The final option *Book Rating* prompts the user to key in a valid ISBN (International Standard Book Number) [18] including a rating $R \in \mathbb{N}$ for it. Again, the rating is stored in both files Q_u and Q_Z.

Seller Search Routine. Pursuant to Algorithm 3, the GUI provides the three options *Rating Popularity, Demand Popularity,* and *Genre Popularity* to successfully authenticated sellers. Reader's Choice again stores the clicked menu item in the variable $o \in \{\text{'D', 'G', 'R'}\}$ before it continues with the chosen database query.

4   Performance Analysis

Users of recommender systems expect to receive ordered lists of recommendations, usually, from best to worst. In some cases, they do not especially bother about an exact ordering and already feel lucky with a set of few *good* recommendations. The scholarly literature knows three widely used metrics to evaluate the utility of recommendations produced by a recommender system: precision, recall [22], and the F-score (a.k.a. F_1 score or F-measure) [12].

The precision connotes the proportion of *good* recommendations [2]. A perfect precision score of 1.0 refers to a result of mere *good* recommendations. Although, it does not necessarily mean that this result encloses all possible *good* recommendations [6].

The recall characterizes the completeness ratio of *good* recommendations. A perfect recall score of 1.0 means that a result encompasses all *good* recommendations. However, it disregards the number of *bad* recommendations in the result.

The F-score measures the accuracy of any statistical test [20]. The score incorporates the values of both precision and recall. It can be interpreted as a weighted average of precision and recall. The F-score takes a value between 1 as the optimum and 0 as the pessimum. In the case of recommender systems, it indicates the overall utility of their results.

Table 2 exhibits the evaluation of the three aforementioned metrics.

True positives indicate correct, positive predictions, false positives correspond to incorrect, positive predictions, and false negatives conform to incorrect, negative indications [7].

Algorithm 3. Seller Search Routine

```
 1: loop
Require:    o ∈ {'D', 'G', 'R'}
 2:    if o = 'R' then {Rating popularity}
 3:        I = {}
 4:        for all q_Z ∈ Q_Z|S_Z = Z do
 5:            if q_{Z_I} ∉ I then
 6:                I = I ∪ q_{Z_I}; R_{q_{Z_I}} = q_{Z_R}; r_{q_{Z_I}} = 1
 7:            else
 8:                R_{q_{Z_I}} ← R_{q_{Z_I}} + q_{Z_R}; r_{q_{Z_I}} = r_{q_{Z_I}} + 1
 9:            end if
10:        end for
11:        for all i ∈ I do
12:            return i : R_i/r_i
13:        end for
14:    else if o = 'D' then {Demand popularity}
15:        I = {}
16:        for all q_Z ∈ Q_Z|S_Z = Z do
17:            if q_{Z_I} ∉ I then
18:                I = I ∪ q_{Z_I}; i_{q_{Z_I}} = 1
19:            else
20:                i_{q_{Z_I}} = i_{q_{Z_I}} + 1
21:            end if
22:        end for
23:        for all i ∈ I do
24:            return i : i_i
25:        end for
26:    else if o = 'G' then {Genre popularity}
27:        G = {}
28:        for all q_Z ∈ Q_Z|S_Z = Z do
29:            if b_G ∉ G|b_I = q_{Z_I} then
30:                G = G ∪ b_G|b_I = q_{Z_I}; g_{b_G} = 1|b_I = q_{Z_I}
31:            else
32:                g_{b_G} = g_{b_G} + 1|b_I = q_{Z_I}
33:            end if
34:        end for
35:        for all g ∈ G do
36:            return g : g_g
37:        end for
38:    else
39:        return Invalid option
40:    end if
41: end loop
```

A thorough performance analysis compared the recently mentioned metrics of Amazon Book Services, Google Books, Goodreads, and Reader's Choice. The comparison incorporated the three book titles *Italian Fascist Activities in the*

Table 2. Information retrieval metrics

Metric	Evaluation
Precision	$\frac{\text{True Positive}}{\text{True Positive}+\text{False Positive}}$
Recall	$\frac{\text{True Positive}}{\text{True Positive}+\text{False Negative}}$
F-score	$2 * \frac{\text{Precision} * \text{Recall}}{\text{Precision}+\text{Recall}}$

United States, The United States and Fascist Italy, and *The United States and Fascist Italy: The Rise of American Finance in Europe.*

The titles served as search phrases in all the four systems. The succeeding analysis classified the top hundred recommendations of each recommender system into *good* and *bad* ones. Thereafter, it calculated the three metrics for the four systems in compliance with the formulas in Table 2.

It can be observed in Table 3 that Reader's Choice outscored Amazon Book Services and Google Books in all metrics. In addition, it measured up to the recall value of Goodreads. At first view, Reader's Choice lags behind Goodreads due to worse precision and F-score. In fact, the sophisticated features of Reader's Choice outweigh and even overcompensate both inferior metrics. In contrast to its competitors, Reader's Choice proffers user recommendations based on their professions and seller recommendations based on user popularity and demand, as explained in Sect. 3.

Table 3. Comparative performance analysis

System	Precision	Recall	F-score
Amazon Book Services	0.12	0.28	0.17
Google Books	0.10	0.10	0.10
Goodreads	1.00	0.71	0.83
Reader's Choice	0.13	0.71	0.21

5   Conclusion

In a sense, any good recommendation system collects data from its users and collates such data to deduce meaning from them. The quality of any recommendation system depends on the number of correctly made predictions, i.e. on its classification accuracy [13]. Continuous evaluation and training is important for improving the accuracy [16]. The novel recommendation system Reader's Choice accurately collects search data to correctly gauge popular books in a region and, therewith, inures to the benefit of both users and sellers. Reader's Choice collection of continuous feedback from its users also conduces to an improvement of this benefit.

In spite of the novelty of Reader's Choice, the explored, superior precision and F-score of Goodreads legitimately calls for amelioration of Reader's Choice.

On account of that, worthwhile future work should preferably take advantage of the virtues of both recommender systems and construct an improved successor by yoking them.

Acknowledgments. Many thanks to Bettina Baumgartner from the University of Vienna for proofreading this paper!

References

1. Adomavicius, G., Tuzhilin, A.: Context-aware recommender systems. In: Ricci, F., Rokach, L., Shapira, B., Kantor, P.B. (eds.) Recommender Systems Handbook, pp. 217–253. Springer US, Boston (2010). https://dx.doi.org/10.1007/978-0-387-85820-3_7

2. Buckland, M., Gey, F.: The relationship between recall and precision. J. Am. Soc. Inf. Sci. **45**(1), 12–19 (1994). https://dx.doi.org/10.1002/(SICI)1097-4571(199401)45:1⟨12::AID-ASI2⟩3.0.CO;2-L

3. Calazans, J., Cavalcanti, G., Lucian, R.: Social media as cultural products promotion platform: the GetGlue case. Animus **12**(24), 202–218 (2013). https://dx.doi.org/10.5902/217549778528

4. Chandak, M., Girase, S., Mukhopadhyay, D.: Introducing hybrid technique for optimization of book recommender system. Procedia Comput. Sci. **45**, 23–31 (2015). https://dx.doi.org/10.1016/j.procs.2015.03.075

5. Cook, J.: Shelfari an online meeting place for bibliophiles (2006). https://www.seattlepi.com/business/article/Shelfari-an-online-meeting-place-for-bibliophiles-1216875.php

6. Davis, J., Goadrich, M.: The relationship between precision-recall and ROC curves. In: Proceedings of the 23rd International Conference on Machine Learning, ICML 2006, pp. 233–240. ACM, June 2006. https://dx.doi.org/10.1145/1143844.1143874

7. Fletcher, R.H., Fletcher, S.W., Fletcher, G.S.: Clinical Epidemiology: The Essentials. Lippincott Williams & Wilkins, Philadelphia (2012)

8. Haupt, J.: Last.fm: people-powered online radio. Music Ref. Serv. Q. **12**(1–2), 23–24 (2009). https://dx.doi.org/10.1080/10588160902816702

9. Hvass, A.: Cataloguing with librarything: as easy as 1,2,3!. Libr. Hi Tech News **25**(10), 5–7 (2008). https://doi.org/10.1108/07419050810949995

10. Layton, J.: How Pandora radio works, May 2006. https://computer.howstuffworks.com/internet/basics/pandora.htm

11. Linden, G., Smith, B., York, J.: Amazon.com recommendations: item-to-item collaborative filtering. IEEE Internet Comput. **7**(1), 76–80 (2003). https://dx.doi.org/10.1109/MIC.2003.1167344

12. Magdy, W., Jones, G.J.F.: PRES: a score metric for evaluating recall-oriented information retrieval applications. In: Proceedings of the 33rd International ACM SIGIR Conference on Research and Development in Information Retrieval, SIGIR 2010, pp. 611–618. ACM, New York, July 2010. https://dx.doi.org/10.1145/1835449.1835551

13. Mooney, R.J., Roy, L.: Content-based book recommending using learning for text categorization. In: Proceedings of the Fifth ACM Conference on Digital Libraries, DL 2000, pp. 195–204. ACM, New York, June 2000. https://dx.doi.org/10.1145/336597.336662

14. Naik, Y., Trott, B.: Finding good reads on goodreads: readers take RA into their own hands. Ref. User Serv. Q. **51**(4), 319–323 (2012). https://dx.doi.org/10.5860/rusq.51n4.319

15. Nirwan, H., Verma, O.P., Kanojia, A.: Personalized hybrid book recommender system using neural network. In: 2016 3rd International Conference on Computing for Sustainable Global Development (INDIACom), pp. 1281–1288, March 2016

16. Pazzani, M.J., Billsus, D.: Content-based recommendation systems. In: Brusilovsky, P., Kobsa, A., Nejdl, W. (eds.) The Adaptive Web. LNCS, vol. 4321, pp. 325 341. Springer, Heidelberg (2007). doi:10.1007/978-3-540-72079-9_10

17. Ricci, F., Rokach, L., Shapira, B.: Recommender systems: introduction and challenges. In: Ricci, F., Rokach, L., Shapira, B. (eds.) Recommender Systems Handbook. Springer US, Boston (2015). https://dx.doi.org/10.1007/978-1-4899-7637-6

18. Schmierer, H.F., Pasternack, H.: Study of current and potential uses of international standard book number in united states libraries. Technical report, Committee for the Coordination of National Bibliographic Control, Washington, DC, March 1977. http://files.eric.ed.gov/fulltext/ED174264.pdf

19. Sharma, S.: Big data landscape. Int. J. Sci. Res. Publ. **3**(6) (2013). http://www.ijsrp.org/research-paper-0613.php

20. Sokolova, M., Japkowicz, N., Szpakowicz, S.: Beyond accuracy, F-score and ROC: a family of discriminant measures for performance evaluation. In: Sattar, A., Kang, B. (eds.) AI 2006. LNCS (LNAI), vol. 4304, pp. 1015–1021. Springer, Heidelberg (2006). doi:10.1007/11941439_114

21. Thusoo, A., Sarma, J.S., Jain, N., Shao, Z., Chakka, P., Anthony, S., Liu, H., Wyckoff, P., Murthy, R.: Hive: a warehousing solution over a map-reduce framework. Proc. VLDB Endow. **2**(2), 1626–1629 (2009). https://dx.doi.org/10.14778/1687553.1687609

22. Ting, K.M.: Precision and recall. In: Sammut, C., Webb, G.I. (eds.) Encyclopedia of Machine Learning, p. 781. Springer US, Boston (2010). https://dx.doi.org/10.1007/978-0-387-30164-8_652

23. Zuckerman, E.: Ethan Zuckerman quotes, January 2017. http://www.azquotes.com/quote/1216776

Accelerating Convolutional Neural Networks Using Fine-Tuned Backpropagation Progress

Yulong Li[1], Zhenhong Chen[1], Yi Cai[1(✉)], Dongping Huang[1], and Qing Li[2]

[1] School of Software Engineering, South China University of Technology,
Guangzhou, China
ycai@scut.edu.cn
[2] Department of Computer Science, City University of Hong Kong,
Kowloon Tong, Hong Kong

Abstract. In computer vision many tasks have achieved state-of-the-art performance using convolutional neural networks (CNNs) [11], typically at the cost of massive computational complexity. A key problem of the training is the low speed of the progress. It may cost much time especially when computational resources are limited. The focus of this paper is speeding up the training progress based on fine-tuned backpropagation progress. More specifically, we train the CNNs with standard backpropagation firstly. When the feature extraction layers got better features, then we start to block the standard backpropagation in the whole layers, the loss function values only back propagates between fully connected layers. So it can not only save time but also pay more attention to train the classifier to get the same or better result compared with training with standard backpropagation all the time. Comprehensive experiments on JD (https://www.jd.com/) datasets demonstrate significant reduction in computational time, at the cost of negligible loss in accuracy.

Keywords: Convolutional Neural Networks · Backpropagation · Acceleration

1 Introduction

The accuracy of convolutional neural networks has been continuously improving, but the computational cost of these networks also increases significantly. It is obvious that training a modestly sized CNNs model is still time consuming [15]. It is the fact that neural networks have a large number of model parameters that must be assigned suitable values through the training progress. It thus of practical importance to accelerate training performance of CNNs.

There have been a lot of studies on accelerating deep CNNs. Some common methods have been used in the practice, for example, Graphic Process Units (GPU) can parallel compute to speed up the training progress, data parallel and model parallel also get good performance in some works. These methods have shown promising speedup ratios and accuracy on deep models [2].

Z. Bao et al. (Eds.): DASFAA 2017 Workshops, LNCS 10179, pp. 256–266, 2017.
DOI: 10.1007/978-3-319-55705-2_20

A common focus of these methods is parallel computation with hardware or software. Considering the whole model structure and the training progress of CNNs, stochastic gradient decent (SGD) and the activation function are the two main factors that effect the speed of the training [7].

Stochastic Gradient Decent. We use the stochastic gradient decent to adjust the parameters and biases of deep model and decrese the loss function values. The mothod's solution is the minimal value when the object function do not have the global optimal solution. The speed of the gradient decent is not efficient.

Activation Function. If the activation function is sigmoid or the analogue of sigmoid, it brings a large number of computation, especially the gradient computation of backpropagation because of referring to division. In addition, the sigmoid activation function will gradually slow down the training progress, because the gradient value of the function would be too small when the input reach to a value of zero or one.

In this paper, we propose an accelerating method that is effective for deep models. We first review the backpropagation and then propose a fine-tuned progress to speed up the training. To demonstrate the fine-tuned progress, We controlled experiments of AlexNet [11] model on clothing image classification [4].

2   Related Works

In the progress of training of CNNs, stochastic gradient decent is a commonly used method [3]. However, in order to train the networks, we commonly need to use some tricks, for example, the choice of model depth, the settings of neurons' number, the initialization, the adjustment of learning rate, the controlling of Mini-batch and so on [1]. Unfortunately, even we know all these tricks, we need more and more try these tricks in practice. In addition, SGD is essentially a serial method. Due to the massive computation and enormous training data of deep neural networks, the time would be long. Now There are only a few general speedup methods for CNNs.

2.1   Graphic Process Units

The many-core structure system of Graphic Process Units contains thousands of stream processors, which can parallel executes computational operation, greatly shorten the operation time of the model. General-Purposed GPU (GPGPU) has become an important means to accelerate to parallel applications [5]. Benefited from the GPU many-core structure system, the speed of applications on GPU system tends to promote dozens of times and even thousands of times compared with on single core CPU. Using GPU to train deep neural networks can bring thousands of computing core efficient parallel ability's superiority into full play.

2.2 Data Parallel

Data parallel refers to divide the training data into N parts, then the N parts of training data are assigned to N workers to parallel train. With all the parallel training done, the gradient of the model is the average of the all N parts of training data, it can use the average gradient to update the parameter of the model [10]. Finally, return the updated parameter to all N workers for another iteration. In the progress of training, all training progress of the N workers are independent to each other [13]. Data parallel also has own shortcomings, when there is too many training data, data parallel have to reduce the learning rate to ensure the smooth of the training process.

2.3 Model Parallel

Model parallel means the model would be spilt into several units, each layer of deep neural networks can be broken up and hold by different training units to complete the training together [13]. When the input of a cell of unit came from the output of another cell of unit, computation overhead would occur between these two different units. Therefore, if the number of parallel split units is too much, the communication overhead and synchronization consumption of model parallel would more than data parallel [8]. But for the large model that cannot accommodate in a single computer memory, model parallel provides a good choice. At the same time, we also can combine data parallel and model parallel to generate the hybrid architecture.

Note, the methods we proposed are not specific to any processing architecture and can be combined with many of the other speedup methods given above.

3 Backpropagation Algorithm

A convolutional neural network is a type of feedforword neural network using convolution instead of general multiplication, inspired by biological processing and firstly introduced by Kunihiko Fukushima in 1980. A CNN consists of one or more convolutional layers and followed by one or more fully connected layers as in one standard Multi-Layer Perception (MLP) structure. It turns out that backpropagation, one of the simplest algorthms of training neural networks, is also the best choice to convolutional neural networks.

3.1 Backpropagation

Backpropagation, at its core, simple consists of repeatedly applying the chain rule through all of the possible paths in our network. However, there are a large number of directed paths from the input to the output [14]. Backpropagetion's real power arises in the form of dynamic programming, where we reuse intermediate results to calculate the gradient. We transmit intermediate errors backwards through a network, thus leading to the name backpropagation.

In fact, backpropagation is closely related to forward propagation, but instead of propagating the inputs forward through the network, we propagate the error backwards.

Backpropagation is a technique to propagates errors in the neural network back through the feedforward architecture and to adapt the weights [17]. Training a neural network with backpropagation is composed of two simple steps the feedforward and the backpropagation step. In the feedforward step a training case is classified using the current neural network. In the backpropagation step a classification error is computed and propagated back through the neural network [12]. The weights are updated based on the error, learning rate and gradient of the activation. Backpropagation is the method used to achieve gradient decent in neural networks.

Output layer error:

$$\delta_j^{(l)} = a_j^{(l)} - y_j. \tag{1}$$

Hidden layer error signal:

$$\delta^{(i)} = (\theta^{(i)})^\tau \delta^{(i+1)} . * \Delta a^{(i)}. \tag{2}$$

where $\theta^{(i)}$ weights to layer i in the neural network. $\delta^{(i)}$ back-propagated error signal used to update the activation values in layer i. $\Delta a^{(i)}$ gradients of the activation function in layer i.

3.2   Vanishing Gradient Problem

The method of training a neural network is important, as a neural network is a complex system with many local minima. The training through backpropagation is largely effected by the complex interactions between the neurons of the neural network. A key issue with backpropagation is the diminishing gradient flow, which essential makes it harder to train the lower layers of a multi layer neural network [6]. Neurons in lower layers of a neural network get their errors passed down from neurons in the higher layers. In the highest layer weights get updated directly based on the error in the output layer and the gradient of the activation function. The weights in the next layer get updated based on the error in the higher level layers. This causes either the error rate decays or explodes exponentially to the lower layers.

The effect of vanishing gradients can be illustrated using the sigmoid activation function, see Fig. 1. A neuron with the sigmoid activation function may get saturated when nearing an activation value of zero or one. In a saturated neuron the gradient of the activation function ($\Delta a^{(i)}$) approximates zero, it diminishes the gradient flow to the lower layers in the neural network. Lowering the error signal ($\delta^{(i)}$) passed by gradient decent to layer i-1. The problem of vanishing gradient flow has motivated many design decisions in deep feedforword neural network architectures.

Algorithm 1. The backpropagation algorithm.

Input: Training datasets:$(x^{(i)}, y^{(i)})$, i = 1, ..., N, Max iteration: T
Output: W, b

1   STEP 1: W, b initialization;
2   STEP 2:
3   **for** t = 1 ... T
4   **do**
5   **for** i = 1 ... N
6   **do**
7   (1) Feedforward compute the state and activation of each layer;
8   (2) Backpropagation to compute the error $\delta^{(i)}$ of each layer;
9   (3) Compute the gradient of each layer;
10   **do**
11                            $\frac{\partial J(W,b;x^{(i)},y^{(i)})}{\partial W^l} = \delta^{(i)}(a^{(l-1)})^T.$
12
13                            $\frac{\partial J(W,b;x^{(i)},y^{(i)})}{\partial b^l} = \delta^{(i)}.$
14   (4) Update parameters;
15   **do**
16                            $W^{(l)} = W^{(l)} - a\sum_{i=1}^{N}(\frac{\partial J(W,b;x^{(i)},y^{(i)})}{\partial W^l}) - \lambda W.$
17
18                            $b^{(l)} = b^{(l)} - a\sum_{i=1}^{N}(\frac{\partial J(W,b;x^{(i)},y^{(i)})}{\partial b^l}).$
19
20   **End**
21   **End**

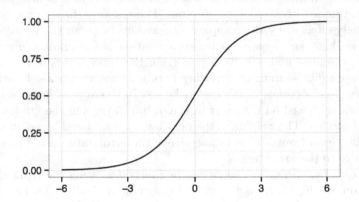

Fig. 1. The sigmoid activation function.

4   Fine-Tuned Backpropagation Progress

4.1   Training Progress Analysis of CNNs

The traditional neural networks are fully-connected, in other words, every neuron of the previous layer is connected to every neuron of the next layer. It would lead to a huge number of parameters, consume too much time to train. However, convolutional neural network is a special kind of deep neural network, Its particularity embodies in two parts, on the one hand, the connection of the neurons of different layers is not fully-connected. On the other hand, the weight of some neurons from the same layer is shared [16].

A typical structure of convolutional neural network is shown in Fig. 2. All convolution layers in front of the model can be seen as a feature extraction, which can generate a hierarchic features. The characteristics of the hierarchic features is the features of a layer is the convolution operation on the previous layer. Then the features will send to a fully-connected classifier. The features send to the classifier is not all the hierarchic features, are only the last layer of the feature extraction.

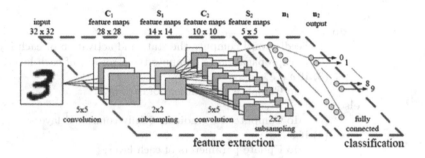

Fig. 2. A typical structure of convolutional neural network.

The feature extraction and the classifier of the model are trained by gradient decent at the same time. In the process of the gradient decent, the loss function values would back propagate to the previous layer, and so on [18].

Convergence of Feature Extraction C_f: The feature extraction get the best hierarchic features.

Convergence of Model C_m: The model get the highest accuracy.

In the training, the classifier is dealing with dynamic input features generated by feature extraction. Dynamic means the features is adjusted from undiscriminating to discriminating. Due to the dynamic characteristic, the classifier of the model cannot reach the best effect. When the feature extraction reached to C_f, the model reach to C_m at the same time.

Because of using the stochastic gradient descent algorithm, on the one hand, it can prevent the model from trapping in a local optimum, on the other hand, the convergence speed of model would be low before the end of the training. The model's direction of parameter adjustment is dynamic.

4.2   Fine-Tuned Progress

Our proposed method is, when the feature extraction tend to convergence, the feedforward remains the same and block the backpropagation to the feature extraction. The stochastic gradient descent algorithm is only operated between fully-connected layers, because the number of fully-connected layers is less than the number of the whole model, so the convergence of the model would be faster.

Algorithm 2. The fine-tuned progress.

Input: Training datasets:$(x^{(i)}, y^{(i)})$, $i = 1, ..., N$, Max iteration: T, Threshold value: τ
Output: W, b

```
1   STEP 1: W,b initialization;
2   STEP 2:
3   for t = 1 ... T
4       do
5           for i = 1 ... N
6               do
7                   (1) Feedforward compute the state and activation of each layer;
8                   (2) Backpropagation to compute the error δ^(i) of each layer;
9                   (3) if δ_Acc ≥ τ
10                      do Compute the gradient of each layer;
11              else
12                      do Compute the gradient of fully-connected layers;
13                  (4) if δ_Acc ≥ τ
14                      do Update parameters of each layer;
15              else
16                      do Update parameters of fully-connected layers;
17              End
18  End
```

---

5   Experiments

Next, we conduct experiments to support our analysis and verify the effectiveness of our proposed method. We use the AlexNet [11] to classify the images of clothing as the benchmark in our experiments.

5.1   Datasets

We self-build a datasets of clothing images. All of the images are crawled from JD, an e-commerce web site of China. There are 18 categories of different clothing images, the details of the datasets is as following, Table 1. Each image size of JD datasets is 256*256.

Table 1. 18 categories of image datasets

No.	Name	Num	No.	Name	Num
1	Woolen overcoat	16694	10	Waistcoat	12240
2	Down jacket	15993	11	Shirt	18667
3	Dress	18850	12	Men's suit	16105
4	Blouse	14864	13	Fur	13123
5	Jeans	15832	14	Man's leather garment	12583
6	Wedding dress	9340	15	Skirt	15316
7	Leggings	14517	16	T-shirt	14480
8	Shorts	14561	17	Full dress	10945
9	Small suit	13549	18	Suit pants	12576

5.2 Classification

We choose a subset of JD datasets of 8 categories, including woolen overcoat, down jacket, dress, blouse, jeans, wedding dress, shorts, T-shirt. Every category contains 1000 images, including trainingset of 800 images and validset of 200 images. We trained AlexNet [11] model using stochastic gradient decent with momentum of 0.9, and weight decay of 0.005. We initialized the weights in each layer from a zero-mean Gaussian distribution with standard deviation 0.01. We trained the network using Caffe [9] on a NVIDIA GTX 980 Ti 6GB GPU.

(1) Our accuracy and loss result on subset of JD datasets is as following, Fig. 3. After 1500 iterations, the model reached convergence and get the highest accuracy of 82.125%.
(2) The detail of training time is as following, Table 2. At the beginning of the training, the time used to training between the same δ_{Acc} is not too long, but gradually increase.

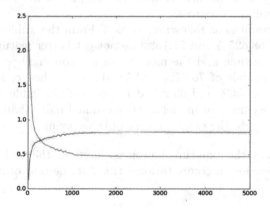

Fig. 3. A typical structure of convolutional neural network.

Table 2. The detail of training time

δ_{Acc}	Accuracy	Time
5%	47.0625–53.3125%	8 s
5%	53.3125–58.375%	15 s
5%	58.375–63.0625%	50 s
5%	63.0625–68.625%	80 s
5%	68.625–73.5625%	225 s
5%	73.5625–78.0625%	422 s
1%	78.0625–79.625%	315 s
1%	79.625–80.6875%	602 s
1%	80.6875–81.75%	1400 s
1%	81.75–82.125% (convergence)	730 s

5.3   Acceleration

We still use the same model and datasets to research the acceleration of the fine-tuned progress. The main progress is as following.

(1) The data and model are the same;
(2) Save the current model every 20 iterations during the training;
(3) Sample five models of 68.625%, 71.75%, 76.25%, 78.8125%, 81.0625% accuracy, adjust the settings of the model to block the error propagate to feature extraction layers, only propagate between the fully-connected layers;
(4) Train each of the five models to convergence, save the details of training and the five models.

5.4   Result and Analysis

(1) Sample result is as following, Table 3. We sample the five models in different stage of the training progress.
(2) Acceleration result is as following, Table 4. From the Table we can see that the models of 68.625% and 71.75% are not get better features, so X means we can not save time and the models are not convergence after fine-tuned progress, the models of 76.25% and 78.8125% reached convergence, so we saved the time of 56% and 65.2% at the cost of negligible loss in accuracy. Y means the accuracy of model after fine-tuned more than before and it is difficult to compute the exact percent of time saving.

We can see that the fine-tuned progress used in the end of the training get significant reduction in computational time, at the cost of negligible loss in accuracy.

Table 3. Sample result.

No.	Time	Acc
1	177 s	68.625%
2	270 s	71.75%
3	572 s	76.25%
4	930 s	78.8125%
5	1733 s	81.0625%
6	4032 s	82.125% (convergence)

Table 4. Topical knowledge of the example

No.	From Acc	To Acc	Time a	Time b	Delta percente
1	68.625%	76.125%	400 s	2100 s	X
2	71.75%	77.1875%	304 s	1785 s	X
3	76.25%	81.0625%	1733 s	721 s	56%
4	78.8125%	82.125%	3102 s	1756 s	65.2%
5	81.0625%	83.125%	2299 s	1879 s	Y

6 Conclusions

The objective of our research is to speed up the CNNs using fine-tuned back-propagation progress. We reviewed the other methods used to accelerate the deep models and introduce the fine-tuned backpropagation progress in detail. We demonstrate the method in AlexNet on the JD datasets. Our results show that fine-tuned progress is a effective method to save the time of training, at the cost of negligible loss in accuracy.

Acknowledgement. This work is supported by National Natural Science Foundation of China (project no. 61300137), Science and Technology Planning Project of Guangdong Province, China (No. 2013B010406004), Tip-top Scientific and Technical Innovative Youth Talents of Guangdong special support program (No. 2015TQ01X633) and Science and Technology Planning Major Project of Guangdong Province (No. 2015A070711001).

References

1. Arel, I., Rose, D.C., Karnowski, T.P.: Deep machine learning-a new frontier in artificial intelligence research [research frontier]. IEEE Comput. Intell. Mag. 5(4), 13–18 (2010)
2. Bengio, Y.: Learning deep architectures for AI. Found. Trends Mach. Learn. 2(1), 1–127 (2009)
3. Bengio, Y.: Deep learning of representations: looking forward. In: Dediu, A.-H., Martín-Vide, C., Mitkov, R., Truthe, B. (eds.) SLSP 2013. LNCS (LNAI), vol. 7978, pp. 1–37. Springer, Heidelberg (2013). doi:10.1007/978-3-642-39593-2_1

4. Di, W., Wah, C., Bhardwaj, A., Piramuthu, R., Sundaresan, N.: Style finder: fine-grained clothing style detection and retrieval. In: Proceedings of the IEEE Conference on Computer Vision and Pattern Recognition Workshops, pp. 8–13 (2013)
5. Dong, C., Loy, C.C., Tang, X.: Accelerating the super-resolution convolutional neural network. In: Leibe, B., Matas, J., Sebe, N., Welling, M. (eds.) ECCV 2016. LNCS, vol. 9906, pp. 391–407. Springer, Heidelberg (2016). doi:10.1007/978-3-319-46475-6_25
6. Glorot, X., Bengio, Y.: Understanding the difficulty of training deep feedforward neural networks. Aistats **9**, 249–256 (2010)
7. Hinton, G.E., Osindero, S., Teh, Y.-W.: A fast learning algorithm for deep belief nets. Neural Comput. **18**(7), 1527–1554 (2006)
8. Jaderberg, M., Vedaldi, A., Zisserman, A.: Speeding up convolutional neural networks with low rank expansions. arXiv preprint arXiv:1405.3866 (2014)
9. Jia, Y., Shelhamer, E., Donahue, J., Karayev, S., Long, J., Girshick, R., Guadarrama, S., Darrell, T.: Caffe: Convolutional architecture for fast feature embedding. In: Proceedings of the 22nd ACM International Conference on Multimedia, pp. 675–678. ACM (2014)
10. Jiang, X., Pang, Y., Li, X., Pan, J.: Speed up deep neural network based pedestrian detection by sharing features across multi-scale models. Neurocomputing **185**, 163–170 (2016)
11. Krizhevsky, A., Sutskever, I., Hinton, G.E.: Imagenet classification with deep convolutional neural networks. In: Advances in Neural Information Processing Systems, pp. 1097–1105 (2012)
12. Le Cun, B.B., Denker, J.S., Henderson, D., Howard, R.E., Hubbard, W., Jackel, L.D.: Handwritten digit recognition with a back-propagation network. In: Advances in Neural Information Processing Systems. Citeseer (1990)
13. Lebedev, V., Ganin, Y., Rakhuba, M., Oseledets, I., Lempitsky, V.: Speeding-up convolutional neural networks using fine-tuned CP-decomposition. arXiv preprint arXiv:1412.6553 (2014)
14. LeCun, Y., Bottou, L., Bengio, Y., Haffner, P.: Gradient-based learning applied to document recognition. Proc. IEEE **86**(11), 2278–2324 (1998)
15. LeCun, Y., Kavukcuoglu, K., Farabet, C., et al.: Convolutional networks and applications in vision. In: ISCAS, pp. 253–256 (2010)
16. Sánchez, J., Perronnin, F.: High-dimensional signature compression for large-scale image classification. In: 2011 IEEE Conference on Computer Vision and Pattern Recognition (CVPR), pp. 1665–1672. IEEE (2011)
17. van Doorn, J.: Analysis of deep convolutional neural network architectures (2014)
18. Wang, L., Yang, Y., Min, M.R., Chakradhar, S.: Accelerating deep neural network training with inconsistent stochastic gradient descent. arXiv preprint arXiv:1603.05544 (2016)

A Personalized Learning Strategy Recommendation Approach for Programming Learning

Peipei Gu[1], Junxia Ma[1(✉)], Wei Chen[2], Lujuan Deng[1], and Lan Jiang[3]

[1] Software Engineering College, Zhengzhou University of Light Industry,
Zhengzhou 450000, China
piaopiaogu@163.com, jxma@zzuli.edu.cn, lujuandeng@163.com
[2] Agricultural Information Institute, Chinese Academy of Agricultural Sciences,
Beijing 100081, China
chenwei@caas.cn
[3] College of Computer and Information Engineering, Xinxiang University,
Xinxiang 453000, China
faye_lanjiang@outlook.com

Abstract. Nowadays, it has been a significant problem to recommend learning strategy for different learners in programming learning projects. This paper discusses a personalized learning strategy recommendation approach to aid programming learning. In this paper, an improved design method of model learner strategies and programming learning strategy recommendation approach are presented. A reward factor is adopted to help to construct a learning strategy recommendation mechanism adaptively. The programming learning strategy recommendation system (ZZULI-PLS) is proposed based on those models to help learners learning in programming according to the actual progresses of learners. Usability tests are conducted to validate the recommendation efficiency in ZZULI-PLS system.

Keywords: Intelligent tutoring system · Learning strategy recommendation system · Learning strategy · Learning strategy model

1 Background and Motivations

Along with rapid development of computer technology and network education, online learning technology applies in many universities and training institution [1]. The various data is available on the web and has increased considerably [2]. The relevant and correct data should be mining and used as the information and knowledge to provide a certain service [3]. It becomes an important issue to study on the tutoring systems that how to serve the needs of tutees adaptively [4].

In the programming learning systems, learners have different interests, backgrounds and various ways to get the knowledge to fulfill themselves [5]. It is important to solve the problem how to help learners adopting appropriate behaviors and thoughts during programming learning process. Though there are many definitions of learning strategies, the common idea of learning strategies is the behaviors and thoughts that a learner takes

© Springer International Publishing AG 2017
Z. Bao et al. (Eds.): DASFAA 2017 Workshops, LNCS 10179, pp. 267–274, 2017.
DOI: 10.1007/978-3-319-55705-2_21

through the study progress [6]. Learning strategy should be chosen individually based on the actual level and situation for a learning strategy could be useful for one learner but not for another.

Many researchers concentrate on recommending learning strategies adaptively based on tutoring learners' personal needs. Some researchers focus on the adaptive e-learning system which builds an experimental system on programming learning [7–9]. Some researches help learners on practice [10, 14]. Motivation would aid learners to acquire higher levels of performance and persistence [12]. It is an important issue studying on the relationship between learning and motivation in higher education [14]. Motivation is a more critical element in programming learning for the learners. Less work concentrate on incorporating motivation into learning strategies recommended in the programming learning project, which is composed of learning and practicing resources.

In this paper, based on our previous work [14, 15], we propose an improved model to formulate learning strategies and a personalized programming learning strategy recommendation approach. We adopt reward factors to motivate learners to acquire knowledge more effectively. And a programming learning strategy recommendation system (ZZULI-PLS) is constructed to help learners choosing appropriate learning strategies to fulfill their learning targets. Usability tests are conducted to validate the recommendation efficiency in ZZULI-PLS system.

The rest of the paper is organized as follows. Section 2 presents an improved model that formulates learning strategies. Section 3 presents programming learning strategy recommendation approach. Section 4 focuses on experiments and evaluation. Section 5 draws the conclusion of this paper and presents the future work.

2   Learning Strategy Model

Scarcella & Oxford have defined learning strategies as "specific actions, behaviors, steps, or techniques – such as seeking out conventional partners, or giving oneself encouragement to tackle a difficult language task – used by learners to enhance their own learning" [16].

In the programming learning process, there are various factors influence the decision which one of learning strategies would be adopted. As we proposed in our previous work, a learner model should include five parts of information concerning learners' static properties, dynamic properties, affective information, history of learning strategy choosing, and test results of all tests [14, 15]. In order to construct programming learning strategy model for ZZULI-PLS that could aid learner to choose the learning strategy more appropriately in programming learning process, an improved learning strategy model representation in ZZULI-PLS is shown in Fig. 1.

There are five groups of learning strategies in ZZULI-PLS:

1. Metacognitive strategies can support the learners to control and manage the process of programming learning. Such as making a plan for next learning stage and summarize the previous learning part.

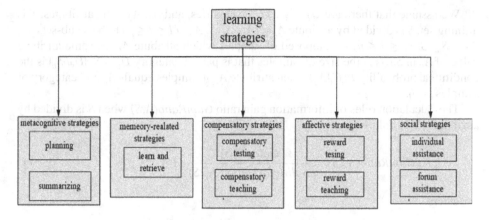

Fig. 1. Learning strategy model representation

2. Memory-related strategies help the learners using the memory-related method to improve the learning process. Such as learning and regular reviewing can enable learners to acquire the knowledge which has learned earlier solidified.
3. Compensatory strategies offer compensatory teaching and testing to help the learners. In the procedure of knowledge acquisition, there always some missing or forgotten knowledge we should learn time after time. Compensatory teaching offers more non-mastered knowledge to learn. And compensatory testing provides more tests cover non-mastered knowledge to test.
4. Affective strategies represent mental strategies. They can help learners re-construct confidence and motivation to learn. Reward testing enables learners to get the reward for rebuilding the learning mood when learners finished the tests. It can attract the learners' interests. Reward teaching helps learners learning knowledge more passionate.
5. Social strategies enable learners to seek help from individuals or forum. They can ask their friends for help openly in the programming learning system or the forum.

3   A Personalized Programming Learning Strategy Recommendation Approach

Based on the previous work of the general learning strategy recommendation algorithm [15], we propose a personalized programming learning strategy recommendation approach (PPLSRA). Learning strategy evaluation factor is incorporated into the recommendation controller to decide the final learning strategy recommendation list for learners.

PPLSRA includes three aspects: a learning strategy classifier, monitor and a recommendation controller.

Learning strategy classifier is constructed based on C4.5 which is proposed by Quinlan [17].

We assume that there are U_1, U_2..., U_m categories, and A_1, A_2... A_p attributes. S is training set. S is divided by attribute $A_t = \{A_{t1}, A_{t2}... A_{tn}\}$ $(1 \leq t \leq p)$ to be n sub-sets S_{t1}, $S_{t2}... S_{tm}$. S_{tj} $(1 \leq j \leq n)$ is composed of samples which attribute A_t has same attribute value of A_{tj} in S. $|U_i|$ is the size of samples that is part of category U_i in S. $|U_i|/|S_{tj}|$ is the conditional probability $P(U_i|A_{tj})$ when attribute A_t of samples equals A_{tj} and category of samples is U_i.

The calculation rules of information gain ratio $GainRatio(A,S)$ when S is divided by attribute A is as follows:

$$GainRatio(A_t, S) = \frac{Gain(A_t, S)}{SplitInformation(A_t, S)} \quad (1 \leq t \leq p) \tag{1}$$

where

$$SplitInformation(A_t, S) = -\sum_{j=1}^{n} \frac{|S_t|}{|S|} \times \log_2 \frac{|S_{tj}|}{|S|} (1 \leq t \leq p) \tag{2}$$

and

$$Gain(A_t, S) = -\sum_{i=1}^{m} \frac{|U_i|}{|S|} \times \log_2 \frac{|U_i|}{|S|} - \sum_{j=1}^{n} \frac{|S_{tj}|}{|S|} \times \sum_{i=1}^{m} \frac{|U_i|}{|S_{tj}|} \log_2 \frac{|S_{tj}|}{|U_i|} \tag{3}$$

The process of construction of learning strategy decision tree (learning strategy classifier) is as follows:

Input: training set S, attributes A_1, A_2... A_p and categories U_1, U_2..., U_m.

(a) If stop criterion is satisfied or set is indivisible, mark it as a leaf node.
(b) Go to step (h) if all the training sub-sets is marked.
(c) For every attribute A_t,$(1 \leq t \leq p)$, go to (d)–(f).
(d) Get the calculation of information gain ratio for attributes according to (1), (2), and (3).
(f) Select attributes A_t, $(1 \leq t \leq p)$ which has best information gain ratio, and divide training set S into sub-sets S_{t1}, $S_{t2}... S_{tm}$. S_{tj} $(1 \leq j \leq n)$ according to A_t, $(1 \leq t \leq p)$.
(g) Repeat step (a)–(f) for new sub-sets S_{t1}, $S_{t2}... S_{tm}$. S_{tj} $(1 \leq j \leq n)$.
(h) Output the final learning strategy decision tree.

Monitor is responsible for selecting the learners' information such as learning strategies history, learning strategies evaluation history, learning tracks and current emotional state. It would identify the types of learners, do previous preparation for the recommendation controller and follow the feedback when a learner selects one learning strategy to aid the learning process.

ZZULI-PLS serves three kinds of learners: (1) novice who has no idea about the system. (2) Learners who use the system but have no learning strategy selection history. (3) Learners who have adopted the recommended programming learning strategy in the system.

For the novice ones, recommendation controller recommends the programming learning strategies ranking by popularity. For the second type of learners, recommendation controller recommends the programming learning strategies according to the learning strategy classifier. For the third type of learners, the programming learning strategies in the learning strategies selection history would have a recommendation weight value which is calculated according to the (4), and (5). And recommendation controller recommends the programming learning strategies according to the calculation of weight value of learning strategies.

In this paper, we incorporate w_r that represents the recommendation weight value of a candidate strategy r. m is the whole size of candidate strategy set in the learning strategies recommendation history. s is the frequency of this strategy which appears in history. e is learning strategy evaluation factor which represents the average evaluation score of one learning strategy in the evaluation history. n is the maximum score for evaluation score. w_0 is the additional weight value for strategy selected in the learning strategy decision tree. The tutor can assign w_0 as a fixed value (we set w_0 to 0.2 in ZZULI-PLS).

If r is a strategy which is recommended by the decision tree, we get the calculation of w_r by using (4). If r is a strategy which appears in recommendation history, we get calculation of w_r by using (5). The calculation rules of w_r for a strategy are as follows:

$$w_r = w_0 + e/n + s/m \tag{4}$$

$$w_r = e/n + s/m \tag{5}$$

4   Experiments and Evaluation

ZZULI-PLS is constructed in Zhengzhou University of Light Industry to evaluate the efficiency of the proposed strategy recommendation approach and learning strategy model. The recommendation process of ZZULI-PLS is shown in Fig. 2.

The monitor in PPLSRA selects the learners' profile and following the feedback. The learning strategy classifier constructs a learning strategy decision tree according to the experiments conducted by Beijing Institute of Technology [15]. And recommendation controller recommends the appropriate programming learning strategy according to the learning strategy classifier and the calculation of weight value of learning strategies. And finally, learners would get a candidate set of learning strategies. Learners can select three of them for further consideration. Then choose one specific programming learning strategy to help the learning process. And at last, learners evaluate the three of them.

PPLSRA can provide a most appropriate learning strategy for learners to know how to learn. Then learner can make the adjustment and decide what to do in the next study process.

A series experiments are conducted on students who major in software engineering in Software Engineering College. Learners gave relatively high appraisals for the incorporation of reward factor.

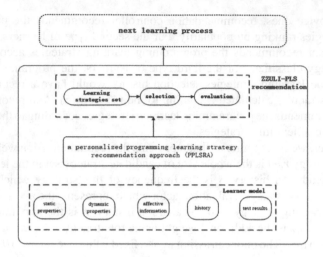

Fig. 2. ZZULI-PLS recommendation process

The programming learning strategy recommendation candidate list for Liye who is a learner in ZZULI-PLS is as Fig. 3 shows. She chooses three of them as the favorite strategies and would have the further specific strategies, as shown in Fig. 4. And she prefers to select affective strategies when she has some learning records in the system. When she finished the reward testing, she would receive a chance to get reward points randomly as shown in Fig. 5. And at last, she would evaluate the three of the selections in programming learning strategy recommendation candidate list. As shown in Fig. 6 Liye marks *3* to social strategy, *4* to compensatory strategy, and *5* to affective strategy.

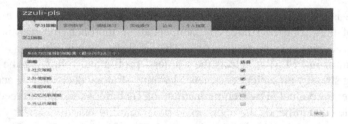

Fig. 3. Programming learning strategy recommendation list for Liye

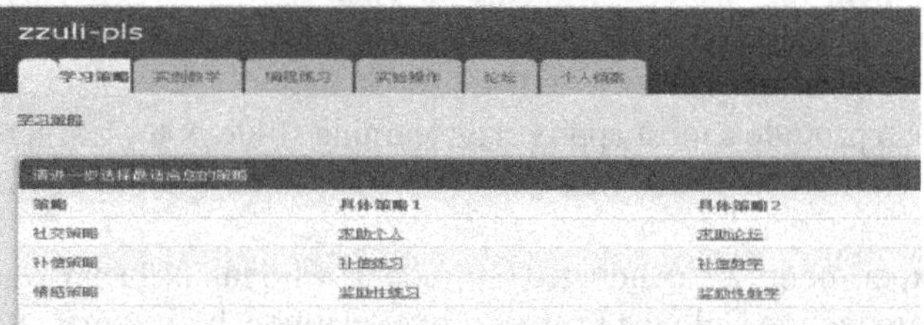

Fig. 4. Recommendation selection and further specific strategies

Fig. 5. Reward points

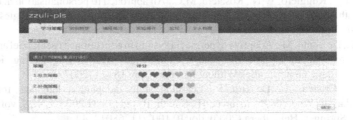

Fig. 6. Evaluation for the three selections

5 Conclusions

In this paper, an improved design method of model learner strategies and programming learning strategy recommendation approach are proposed. A programming learning strategy recommendation system is constructed to help learners find a good method to learn.

In our future work, we will expand the scope of experiments, and consider making adjustments for programming learning strategy group in real applications.

Acknowledgments. This work was supported by the Educational Commission of Henan Province of China (grant no. 15A520030), the 3rd Young teachers teaching reform and research project of Zhengzhou University of Light Industry (project no. 31), and 11th teachers teaching reform and research project of Zhengzhou University of Light Industry (project no. 3).

References

1. Zou, D., Xie, H., Wong, T.-L., Wang, F.L., Wu, Q.: The augmented hybrid graph framework for multi-level e-learning applications. In: Cheung, S.K.S., Kwok, L., Shang, J., Wang, A., Kwan, R. (eds.) ICBL 2016. LNCS, vol. 9757, pp. 360–370. Springer, Heidelberg (2016). doi:10.1007/978-3-319-41165-1_32
2. Xie, H., Zou, D., Lau, R.Y.K., Wang, F.L.: Generating incidental word-learning tasks via topic-based and load-based profiles. IEEE Multimed. **23**(1), 60–70 (2016)

3. De Paula, L.C., de Oliveira Fassbinder, A.G., Barbosa, E.F.: A recommendation system to support the students performance in programming contests. In: Frontiers in Education Conference (FIE). IEEE (2014)
4. Zou, D., et al.: A comparative study on various vocabulary knowledge scales for predicting vocabulary pre-knowledge. Int. J. Distance Educ. Technol. (IJDET) 15(1), 69–81 (2017)
5. Curilem, S.G., de Azevedo, F.M., Barbosa, A.R.: Adaptive interface methodology for intelligent tutoring systems. In: Lester, J.C., Vicari, R.M., Paraguaçu, F. (eds.) ITS 2004. LNCS, vol. 3220, pp. 741–750. Springer, Heidelberg (2004). doi:10.1007/978-3-540-30139-4_70
6. Selçuk, G.S., Çalişkan, S., Erol, M.: Learning strategies of physics teacher candidates: relationships with physics achievement and class level. In: Sixth International Conference of the Balkan Physical Union, pp. 511–512 (2007)
7. Stiubiener, I., Ruggiero, W.N., Rosatelli, M.C.: An approach to personalisation in e-learning. In: Seventh IEEE International Conference on Advanced Learning Technologies (ICALT), pp. 189–193 (2007)
8. Colace, F., De Santo, M.: Adaptive hypermedia system in education: a user model and tracking strategy proposal. In: Frontiers Education Conference - Global Engineering: Knowledge Without Borders, Opportunities Without Passports, pp. 18–23 (2007)
9. Stash, N., Cristea, A., De Bra, P.: Learning styles adaptation language for adaptive hypermedia. In: Wade, V.P., Ashman, H., Smyth, B. (eds.) AH 2006. LNCS, vol. 4018, pp. 323–327. Springer, Heidelberg (2006). doi:10.1007/11768012_42
10. Chang, K.-E., Chiao, B.-C., Chen, S.-W., Hsiao, R.-S.: Programming learning system for beginners - a completion strategy approach. IEEE Trans. Educ. 43(2), 211–220 (2000)
11. Yoneyama, Y., Matsushita, K., Mackin, K.J., Ohshiro, M., Yamasaki, K., Nunohiro, E.: Puzzle based programming learning support system with learning history management. In: Proceedings - ICCE 2008: 16th International Conference on Computers in Education, pp. 623–627 (2008)
12. De Paula, L.C., De Oliveira Fassbinder, A.G., Barbosa, E.F.: A recommendation system to support the students performance in programming contests. In: 2014 Frontiers in Education Conference, vol. 2015 (2014)
13. Tony, J.: The motivation of students of programming. In: Proceedings of the Conference on Integrating Technology into Computer Science Education, ITiCSE, pp. 53–56 (2001)
14. Zhang, W.S.: Research on Learning Strategy-Oriented Service model in E-Learning. Beijing Institute of Technology, Beijing (2012)
15. Niu, Z., Gu, P., Zhang, W., Chen, W.: Learning strategy recommendation agent. In: Uden, L., Corchado Rodríguez, E.S., De Paz Santana, J.F., De la Prieta, F. (eds.) Workshop on Learning Technology for Education in Cloud (LTEC 2012), vol. 173, pp. 205–216. Springer, Heidelberg (2012)
16. Scarcella, R., Oxford, R.: The Tapestry of Language Learning the Individual in the Communicative Classroom, p. 63. Heinle & Heinle, Boston (1992)
17. Quinlan, J.R.: Programs for Machine Learning. Morgan Kauffman publishers, Burlington (1993)

Wikipedia Based Short Text Classification Method

Junze Li[1], Yi Cai[1(✉)], Zhiwei Cai[1], Hofung Leung[2], and Kai Yang[1]

[1] School of Software Engineering, South China University of Technology,
Guangzhou, China
ycai@scut.edu.cn
[2] Department of Computer Science and Engineering,
The Chinese University of Hong Kong, Sha Tin, China

Abstract. Short text is usually expressed in refined slightly, insufficient information, which makes text classification difficult. But we can try to introduce some information from the existing knowledge base to strengthen the performance of short text classification. Wikipedia [2, 13, 15] is now the largest human-edited knowledge base of high quality. It would benefit to short text classification if we can make full use of Wikipedia information in short text classification. This paper presents a new concept based [22] on Wikipedia short text representation method, by identifying the concept of Wikipedia mentioned in short text, and then expand the concept of wiki correlation and short text messages to the feature vector representation.

Keywords: Short text classification · Concept · Wikipedia

1 Introduction

With the advent of the era of big data, the information in the Internet are explosive growth, and the user's own electronic documents generated gradually occupied the main body of the network information resources. Since short text like microblog and all kinds of user reviews have emerged within the Internet, short text occupied the main body of new information. How to make the management and use of these information resources, and help people find what they need, has become a big challenge of information processing technology.

Text classification technology is the key technology of processing huge amounts of documents data which usually include the text pretreatment, the representation of a text and classifier selection and training. And text classification methods are mostly for long text classification. Different from long text, short texts feature is sparse, incomplete, and fuzzy semantic information which makes the traditional text classification methods always cannot reach a good result compare with long text [1,5].

In this paper, we described research background and research significance of short text classification in detail based on the concept. And introduces the

© Springer International Publishing AG 2017
Z. Bao et al. (Eds.): DASFAA 2017 Workshops, LNCS 10179, pp. 275–286, 2017.
DOI: 10.1007/978-3-319-55705-2_22

processing method which is generally used in text categorization, including word segmentation, feature representation, classifier, etc., and compares the present method of classification based on the concept.

We also proposed a method called WBST (Wikipedia Based Short Text Classification Method) which combine the Wikipedia and concept dimension extension to make the short text classification better. Text feature expression method used concepts of Wikipedia, which is mainly includes the concept of dimension expansion, using the concept of information, using the concept of relationship, etc., and introduces the short text classification process based on concepts on Wikipedia. The experiment results will prove that the words and concepts based on the combination of the concept of Wikipedia dimension text representation method in short text has certain advantage in this classification. Details will be introduced later.

Section 2 will introduce related works which are Text Vector Space Model and some popular classification methods. Section 3 will introduce the representation of text characteristic and the short text classification method based on concept. Section 4 will show the experiment.

2   Relative Works

2.1   Text Vector Space Model

Up to now, the most commonly feature representation model method used in text categorization problem is Vector Space Model (VSM) [12,14]. The basic assumption is that every word is independent of one dimension in Vector space model [20], and then, the text will representation into a vector of word forms. Text corresponding to each word will have specific weight, as shown in Formula 1.

$$q_i = < t_{i1} : w_{i1}, t_{i2} : w_{i2}, \ldots, t_{in} : w_{in} > \tag{1}$$

In the process, the weight w_{ij} of corresponding word has a different calculation method. One of the most commonly used method is Term Frequency - Inverse Documentation Frequency (TFIDF) [9]. Where t_{ij} is corresponded to the words have appeared in the data set, w_{ij} is the word t_{ij}s weight in the document, and w_{ij} can be calculated by the TFIDE value of corresponding document. The calculation formula is as follow:

$$w_{ij} = tf_{ij} * idf_j = tf_{ij} * log(N/n_j) \tag{2}$$

The most basic assumptions of the relationship between the words were independent of each other in vector space model, which is ignoring the correlation between the words. However, short text can only express less semantic information, and it needs more of the correlation between the words. Although the traditional long text categorization rarely use the fuzziness, polysemy and synonyms of natural language, these are very common in the short text. Hence, traditional classification methods often cannot achieve the desired accuracy in short text classification problem. In vector space model, ordinarily, every word is as one dimensional independently, and is unrelated between different dimensions, but this is not conformed to the actual characteristics of text data, because of the word is not independent completely, there are a lot of synonyms and polysemy between words, which makes different samples in the process of short text classification almost have no common non-zero dimension, so, it is difficult to get the satisfactory accuracy of classification. In this paper, we are trying to identify some concepts like Obama in short text, and then combine with some existing knowledge based on Wikipedia which is as a knowledge base, and we combine the information from the knowledge base into the characteristics of short text. Hence, in the following steps, we will identify the concept of the text based on Wikipedia, and attempt to combine the corresponding information added in short text, so as to improve the classification accuracy of short text.

2.2 Methods for Short Text Classification

Entity_ESA. This is a method based on the physical method of semantic [8]. Many works based on the degree of Wikipedia has been committed to measuring semantically related degree [10,11,16]. In order to deal with the problem of semantically related but the dimensions of the word do not match, the method ESA related all the word in Wikipedia pages to Wikipedia entry (concept) according to the TFIDF value as correlation matching, to classify every word in this short text according to the mapping relationship to the concept of dimension, and then classify text in the dimension of the concept. It is important to note that this method is different with what we do, this method does not use entity recognition technology, it is just simply map according to the correlation of word and the concept. It does not consider whether the text actually have mentioned a concept and just keep only concept dimension, and this mapping progress may lose a part of information. After representing as vector, this method will use the SVM classifier to classify.

Probase BocSTC. This method is based on a knowledge base called Probase [23]. The knowledge base is extracted entity relationship and related information from a large number of Internet data by Microsoft, and we mainly use the isA relationship of them to help short text classifier. This relationship is between the corresponding word or phrase text to the concept. Through the relational tables to match data in the short text, the short text will be mapping to the Probase concept. And then process the concept follow a series steps like disambiguation, weight calculation, etc., and classify according to the text similarity with the class. This method also only considers of concept dimension, and this mapping progress may lose a part of information, which makes recall value may be very low. Besides, what is different from Wikipedia is that Probase itself is a knowledge base according to the algorithm of automatic extraction, and isA relationship matching may have some mistakes (the accuracy is about 92.8%), which can also extend to the process of text classification.

3 Short Text Classification Method Based on Concept

3.1 Extension of the Text Characteristic

This part will be shown as an example in Table 1 of the following short text. It is a real twitter data, and the underline parts were identified as two concepts, Emmy nominations recognition for concept $Emmy\_Award$, primetime identified as $Prime\_time$.

Table 1. An example of short text

Example
Emmy nominations highlight new shows: the 2011 Emmy
nominations for primetime television were released Thursday

This text will be expressed as the vector space model according to TFIDF value firstly, as shown in Table 4-1. In this table, we only show a part of the dimensions, each dimensions data is a TFIDF value.

Dimension Expansion. In this paper, in order to reflects the characteristics of the concept in the text, after we recognized the concept, which is similar to the current vector space model, we add some dimensions of the concept on the basis of the dimension of the word. The concept of the dimension value is:

$$W(c_i) = W_{st}(t_i) * P(c_i|t_i) \tag{3}$$

where the subscript ST is the abbreviations of Short Text, which is in order to tell the difference between the weight of Wikipedia and the weight of the short text. And $W_{st}(t_i)$ is the weight of the word in short text, in this paper, we use the TFIDF value as the weight value. $P(c_i|t_i)$ is corresponded to the concept of probability value which we get in the process of concept identification of the current word.

After training, useful features will be given a higher weighting, not useful features will be given a low weight, small or even zero weight, so the characteristics of redundant information will have less negative effect for the classification effect. But missing information's influence on the classification effect is high, so, enough of the characters in the text will help developing classification effect (Table 2).

Table 2. Representation of text vector space model

Emmy	Nomination	Highlight	$Concept_{Emmy Award}$	$Concept_{primetime}$
0.281	0.276	0.081	0.243	0.121

Using the Relation of Concept to Extension. In the previous step, we added the dimension of the concept, but the concept of dimension is still very sparse [17], the vast majority of the dimensions of the concept of the value is 0. In order to make the dimension of the concept of more useful, we used the concept of similarity between into other dimensions of the concept of value. After joining concept dimension, the characteristics still have the weakness of vector space model, and concepts are independent with each other, any two may have very similar concepts will be treat as two completely different concepts, this kind of text representation is not enough to express the characteristics of short text. And in the real-world data set, the number of training set samples is usually less, this kind of sparse features almost won't bring any help to text classification accuracy, some characteristics even has only one sample as a nonzero value. In order to make full use of these concepts, this paper tries to use the similarity between the concepts, and connect the similar characteristics with each other. In this paper, we use the cos of concept of feature vector similarity $cos(c_i, c_j)$ and the distance between concept and corresponding wiki category tree $sim(c_i, c_j)$:

$$sim(c_i, c_j) = (cos(c_i, c_j) + dis(c_i, c_j))/2 \qquad (4)$$

where similarity cos means cosine similarity, and what we use is the angle between two vectors in the vector space determine to judge the size of the similarity of two vectors. The most common application is computing text similarity.

According to their word, establish the two text to two vectors, computing the cosine values of the two vectors, and we can know their similarity in statistical methods. With practice, we can prove that this is a very effective method. The specific formula is as below:

$$cos(c_i, c_j) = \frac{\sum_1^n (A_i \times B_i)}{\sqrt{\sum_1^n A_i^2} \times \sqrt{\sum_1^n B_i^2}} \qquad (5)$$

where A_i and B_i are the feature vector of c_i and c_j respectively.

After getting the similarity value, the value of the concept will be calculated again as:

$$W\prime(c_i) = \Sigma sim(c_i, c_j) \times W(c_i) \qquad (6)$$

The new calculation method of weight is based on the similarity between the concept and the concepts weight in the text. In the text, when the concepts weight is higher, and is very similar to another concept, the more the weight of another concept should also increase. In this way, the similar concepts in different text weight will increase at the same time. When we measure the similarity, the concept of dimension will increase the mention of the concept of similar text similarity, which makes the characteristics of short text performance increased, and data sparse will greatly abate.

Using the Feature of Concept to Extension. When we know that a piece of text is related to a concept, this concept of information is to be able to help this paragraph of text classification. In this example, *Emmy_Award* is an award of a TV series, its corresponding wiki page is likely to appear a lot of entertainment aspects of the word, and it is benefit to classifier when these words are added in this short text.

According to this idea, we add the characteristics of the concept of the words to the text, which is increasing the short text eigenvector corresponding word in weight. The weight will be calculated again as below:

$$W_{st}(w_i) = W_{st}(w_i) + W(c) \times W_{wiki}(c, W_i) \tag{7}$$

In this formula,

$$W_{wiki}(c, w_i) = \frac{TFIDF(w_i)}{\Sigma_{w_j \in c} TFIDF(w_j)} \tag{8}$$

Wikipedia concepts of key concepts in the set is the last chapter of extraction process to extract the TFIDF maximum 100 words per page, and here use TFIDF value according to the word in the wiki page accounts for the ratios of the concept of the weight to redistribute word space dimensions. TFIDF value is calculated by the word or word occurrences in the text and the emergence of probability between different documents, which can effectively measure the importance of a word. So, we use words TFIDF value as the importance of text. When we want to represent feature words of concepts in short text, we hope that the more important the concept is, the higher weight the corresponding word has. Whats more, when the correlation of the concept and short text are higher, the feature word of the concept should also get the higher weights in the text.

Comprehensive above, we add the product of the feature word of concepts proportion in all feature words TFIDF value as well as the concepts weight in the text to the feature words weight in short text.

3.2 Process of the Characteristics of Short Text Based on the Concept

– **Step 1:** The short text preprocessing is the preprocessing to the text mainly about dealing with the word segmentation, delete stop words, morphological reduction and so on.
– **Step 2:** The process of Vector space model is to traverse document to calculate the IDF value of words, and then calculate the TF value of each word in a document, finally calculate the TFIDF value based on the words TF value and IDF value as the weight of the word in a document. All the words weight will get together as the document feature vectors.
– **Step 3:** The concept recognition and the concept dimension extension features, are mainly identified the concept of Wikipedia according to the concept of matching and the method of ambiguity elimination, and then add the concept to the text feature dimension.

– **Step 4:** Concept relations and information supplement, is extensive based
on the similarity between the concepts, then extension based on information
according to the concept itself, and we can get the short text feature based on
concept.

4 Experiments

4.1 Experiment Dataset

This paper uses the Wikipedia data released on January 12, 2015 version
at "https://dumps.wikimedia.org/enwiki" which is Wikidump format[1], size of
48.81 GB. We only care about the body parts in the data. The example in Fig. 1
is part of the *Barack_Obama* pages of text data. This part of corresponding to
the first half of the data as the following wiki page in Fig. 2.

> Obama was born on August 4, 1961,<ref name="biography">{{ci
> te web|year=2008|title=President Barack Obama|location=Washington, D.C.|publisher=T
> he White House|url=http://www.whitehouse.gov/administration/president-obama|accessdate
> =December 12, 2008}}</ref> at Kapi{{okina}}olani Maternity & Gynecologic
> al Hospital (now [[Kapiolani Medical Center for Women and Children|Kapi{{okina}}ol
> ani Medical Center for Women and Children]]) in [[Honolulu]], Hawaii,

Fig. 1. The example of dataset

Obama was born on August 4, 1961,[4] at Kapi'olani Maternity & Gynecological Hospital (now Kapi'olani Medical
Center for Women and Children) in Honolulu, Hawaii.[5][6][7] and would become the first President to have been
born in Hawaii.[8] His mother, Stanley Ann Dunham, born in Wichita, Kansas, was of mostly English ancestry.[9] His

Fig. 2. The example of wiki page

As in this paper, we only care about the body of the text as well as what
are mentioned in the text to other wiki page link, other display format, like bold
and reference will be directly removed. In wikidump format, the pages point to
other pages of text will be marked as [[A]] or [[A | B]], in the form of A text
point to A wiki page A, or text to wiki page B respectively.

Because this paper considers the wikipedia between concepts in the similarity
in the directory structure, we need to establish the dimensional basc concept tree
structure, and find out the category which is the concept belonging to.

This paper adopted the DBpedia ("http://wiki.dbpedia.org") which is provided a wiki category of the structure of the data set. This paper based on the
two data sets to measure short text classification accuracy. The first data set is

---

[1] https://en.wikipedia.org/wiki/Wikipedia:Database_download.

from Twitter, which are send from the real users and have 6422 in total, artificial markers for seven categories. Second data set is from Snippet, which are abstract from search engine [3, 19, 21] and have 12340 in total, artificial markers for eight categories. Because the Twitter data almost from people's daily life of daily status or news brief, so they don't pay attention to the format of the words, which makes there will be some spelling errors and data quality is relatively poor. But things get much better in Snippet data, since the data is in the Google search engine page summary, and for some specific commercial stronger category such as mechanical, this kind of website will describe in precision according to your own business types, so its data quality is much more better compared to Twitter.

4.2   Text Data Preprocessing

This paper uses the Natural Language processing library (Natural Language Toolkit [4]) to preprocess the text data. Me is the most commonly used treatment of natural language toolkit by far, based on the Python program. It provides interface and vocabulary of more than 50 corpus resources, as well as the process of text processing including classification, tag, prevent, labels, parsing, and the encapsulation of semantic reasoning and provide friendly interface which is easy to use. In this paper, we mainly use its sentence segmentation (*nltk.tokenize.sent_tokenize*), word segmentation (nltk.tokenize), morphological reduction based on WordNet (nltk.stem.wordnet.WordNetLemmatizer), and stop word table (nltk.corpus.stopwords).

4.3   Classification Method and Evaluation Indicators

Classification Method. In this experiments, we use 10-Fold verification method, which is used the average random divided into 10 portions, take turns to take one of them as the test data, the other 9 together as the training data, repeat 10 times, and take the average result of the ten results as final results. This paper use the SVM classifier realized by LibLinear [6], all parameters are default. LibLinear is developed by a team of scientists at the university of Taiwan's national which is an open source and suitable for the large-scale data set of linear classifier. With SVM, compared to the realization of the most commonly used LibSVM, LibLinear has a faster speed on training and prediction, but does not support the definition of kernel function, and cannot be achieved through the kernel function to a non-linear classifier.

For text classification problem, the vector dimension is bigger, and often have more than a dozen ten thousand d, and each vector has only a few of the dozens of d and the value is 0, so the data of nonlinear kernel function for classification can increase less on effect and easy over-fitting, time-consuming, and high dimension also makes the classifier training a long time, so LibLinear is more suitable for short text classification problem.

Evaluation Indicators. We have three evaluation indicators for this experiment, AVG-Precision, AVG-Recall and AVG-F1 value. AVG-Precision means the average classification accuracy. The calculation formula is as follows:

$$AVG - Precision = \frac{1}{N} \sum_{i=1}^{N} Precision_i \tag{9}$$

$$Precision_i = \frac{TP_i}{TP_i + FP_i} \tag{10}$$

AVG - average Recall means the recalling rate of each category. The calculation formula is as follows:

$$AVG - Recall = \frac{1}{N} \sum_{i=1}^{N} Recall_i \tag{11}$$

$$Recall_i = \frac{TP_i}{TP_i + FN_i} \tag{12}$$

TP_i, FP_i, and FN_i in the above formula represent of true positive values, false positive and false Yin value in the classification respectively. These three values the final classification of correct samples which were classified to correct result, the final classification of correct samples which were classified to wrong result, the final classification of error samples which were classified to wrong result respectively. AVG-F1 is the average value for each category of F1 value, F1 value can result in mediate the value relationship between the accuracy and recall rate. So, we concluded that a comprehensive evaluation model of a classification index formula as follows:

$$AVG - F1 = \frac{1}{N} \sum_{i=1}^{N} F1_i \tag{13}$$

$$labelEquation: f1_i F1_i = 2 \times \frac{Precision_i \times Recall_i}{Precision_i + Recall_i} \tag{14}$$

4.4 The Short Text Classification Based on the Wiki Concept Performance Evaluation

In this paper, the short text classification experiment is for two different short text data sets which are Twitter and Snippets, and then we compare the final AVG - F1 value. We compare the performance of VSM [7,18], $Entity_E SA$, Probase BocSTC and WBST (Wikipedia Based Short Text Classification Method). WBST is the method we used in this paper. Experimental results are shown in Table 3 and Fig. 3.

Table 3. Different classification method under the two datasets of AVG - F1 value

	VSM	ESA	BocSTC	WBSTC
Twitter	0.7951	0.7743	0.7687	0.8254
Snippet	0.6722	0.6921	0.7033	0.7226

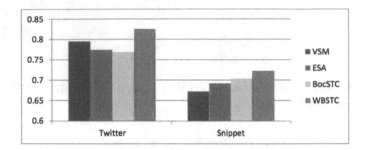

Fig. 3. Different classification method under the two datasets of AVG - F1 value

In these experiments, two data set in the threshold of 0.3 and 0.4 respectively reached the highest classification accuracy, which means that this is a balance point of using the relevant information and avoiding introducing too much uncorrelated information at the same time. When the threshold is too low, the irrelevant information is too much, so the accuracy will fall. When the threshold is too high, there are little concept can be identified, so the accuracy is very close to VSM method.

5   Conclusion

This paper mainly introduced the method of text feature expression method used concepts of Wikipedia, which is mainly includes the concept of dimension expansion, using the concept of information, using the concept of relationship, etc., and introduces the short text classification process based on concepts on Wikipedia. The Wikipedia short text representation method, by identifying the concept of Wikipedia mentioned in short text, and then expand the concept of wiki correlation and short text messages to the feature vector representation. And we can easily see from the experiments results that WBSTC method shows the best performance among VSM, ESA and BocSTC.

Acknowledgement. This work is supported by National Natural Science Foundation of China (project no. 61300137), Science and Technology Planning Project of Guangdong Province, China (No. 2013B010406004), Tip-top Scientific and Technical Innovative Youth Talents of Guangdong special support program (No. 2015TQ01X633) and Science and Technology Planning Major Project of Guangdong Province (No. 2015A070711001).

References

1. Cai, Y., Chen, W.-H., Leung, H.-F., Li, Q., Xie, H., Lau, R.Y., Min, H., Wang, F.L.: Context-aware ontologies generation with basic level concepts from collaborative tags. Neurocomputing **208**, 25–38 (2016)
2. Cucerzan, S.: Large-scale named entity disambiguation based on Wikipedia data. EMNLP-CoNLL **7**, 708–716 (2007)
3. Dai, H.K., Zhao, L., Nie, Z., Wen, J.-R., Wang, L., Li, Y.: Detecting online commercial intention (OCI). In: Proceedings of the 15th International Conference on World Wide Web, pp. 829–837. ACM (2006)
4. Davidson, D., Harman, G.: Semantics of Natural Language, vol. 40. Springer Science & Business Media, Netherlands (2012)
5. Du, Q., Xie, H., Cai, Y., Leung, H.-F., Li, Q., Min, H., Wang, F.L.: Folksonomy-based personalized search by hybrid user profiles in multiple levels. Neurocomputing **204**, 142–152 (2016)
6. Fan, R.-E., Chang, K.-W., Hsieh, C.-J., Wang, X.-R., Lin, C.-J.: Liblinear: a library for large linear classification. J. Mach. Learn. Res. **9**, 1871–1874 (2008)
7. Faruqui, M., Dyer, C.: Improving vector space word representations using multilingual correlation. Association for Computational Linguistics (2014)
8. Gabrilovich, E., Markovitch, S.: Computing semantic relatedness using wikipedia-based explicit semantic analysis. IJcAI **7**, 1606–1611 (2007)
9. Guo, A., Yang, T.: Research and improvement of feature words weight based on TFIDF algorithm. In: Information Technology, Networking, Electronic and Automation Control Conference, pp. 415–419. IEEE (2016)
10. Han, X., Zhao, J.: Named entity disambiguation by leveraging Wikipedia semantic knowledge. In: Proceedings of the 18th ACM Conference on Information and Knowledge Management, pp. 215–224. ACM (2009)
11. Hu, X., Zhang, X., Lu, C., Park, E.K., Zhou, X.: Exploiting Wikipedia as external knowledge for document clustering. In: Proceedings of the 15th ACM SIGKDD International Conference on Knowledge Discovery and Data Mining, pp. 389–396. ACM (2009)
12. Kiela, D., Clark, S.: A systematic study of semantic vector space model parameters. In: Proceedings of the 2nd Workshop on Continuous Vector Space Models and Their Compositionality (CVSC) at EACL, pp. 21–30 (2014)
13. Mihalcea, R., Csomai, A.: Wikify!: linking documents to encyclopedic knowledge. In: Proceedings of the Sixteenth ACM Conference on Information and Knowledge Management, pp. 233–242. ACM (2007)
14. Mikolov, T., Chen, K., Corrado, G., Dean, J.: Efficient estimation of word representations in vector space. arXiv preprint arXiv:1301.3781 (2013)
15. Milne, D., Witten, I.H.: Learning to link with Wikipedia. In: Proceedings of the 17th ACM Conference on Information and Knowledge Management, pp. 509–518. ACM (2008)
16. Ni, X., Sun, J.-T., Hu, J., Chen, Z.: Mining multilingual topics from Wikipedia. In: Proceedings of the 18th International Conference on World Wide Web, pp. 1155–1156. ACM (2009)
17. Phan, X.-H., Nguyen, L.-M., Horiguchi, S.: Learning to classify short and sparse text & web with hidden topics from large-scale data collections. In: Proceedings of the 17th International Conference on World Wide Web, pp. 91–100. ACM (2008)
18. Salton, G., Wong, A., Yang, C.-S.: A vector space model for automatic indexing. Commun. ACM **18**(11), 613–620 (1975)

19. Shen, D., Sun, J.-T., Yang, Q., Chen, Z.: Building bridges for web query classification. In: Proceedings of the 29th Annual International ACM SIGIR Conference on Research and Development in Information Retrieval, pp. 131–138. ACM (2006)
20. Sidorov, G., Gelbukh, A., Gómez-Adorno, H., Pinto, D.: Soft similarity and soft cosine measure: similarity of features in vector space model. Comput. Sist. **18**(3), 491–504 (2014)
21. Szpektor, I., Gionis, A., Maarek, Y.: Improving recommendation for long-tail queries via templates. In: Proceedings of the 20th International Conference on World Wide Web, pp. 47–56. ACM (2011)
22. Wang, F., Wang, Z., Li, Z., Wen, J.-R.: Concept-based short text classification and ranking. In: Proceedings of the 23rd ACM International Conference on Information and Knowledge Management, pp. 1069–1078. ACM (2014)
23. Wu, W., Li, H., Wang, H., Zhu, K.Q.: Probase: a probabilistic taxonomy for text understanding. In: Proceedings of the 2012 ACM SIGMOD International Conference on Management of Data, pp. 481–492. ACM (2012)

An Efficient Boolean Expression Index by Compression

Jin Tao, Chenxi Zhang, and Weixiong Rao$^{(\boxtimes)}$

School of Software Engineering, Tongji University, Shanghai, China
wxrao@tongji.edu.cn

Abstract. Boolean expressions (BEs) have been widely used to represent subscriptions and publications in Publish/Subscribe (Pub/Sub) applications. Such applications frequently exhibit high dimensionality caused by the diversity of attributes and values. Given the high dimensionality in Pub/Sub, how to efficiently index BEs (*space optimization*) and next match events against the indexed BEs (*matching efficiency*) becomes a challenging task. Unfortunately, no existing Pub/Sub solutions could meet both objectives without compromising either of them.

In this paper, we proposed a novel approach, namely *BE-Matrix*, to address the above issues. Firstly, we model BE subscriptions as a binary matrix and then encode the binary matrix into lists of encoded numbers, i.e. encoding lists, for lower space cost. Next, by adopting fast bitwise operation AND over encoding lists without a costly decoding phase, we can significantly speedup the matching process. Finally we conduct extensive experiments to evaluate the performance of *BE-Matrix* with the state of art indexing approach BE-Tree [9,10], the results show that BE-Matrix outperforms BE-Tree by using **12.72×** less space cost and **11.38×** faster running speed.

Keywords: Pub/Sub · BE Matrix · Algorithm optimazition

1 Introduction

Boolean expressions (BEs) have been widely used to represent subscriptions and publication events in Pub/Sub applications, such as target advertising systems, information filtering applications and financial news feed applications [2,7,9,10,12].

Pub/Sub applications frequently exhibit high dimensionality caused by the diversity of attributes appearing in BEs. There are two challenges regarding with the matching problem within high dimensionality. Firstly, there is space cost challenge, which address the issue of how to efficiently index BEs with minimum space. Secondly, there is a matching efficiency challenge, which address the issue of figure out a fast matching algorithm which can output match result quickly, and this matters to most time-crunched applications. Unfortunately, most of existing Pub/Sub solutions can optimize only one of the target while compromising the other one.

© Springer International Publishing AG 2017
Z. Bao et al. (Eds.): DASFAA 2017 Workshops, LNCS 10179, pp. 287–300, 2017.
DOI: 10.1007/978-3-319-55705-2_23

In this paper, we proposed *BEMatrix*, to tackle the challenges. On the one hand, we model BE subscriptions as binary matrices. Due to the diversity of BEs, each matrix could be very sparse. We then design a coding scheme, which treats a set of binary elements in matrix as 0/1 bits and encodes them into a single integer number, to optimize the space cost. On the other hand, we propose a matching algorithm, which directly adopt a fast bitwise operation (AND) over the encoded numbers without a costly decoding phase, to achieve high event matching efficiency. As a result, the proposed coding scheme can unify the two optimization objectives without compromising either of them.

Our contribution in this paper is threefold. Firstly, we propose a new encoding scheme by *permuting matrix columns* to further capsule the matrix. In our scheme the encoded numbers with all zero bits are not maintained, so the permutation helps to make the 1-element in the matrix more densely co-clustered and save space cost. By modeling the matrix as a bipartite graph and we propose an efficient algorithm on the bipartite graph to permute matrix columns.

Secondly, we design an efficient data structure to index the (permuted) BE binary matrix and an efficient event matching algorithm based on the indexing structure to process matching problems. The key idea of the matching algorithm is to leverage a fast bitwise operation (AND) on the encoded numbers for the matching between BEs and events.

Lastly, We conduct extensive experiments on synthetic datasets. The results show that our work significantly outperforms the state of art *BE-Tree* [9,10], in terms of index space cost and average matching time. For example, *BE-Matrix* can achieve 12.72× less space cost and 11.38× faster running time than *BE-Tree* in high dimensionality.

The rest of the paper is organized as follows. We present the preliminaries including data model and problem definition in Sect. 2, the coding scheme and the index structure are shown in Sect. 3. Section 4 presents the event matching algorithm, and Sect. 5 reports experimental results. After that, Sect. 6 reviews literature works, and Sect. 7 finally concludes the paper.

2 Preliminaries

In this section, we introduce the data model of BE subscriptions and BE Binary Matrix.

2.1 BE Subscriptions

Definition 1. $be = \{p_1^{\langle attr_1, op_1, vals_1 \rangle} \cap \ldots \cap p_k^{\langle attr_k, op_k, vals_k \rangle}.\}$

We define each BE *be* to be a conjunctions of predicates as in Definition 1. k refers to the BE's size. Here, a predicate $p^{\langle attr, op, vals \rangle}$, in short p, consists of a triple of an attribute *attr*, an operator *op* and a set of attribute values *vals*. The typical operator *op* in BE supports $\in$. Notice that the operator $\in$ subsumes the simple $=$ operator. Thus we can also use BEs to represent publication events $e = \{attr_1 = val_1, \ldots, attr_{k_e} = val_{k_e}\}$, here the event size is k_e.

A subscription BE *be* defines a stateless boolean function $be(\cdot)$ that accepts an event e as an argument. An event e is said to *match* a BE *be*, if $be(e) = true$. For example, the following BE *be* represents a user profile: *gender* $\in$ $\{Male\}, age \in \{20\}, interests \in \{sports, entertainment, game\}$. The event e: $\{gender = Male, age = 20, interests = game\}$ successfully matches the BE *be*.

We note that BEs slightly differ from the traditional data model in Pub/Sub, which defines an expressive set of operators (such as $<, >, =, \neq$) on either discrete or continuous data types. For the predicate in which involves a range of continuous values such as $\{20 \leq age < 50\}$, we follow previous works [12,13] to discretize the data range into multiple intervals. For example, if the given granularity for *age* is 10, the range $20 \leq age < 50$ can be exhaustively enumerated into $age \in \{20, 30, 40\}$. In this way, we can conformably convert predicates, no matter the involving data types, either discrete or continuous, into BEs.

Nevertheless, the discretization of multi-value predicates and events could lead to false positives. For example, the converted predicate $\{26 \leq age < 50\}$ and event $age = 22$ would be a match for granularity set to 10. To fix the issue, we have to validate the output of a BE-based matching algorithm against an event and filter out those incorrect matching result, if needed. While many recent Pub/Sub applications, such as targeted Web advertising, prefer to get the results with false positives for they could find more potential targeted customers. Thus, it makes sense to model subscriptions and events in Pub/Sub by BEs.

As for the discretization of subscriptions and events [12], it is a good way to construct the hierarchical subranges and is of separate interest and not covered in this paper.However, no matter how granularity is tuned, the BE data model and proposed techniques in this paper still work.

2.2 BE Binary Matrix

Given a set of BEs, we first group the BEs by their size. For a specific size say k, we model the grouped BEs (with size of k) as a matrix. Consider that BEs typically involves high dimensionality. We thus decompose the BEs from a high dimensional space into multiple low-dimensional space. The decomposition is similar to the producing of inverted list.

Now, given the above decomposed predicates p_i, we model them as a matrix as follows. In the matrix, each row indicates a discrete attribute and a specific value in its own data range, while each column indicates a BE containing predicates p_i. For simplicity, here we only present one matrix and assume that each BE is associated with a unique ID. The element in matrix is set to 1 if and only if the corresponding BE overlaps with the attribute and value associated in the row.

Example 1. The BEs with $k = 2$ in Table 1 is decomposed as in Table 2 and then modeled into matrix in Table 3.

Table 1. Eight subscription BEs

ID	BE	Size
0	$gender \in \{male\}$, $age \in \{30\}$	2
1	$gender \in \{female\}$, $interests \in \{sports, game\}$	2
2	$dept \in \{sales\}$, $position \in \{staff\}$, $age \in \{20, 30, 40, 50\}$	3
3	$interests \in \{entertainment, driving\}$, $age \in \{20, 30, 40\}$	2
4	$gender \in \{male\}$, $dept \in \{sales\}$, $age \in \{30, 40\}$	3
5	$gender \in \{male\}$, $dept \in \{sales\}$	2
6	$interests \in \{sports, driving\}$, $age \in \{20\}$	2
7	$dept \in \{finance\}$, $age \in \{30\}$	2

Table 2. Decomposed predicates on 4 dimensional space ($k = 2$)

Attribute	Predicate
$gender$	$be_0 : gender \in \{male\}$
	$be_1 : gender \in \{female\}$
	$be_5 : gender \in \{male\}$
age	$be_1 : age \in \{30\}$
	$be_3 : age \in \{20, 30, 40\}$
	$be_6 : age \in \{20\}$
	$be_7 : age \in \{30\}$
$interests$	$be_1 : interests \in \{sports, game\}$
	$be_3 : interests \in \{entertainment, driving\}$
	$be_6 : interests \in \{sports, driving\}$
$dept$	$be_6 : dept \in \{sales\}$
	$be_7 : dept \in \{finance\}$

Table 3. BE binary matrix of size 2. By base number $b = 2$, we divide the columns into 3 groups.

ID	0	1	3	5	6	7
gender:male	1	0	0	1	0	0
gender:female	0	1	0	0	0	0
age:20	0	0	1	0	1	0
age:30	1	0	1	0	0	1
age:40	0	0	1	0	0	0
interests:sports	0	1	0	0	1	0
interests:game	0	1	0	0	0	0
interests:entertainment	0	0	1	0	0	0
interests:driving	0	0	1	0	1	0
dept:sales	0	0	0	1	0	0
dept:finance	0	0	0	0	0	1

3   BE Index

In this section, we give the detail to index BEs. Section 3.1 first focuses on the encoding of binary matrices, and Sect. 3.2 then presents the BE index.

3.1   Matrix Encoding Scheme

Intuition. The key idea of encoding scheme is: when using a base number b to divide each row in matrix into $\lceil n/b \rceil$ groups and encode bits in each group as a number, we could save the space cost to $\frac{1}{\lceil n/b \rceil}$ of its origin size. For lower space cost the encoded numbers equal to 0 are not maintained. By tuning base number b we can get different space cost as in Example 2. And it's obvious for all that if the 1-elements in a row are clustered, the space cost will be less. So in the next part, we will discuss how to permute the columns of $B_{m \times n}$ to minimize the space cost in detail.

Example 2. By $b = 2$, we divide the binary matrix $B_{11 \times 6}$ as in Table 3, by encoding each group we can get 17 non-zero encoded numbers and 16 zero encoded numbers. If $b = 4$, we have 15 non-zero encoded numbers and 7 zero encoded numbers. However, if we permute the matrix as in Table 4, we can get 16 non-zero encoded numbers and 17 zero encoded numbers if $b = 2$, 13 non-zero encoded numbers and 9 zero encoded numbers if $b = 4$. Notice that after this permutation the space cost become lower no matter $b = 2$ or $b = 4$.

Table 4. Binary matrix of size 2 with the permutation of columns. By $b = 4$, we divide the 6 columns into 2 groups

origin ID	3	1	6	5	0	7
new ID	0	1	2	3	4	5
gender:male	0	0	0	1	1	0
gender:female	0	1	0	0	0	0
age:20	1	0	1	0	0	0
age:30	1	0	0	0	1	1
age:40	1	0	0	0	0	0
interests:sports	0	1	1	0	0	0
interests:game	0	1	0	0	0	0
interests:entertainment	1	0	0	0	0	0
interests:driving	1	0	1	0	0	0
dept:sales	0	0	0	1	0	0
dept:finance	0	0	0	0	0	1

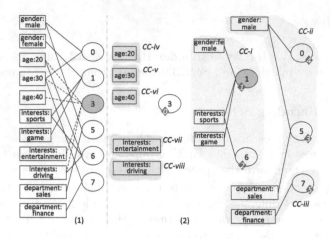

Fig. 1. Running example of Algorithm 1

Permutation. Following [4], we know that the permutation problem, reducible from Travelling Salesman Problem (TSP) [3], is NP-hard, and no efficient solution can solve the problem. To enable a sub-optimal solution, some previous works [5,6] transform the above problem to a co-clustering problem [1] on the binary matrix $B_{m\times n}$, and design classic clustering work such as spectral clustering [8]. While this does not work well if the BE matrix follows a skewed distribution, which appears in most scenario.

To solve problem, we alternatively use a bipartite graph G_B to model the binary matrix B. The vertices in G_B represent the rows and columns of B. The v_r and v_c represent the associated rows and columns in B, respectively. There exists an edge between v_r and v_c if and only if the associated element in B is 1 (Fig. 1(1)). Once the bipartite graph G_B is clustered, the 1-elements in matrix B are then equivalently clustered.

Basic Idea. Instead of directly finding good cut on G_B, we would like to leverage the graph topology structure in G_B. In detail, we first remove *hub vertices (those with high degrees) and associated neighbors*. The removal of hub vertices and their neighbors then decomposes the remaining bipartite graph of G_B into disjoint connected components (CCs). Among such CCs, we call the largest CC to be a giant CC (GCC). *Such disjoint CCs except the GCC then become natural clusters.* The GCC becomes the new bipartite graph G_B to replace the original one. On the new bipartite graph G_B, we repeat the steps of removing hub vertices in the new G_B and decomposing the new G_B into disjoint CCs, until the entire graph is processed.

The Algorithm 1 takes a bipartite graph G_B as an input, and generates an output array L_c. The L_c records the array list of permuted column IDs. In Algorithm 1, line 2 finds the hub vertex v_h (with the largest degree) if existing, and removes v_h (plus its edges) from G_B. After the removal, we have the remaining graph G'_B. Note that v_h could be the vertex with respect to either a row or a column in the corresponding binary matrix. Thus, we add v_h to L_c only if v_h is

---

Algorithm 1. Permutation in a bipartite graph G_B

---

   input : bipartite graph G_B
   output: L_c
1 Create an array L_c;
2 **for** *Remove the largest hub* v_h *from* G_B *to make new graph* G'_B **do**
3   **if** v_h *is a column vertex* v_c **then** Add v_h to L_c;
4   Find all connected components (CCs) in G'_B;
5      $G_B \leftarrow$ the largest giant CC (GCC);
6   Sort all other CCs (except the GCC) by decreasing order of their size;
7   **for** *each vertex* v_d *in every sorted CC (except the GCC)* **do**
8   **if** v_d *is a vertex of* v_c **then** Add v_d to L_c;

9 output the array L_c;

---

a type of a column vertex v_c. In the remaining graph G'_B, we find all disjoint connected components (CCs). Among such CCs, we find the largest GCC and let it become the new graph G_B and repeat the steps above. For the rest of such CCs (except the GCC), we treat them as nature clusters. By sorting them by decreasing order of their size, we add each member vertex in the sorted CC to L_c (in case that the vertex is the type of a column vertex v_c).

Complexity Analysis. The running time of Algorithm 1 depends on the two following aspects: (*i*) The amount of iterations in Algorithm 1. Note that the ratio r_{G_B}, computed by the number of total hub nodes against the total number of vertices in G_B, decides the amount of iterations in Algorithm 1. Obviously, such an amount of iterations can be computed by $O(|V| \times r_{G_B})$. A smaller r_{G_B} leads to faster running time of Algorithm 1. (*ii*) In each iteration, finding the CCs (line 4) takes $O(|E| + |V|)$ and sorting non-GCC vertices (line 6) takes $O(|V| \log |V|)$, leading to the total cost $O(|E| + |V| \log |V|)$. Thus, the overall cost of Algorithm 1 is $O(r_{G_B} \times |V| \times (|E| + |V| \log |V|))$.

Performance Optimization: Given a large bipartite graph G_B, the above algorithm could use high running time. Thus, we address the challenge by the following improvement.

- (*i*) We first preprocess the matrix B directly by the row-level encoding scheme but without column permutation. In this way, we reduce the amount of matrix columns from n to $\lceil \frac{n}{b} \rceil$ and have a new matrix $B_{\lceil \frac{n}{b} \rceil}$. After that, we use Algorithm 1 to process the matrix $B_{\lceil \frac{n}{b} \rceil}$ for higher computation efficiency.
- (*ii*) We avoid running line 5 (used to find CCs) in each iteration of the **for** loop to find a new hub vertex v_h. Instead, we find CCs only when hub vertices with top k highest degrees are found. The batch manner can greatly optimize Algorithm 1 for faster running time.

3.2   BE Index

The BE Index *Idx* consists of 3 components: *row directory, encoding lists* and *Col IDs*. Figure 2 gives the *Idx* to index the binary matrix in Table 4. Here, the Col IDs (which are just the output result of Algorithm 1) correspond the columns of binary matrix in Table 4.

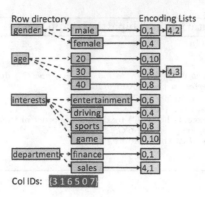

Fig. 2. BE index

Row Directory. *Row directory* maintains the mapping from each row of a BE binary matrix to a specific encoding list. As shown in Fig. 2, the row directory maintains only a full list of attributes instead of the full item. An attribute next points to a set of attribute values. An attribute value finally associate with an encoding list.

Encoding List. The encoding list consists of pairs $\langle cid, eid \rangle$, sorted by ascending order of cid. Specifically, based on the base number b, we can get $s = \lceil n/b \rceil$ blocks for each row, each blocks with at most b elements are encoded into a pair $\langle cid, eid \rangle$. The last block which might encodes less than b bits requires special attention. The number cid represents the ID of the leftmost column ID in the block and eid is a non-zero number to encode the associated binary elements.

4 Event Matching

4.1 Overview

Consider a BE index Idx of size k and an event e consisting of $|e|$ event BEs $attr_i = val_i$ (with $1 \le i \le |e|$). We light the even matching as follows.

First, for each BE $attr_i = val_i$ in event e, we lookup the row directory of Idx to find an associated encoding list, if exists. Among the $|e|$ BEs $attr_i \in \{val_i\}$ in e, it's required to get at least k BEs with non-empty encoding lists to continue the next matching step. Otherwise we just return an empty result.

Second, using at least k encoding lists (denoted by L) found by the first step above and k as input arguments, Algorithm 2 will traverse the encoding lists $L_i \in L$ and output a set S consisting of the pairs $\langle cid, eid \rangle$.

4.2 Retrieving Encoding Lists

While retrieving encoding list in L, the associated overhead is with the traversal of the pairs. If we can avoid redundant traversal of the pairs in L meanwhile without falsely missing correct result, we can accelerate the traversal and optimize the matching performance. So we design Algorithm 2 with the following key points.

Algorithm 2. Event Matching

input : Set L of encoding lists, size k
output: Set S of BE IDs
1 **if** $|L| < k$ **then** return empty;
2 Create a sorted list H and empty set S;
3 add the head pair in every encoding list $L_i \in L$ to H;
4 **while** *at least k encoding lists do not reach the end position* **do**
5 | $eid \leftarrow 0$; sort H by ascending order of $H[i].cid$;
6 | find k' pairs $H[0 \ldots k' - 1]$ with $H[0].cid = \ldots = H[k' - 1].cid$;
7 | **if** $k' < k$ **then** goto line 13;
8 | **for** int $i = 0; i < b; i + +$ **do**
9 | | $cnt \leftarrow 0$;
10 | | **for** int $j = 0; i < k'$ && $cnt < k; j + +$ **do**
11 | | └ **if** $H[j].eid \wedge 1^i > 0$ && $cnt + + == k$ **then** $eid \leftarrow eid \vee 1^i$;

12 | **if** $eid > 0$ **then** add $\langle H[0].cid, eid \rangle$ to S;
13 └   move the position of L_i w.r.t $H[0]$ until $H[0].cid \geq H[1].cid$;

14 return S;

(i) If we do have a BE ID that co-appears in k encoding lists. Since BE IDs are encoded by pairs $\langle cid, eid \rangle$, finding the k pairs having the equal cid values becomes the *necessary condition* of finding the BE ID that co-appear in k encoding lists. The **if** condition in line 7 is used to determines whether we have at least k pairs with the same cid values.

(ii) Consider that we have k' ($\geq k$) pairs having the same cid values. Among such pairs, we need to check whether or not a BE ID co-appear inside k pairs. To this end, we follow an *interleaved* fashion to traverse the encoding lists. We use a sorted list H to maintain the pairs at current position of all encoding lists $L_i \in L$. Now if the pairs $H[0]$ and $H[k' - 1]$ (with $k' > k$) are with the same cid values, we perform a bitwise AND operation $\wedge$ on $H[j].eid$ with a number 1^i (line 11). Here we use 1^i ($0 \leq i \leq b - 1$) to denote a number with the i-th bit set to 1. By the inner **for** loop in lines 10–11, we check whether or not at least k $eids$ are all with 1-bits at the i-th position. If *true*, we set the i-bit of eid to be 1 (by $eid \leftarrow eid \vee 1^i$ in line 11).

(iii) Finally, when the number of encoding lists that do not reach the end position is smaller than k, it is obvious that we could find no more matches as $|L_i| < k$. In such case (i.e.,line 4), we can simply stop the **while** loop and omit the traversal of the remaining encoding lists.

Cost Analysis. Since each pair is visited at most one time, the complexity cost of Algorithm 2 is $O(N)$, where N is the total number of pairs in L. Recall that the permutation algorithm (Algorithm 1) is to co-cluster the 1-elements together. The smaller the number N of pairs in encoding lists L is, the lower of the traversal cost of Algorithm 2. Thus, the permutation can unify the space cost of encoding lists L and traversal overhead of event matching.

5   Evaluation

In this section, we first introduce the experimental setting, and next describe the detail of performance evaluation.

5.1   Experimental Setting

Metrics and Counterparts. We compare the performance of our approach, namely *BE-Matrix*, in terms of both space cost of index *Idx* and the average matching time with *BE-Tree* [9,10]. we implement *BE-Matrix* by C++ and test both *BE-Tree* and *BE-Matrix* on a Linux workstation equipped with 12 cores, 2.4 MHz CPU and 512 GB main memory.

Data Generation. Following the data generator of *BE-Tree*, we make synthetic publication events and BE subscriptions by varying the following parameters given in Table 5.

– *BE subscriptions*: (*1*) the total number of BE subscriptions N_S; (*2*) the maximal size of BE subscriptions K_S; (*3*) the maximal number of discrete attribute values of one predicate in subscriptions V_s.
– *Attribute Space*: (*1*) the total number of attributes N_a (i.e., the dimensionality of full semantic space); (*2*) total number of possible values in each attribute V_a (i.e., the data range for each attribute).
– *Event publications*: the maximal number of predicates in one event K_e (i.e., the size of events).
– *Miscellaneous parameters*: (*1*) Zipf parameter α to generate skewed attribute values in Subs and Pubs, (*2*) base number b to encode a BE binary matrix, and (*3*) granularity g of attribute values to divide the data range of BE Subs. Depending on b, we use a corresponding data type to represent a number. For example $b = 8$, we use the data type `uint8_t` to encode 8 bits.

Table 5. Parameters to generate subscriptions & publications

Parameter	Data range	Default val
N_a: # of attributes (* 10^4)	2–6	2
N_S: # of subs (*10^5)	2–10	2
K_S: size of subs	4–20	4
V_S: # of attr. values in subs	5–100	50
V_a: # of attribute values (* 10^3)	1–10	1
K_e: size of events	20–100	20
α: Zipf parameter	0.2 - 1.0	0.8
b: Base num	8–32	8
g: Granularity of attr. values	1,10	1

Given the above parameters in Fig. 5, we follow the Zipf distribution (by parameter α) to generate attribute space, BE subscriptions and publication events with related parameters. And we totally generate 10,000 events to measure the average time to match an event against indexed BE subscriptions.

5.2   Experimental Result

To evaluate the performance of *BE-Tree* and *BE-Matrix*, we vary the value of the parameters listed in Table 5 and measure their space cost and matching time.

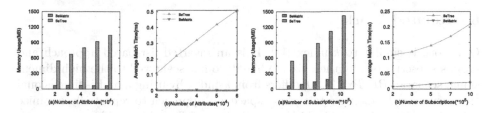

Fig. 3. Num. of attributes BEs (left) and subscriptions (right): (a) Space cost; (b) Matching time

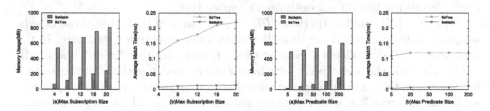

Fig. 4. Size of BEs (left) and num. of discrete predicates (right): (a) Space cost; (b) Matching time

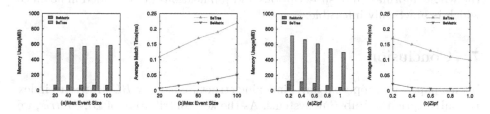

Fig. 5. Size of events (left) and Zipf (right): (a) Space cost; (b) Matching time

From Figs. 3, 4, 5, we can conclude that *BE-Matrix* outperforms *BE-Tree* in any same circumstances. And the space cost basically increases as the parameter grows. And so with the avg matching time. In Fig. 5, the line decrease becuase when the α grows, the generated data is more clustered. From Fig. 6, we can conclude that the base number and granularity both affect the performance.

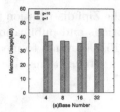 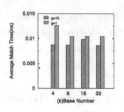

Fig. 6. Base number b: (a) Space cost; (b) Matching time

6   Related Work

Our work is similar to k-index [12]. It is an inverted list-based approach and groups subscriptions by their size. While it do not scale well to the high dimensional scenario. *BE-Tree* [9,10] differs from k-index by using a multi-dimensional tree structure. It adopts a two-phase space cut approach to repeatedly organize subscriptions as a hierarchical index by attribute followed by a value space partitioning. Although *BE-Tree* [9,10] can achieve better performance over k-index, it still incur large space cost and slow matching time in high dimensionality. A recent improvement [11] extends [9,10] and studies the problem of parallel event processing but do not optimize the space cost.

OpIndex [13] could outperform *BE-Tree* in terms of both space cost of subscription index and matching time. It adopts a two-level index structure. In the first level, *OpIndex* selects a *pivot attribute* from each subscription. Those subscriptions with the same pivot attribute are grouped together. In the second level, subscriptions are further partitioned according to their predicate operator, such that subscriptions with the same operator are clustered into segment signatures of predicate lists. The selection of pivot attributes is the key of *OpIndex* to achieve less space cost and faster running time than k-index and BE-Tree. Unfortunately, the selection of pivot attribute in *OpIndex* works if and only if the whole information of subscriptions and publications are provided in advance which is practically infeasible.

7   Conclusion

In this paper, we propose a novel indexing structure namely *BE-Matrix* to index BE subscription in Pub/Sub systems. As the key contribution of *BE-Matrix*, we model BE subscriptions as a binary matrix and propose an algorithm to permute matrix columns. By encoding the permuted matrix, we can achieve shorter encoding lists for lower space cost. Meanwhile, by the fast bitwise operation, the proposed event match algorithm requires less traversal overhead caused by shorter encoding lists. Consequently, *BE-Matrix* can unify the objectives of low space cost and fast event matching time. As future work, (i) we consider that matrix elements could be any numbers (beyond 0/1 values), and are interested in encoding such matrixes, and (ii) we extend our solution to support those BE consisting of many other operators such as ranked BEs [7] and Disjunctive/Conjunctive Normal Forms (DNF/CNF) [2].

Acknowledgment. This work is partially supported by National Natural Science Foundation of China (Grant No. 61572365) and Science and Technology Commission of Shanghai Municipality (Grant No. 15ZR1443000). We also would like to thank the anonymous the reviewers for their valuable comments that helped improve this paper.

References

1. Davis, J.V., Kulis, B., Jain, P., Sra, S., Dhillon, I.S.: Information-theoretic metric learning. In: Machine Learning, Proceedings of the Twenty-Fourth International Conference (ICML 2007), Corvallis, Oregon, USA, 20–24 June 2007, pp. 209–216 (2007). http://doi.acm.org/10.1145/1273496.1273523
2. Fontoura, M., Sadanandan, S., Shanmugasundaram, J., Vassilvitskii, S., Vee, E., Venkatesan, S., Zien, J.Y.: Efficiently evaluating complex boolean expressions. In: Proceedings of the ACM SIGMOD International Conference on Management of Data, SIGMOD 2010, Indianapolis, Indiana, USA, 6–10 June 2010, pp. 3–14 (2010). http://doi.acm.org/10.1145/1807167.1807171
3. Garey, M.R., Johnson, D.S.: Computers and Intractability: A Guide to the Theory of NP-Completeness. WH Freeman, California (1979)
4. Johnson, D.S., Krishnan, S., Chhugani, J., Kumar, S., Venkatasubramanian, S.: Compressing large Boolean matrices using reordering techniques. In: VLDB, pp. 13–23 (2004)
5. Kang, U., Faloutsos, C.: Beyond 'caveman communities': hubs and spokes for graph compression and mining. In: 11th IEEE International Conference on Data Mining, ICDM 2011, Vancouver, BC, Canada, 11–14 December 2011, pp. 300–309 (2011). http://dx.doi.org/10.1109/ICDM.2011.26
6. Lim, Y., Kang, U., Faloutsos, C.: SlashBurn: graph compression and mining beyond caveman communities. IEEE Trans. Knowl. Data Eng. **26**(12), 3077–3089 (2014). doi:10.1109/TKDE.2014.2320716. http://doi.ieeecomputersociety.org
7. Machanavajjhala, A., Vee, E., Garofalakis, M.N., Shanmugasundaram, J.: Scalable ranked publish/subscribe. PVLDB **1**(1), 451–462 (2008). http://www.vldb.org/pvldb/1/1453906.pdf
8. Ng, A.Y., Jordan, M.I., Weiss, Y.: On spectral clustering: analysis and an algorithm. In: Advances in Neural Information Processing Systems 14 Neural Information Processing Systems: Natural and Synthetic, NIPS 3–8, 2001, Vancouver, British Columbia, Canada, pp. 849–856 (2001). http://papers.nips.cc/paper/2092-on-spectral-clustering-analysis-and-an-algorithm
9. Sadoghi, M., Jacobsen, H.: BE-Tree: an index structure to efficiently match boolean expressions over high-dimensional discrete space. In: Proceedings of the ACM SIGMOD International Conference on Management of Data, SIGMOD 2011, Athens, Greece, 12–16 June 2011, pp. 637–648 (2011). http://doi.acm.org/10.1145/1989323.1989390
10. Sadoghi, M., Jacobsen, H.: Analysis and optimization for Boolean expression indexing. ACM Trans. Database Syst. **38**(2), 8 (2013). doi:10.1145/2487259.2487260. http://doi.acm.org

11. Sadoghi, M., Jacobsen, H.: Adaptive parallel compressed event matching. In: IEEE 30th International Conference on Data Engineering, Chicago, ICDE 2014, IL, USA, 31 March–4 April, 2014, pp. 364–375 (2014). http://dx.doi.org/10.1109/ICDE. 2014.6816665
12. Whang, S., Brower, C., Shanmugasundaram, J., Vassilvitskii, S., Vee, E., Yerneni, R., Garcia-Molina, H.: Indexing Boolean expressions. PVLDB **2**(1), 37–48 (2009). http://www.vldb.org/pvldb/2/vldb09-83.pdf
13. Zhang, D., Chan, C., Tan, K.: An efficient publish/subscribe index for ecommerce databases. PVLDB **7**(8), 613–624 (2014). http://www.vldb.org/pvldb/vol7/p613-zhang.pdf

DMMOOC

MOOCon: A Framework for Semi-supervised Concept Extraction from MOOC Content

Zhuoxuan Jiang[(✉)], Yan Zhang, and Xiaoming Li

School of Electronics Engineering and Computer Science, Peking University,
Beijing, China
{jzhx,lxm}@pku.edu.cn, zhy@cis.pku.edu.cn

Abstract. Recent years have witnessed the rapid development of Massive Open Online Courses (MOOCs). MOOC platforms not only offer a one-stop learning setting, but also aggregate a large number of courses with various kinds of textual content, e.g. video subtitles, quizzes and forum content. MOOCs are also regarded as a large-scale 'knowledge base' which covers various domains. However, all the contents generated by instructors and learners are unstructured. In order to process the data to be structured for further knowledge management and mining, the first step could be concept extraction. In this paper, we expect to utilize human knowledge through labeling data, and propose a framework for concept extraction based on machine learning methods. The framework is flexible to support semi-supervised learning, in order to alleviate human effort of labeling training data. Also course-agnostic features are designed for modeling cross-domain data. Experimental results demonstrate that only 10% labeled data can lead to acceptable performance, and the semi-supervised learning method is comparable to the supervised version under the consistent framework. We find the textual contents of various forms, i.e. subtitles, PPTs and questions, should be separately processed due to their formal difference. At last we evaluate a new task: identifying needs of concept comprehension. Our framework can work well in doing identification on forum content while learning a model from subtitles.

Keywords: Concept extraction · MOOC · Semi-supervised · CRF

1 Introduction

With the prosperity of Massive Open Online Courses (MOOCs), tens of millions of students all over the world have been beneficial in recent years. An important characteristic of MOOC is providing a one-stop online learning setting which is composed by: (1) video clips, (2) homework, (3) email notification, (4) discussion forum, and (5) quiz/assignment. As more and more disciplines are offered on MOOC platforms, a massive number of cross-domain knowledge have been aggregated in a form of textual content, e.g. subtitles of videos, questions in quizzes, and posts in forums. However, most of the textual content are

© Springer International Publishing AG 2017
Z. Bao et al. (Eds.): DASFAA 2017 Workshops, LNCS 10179, pp. 303–315, 2017.
DOI: 10.1007/978-3-319-55705-2_24

unstructured and diverse. For example, subtitles are well-organized and formal, since they are generated by instructors. While the contents of posts are written by various learners, thus they are colloquial and informal. So there is an issue that how to make the textual content structured, in order to facilitate subsequential knowledge management and mining.

As MOOCs are in fact an educational and unstructured knowledge base, an intuitive and first-step idea is to extract knowledge points, i.e. concepts, from textual content. Concept extraction from MOOC data may be faced with several difficulties: (1) MOOC data is multi-discipline so the method should be instructor- and course-agnostic, (2) obtaining labeled training dataset is extremely expensive since usually domain expertise is required. However, once concepts are well extracted, many subsequential applications are feasible, e.g. building course- or domain-specific concept map, structured cross-domain concepts management, and even personalized learning.

In this paper, we explore the feasibility of concept extraction from MOOC textual content. We regard this task as a problem of sequence labeling by natural language processing method. Since conditional random fields (CRFs) has been proved successful in many sequence labeling tasks like part of speech (POS), named entity recognition, and word segment [19], we propose a framework called MOOCon by adapting CRFs to a semi-supervised version in order to alleviate the human effort of labeling training dataset. The framework is shown as Fig. 1.

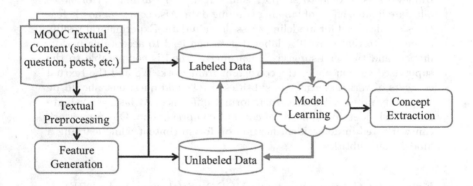

Fig. 1. Framework of semi-supervised concept extraction from MOOC textual content.

As shown in Fig. 1, after collecting raw MOOC textual content, our basic idea is to first preprocess the data with natural language processing tools. We define instructor- and course-agnostic features during feature generation. The thoroughly-considered features should be applied to various courses. Then by obtaining a handful of labeled data, an initial model can be learned. Some unlabeled data are selected based on a measure and can be tagged by the model. The newly-labeled data join the 'Labeled Data', and all labeled data are used again to learn a new model. Cyclically until there is no unlabeled data, the final model can be used to do concept extraction. Note that the data used during the 'Concept Extraction' are usually unobserved samples during 'Model Learning'.

Experimental results demonstrate only 10% labeled data can lead to satisfactory performance. And semi-supervised model is comparable to the supervised version. On the other hand, data in various forms, e.g. subtitle, question and post, should be separately learned, because they have different model ability when being regarded as training data. Especially, an experiment of identifying threads about need of concept comprehension is conducted. The result shows the proposed framework can indirectly run binary classification on MOOC forum content.

2   Related Work

Concept extraction in MOOC settings is a little different from traditional information extraction, e.g. key phrase extraction [7], terminology extraction [4], and named entity recognition (NER) [14,17]. Key phrases (or key words) are usually involved in search engine. Terminology is domain-specific. And named entity is usually person name, location, time and address. Concepts in MOOC settings are not only domain-specific but also cross-domain. For example, the word of *Network* may be a concept in both courses of *Criminal Laws* and *Introduction to Computer Network*. But the two *'Networks'* are actually different concepts.

In the past decades, methods for sequence labeling have been largely studied in the field of information extraction [5]. For example, [10] proposes a rule-based method to extract terms, and [2,6] propose statistical methods. Recently some machine learning methods are proposed for terminology extraction [15,16]. But [15] is designed for software document domain where terminologies are domain-specific proper nouns, e.g. *User ID field*. While [16] only considers nouns when doing extraction. By the way, [15] also proves that directly applying existing methods of NER to terminology extraction will not perform well.

In terms of MOOC data mining, large of studies have been proposed in recent years. For example, [1] studies a badge system to produce incentives for activity and contribution in the forum based on behavior patterns. [8] analyzes the behavior of superposter in 44 MOOCs forums and finds MOOCs forums are mostly healthy. [22] studies the sentiment analysis in MOOCs discussion forums and finds no consistent influence of expressed sentiment to dropout exists. [21] studies the learning gain reflected through forum discussions. [9] conducts an analysis from the perspective of influence in MOOCs forum. In summary the previous studies with MOOCs data focus more on how to improve learning efficacy.

3   Semi-supervised Concept Extraction

In this section, we introduce our framework for concept extraction from MOOC content. Firstly, we introduce the CRFs model which can be applied to our task. Then we state the feature engineering. Then the method of learning a CRF's model is described. At last we introduce the semi-supervised framework which can alleviate human effort for labeling. Note that the framework can be adapted to other probabilistic graphical models, rather than only CRFs.

3.1 Conditional Random Fields Model

As a sequence labeling problem, concept extraction is similar to other sequence labeling tasks, e.g. named entity recognition and part-of-speech annotation. Probabilistic graphical models are the corresponding solution to this kind of tasks, and conditional random fields (CRFs) can obtain the state-of-the art performance [19]. We leverage the CRFs framework to our task.

The problem of sequence labeling can be formally described as solving the conditional probability $P(Y|X)$. The random variable X is features of each sentence which consists of a word sequence $x = \{x_1, x_2, ..., x_T\}$, and the random variable Y is a label sequence of the sentence $y = \{y_1, y_2, ..., y_T\}$.

As to our task, we concern more on the conditional probability of labeling sequence Y, i.e. $p(Y|X)$, rather than their joint probability $p(Y, X)$, so linear chain CRFs framework [11] is the natural choice. The conditional distribution over label sequence y given an observation word sequence x can be defined as:

$$p(y|x) = \frac{1}{Z(x)} \exp \left(\sum_{t=1}^{T} \sum_{k=1}^{K} \lambda_k f_k(y_{t-1}, y_t, x_t) \right) \tag{1}$$

where $Z(x) = \sum_y \exp \left(\sum_{t=1}^{T} \sum_{k=1}^{K} \lambda_k f_k(y_{t-1}, y_t, x_t) \right)$ and $\mathscr{F} = \{f_k(y_{t-1}, y_t, x_t)\}_{k=1}^{K}$ are the set of feature functions defined on given x; $\Theta = \{\lambda_k\} \in \mathfrak{R}^K$ are parameter vector. N is the length of sentence and K is the number of features. Thus in order to fulfil the task of concept extraction, the next steps are defining feature functions $\mathscr{F}$ and learning the model $\Theta = \{\lambda_k\}$.

3.2 Feature Engineering

A crucial part of CRFs framework is the definition of feature functions. Based on our observation, we define five kinds of features which are adapted to our educational data. All the features are course-agnostic and make our framework flexible for scalability.

Text Style Features

- Whether the target word are English.
- Whether the two neighbor words are English.
- Whether the word is the first word in a sentence.
- Whether the word is the last word in a sentence.
- Whether the target word is in a quotation.

Text style features capture the stylistic characteristics. Some concepts usually appear at the beginning or the last of a sentence in instructor's language, e.g. "*Netwrok* means..." or "...This is the definition of *Network*." Since our data are from a Chinese MOOC, we regard whether the word is English as a feature. Obviously, when it comes to English MOOCs, capitalization is the key feature of English concepts. So this kind of features are flexible to different situations.

Structure Features

- Part-of-speech tag of the target word.
- Part-of-speech tag of the previous word.
- Part-of-speech tag of the next word.

We treat the part-of-speech as a feature because fixed combination of part-of-speech, e.g. adjective + noun or noun + noun, may indicate concept phrases. We utilize the Stanford Log-linear Part-Of-Speech Tagger (POS Tagger)[1] [20] to assigns parts-of-speech to each word. Note that as to the corresponding feature functions, we adopt binary value, 0 or 1, to every part-or-speech. For example, there is a function to capture whether target word is a noun or not, and so on.

Context Features

- TF-IDF value of the target word and two neighbor words.
- Normalized uni-gram BM25 score of the target word.
- Normalized bi-gram BM25 score of the target word.
- Normalized bi-gram BM25 score of the two neighbor words

Context features capture the importance of words and word-level information within the whole documents. The training set is partitioned to documents based on video clips. Statistical metric of normalized bi-grams BM25 scores [18] is used to quantify word relevance by default parameters.

Semantic Features

- Semantic similarity of the target word with the previous two words respectively.
- Semantic similarity of the target word with the next two words respectively.

Some frequent-co-occurrence words may be concept phrases. Also close words in the semantic space may be concepts. So by learning the word semantics, features of adjacent words can be captured. The similarity of two adjacent words in semantic space is calculated with the corresponding word vectors trained by Word2Vec[2] [13]. All textual content are used to learn the word embeddings. The corpus size is 145,232 words and the vector dimension is set as 100 by default.

Dictionary Features

- Whether the target word and two neighbor words are in dictionary.
- Whether the two neighbor words are in dictionary.

As in most tasks about natural language processing, a dictionary is useful, we design a run-time dictionary which is just a set of concepts in training dataset.

---

[1] Stanford Log linear Part-Of-Speech Tagger: http://nlp.stanford.edu/software/tagger.shtml.

[2] Word2Vec: https://code.google.com/p/word2vec/.

3.3 Learning and Inference

Given a training dataset, the model $\Theta = \{\lambda_k\}_{k=1}^{K}$ could be learned by Maximum Likelihood Estimation (MLE). To avoid overfitting, we add a regularized term to the function. Then the log-likelihood function of $p(y|x, \lambda)$ based on the Euclidean norm of $\lambda \sim (0, \sigma^2)$ is represented as:

$$L(\Theta) = \sum_{x,y} \log p(y|x, \Theta) - \sum_{k=1}^{K} \frac{\lambda_k^2}{2\sigma^2} \tag{2}$$

So the gradient function is:

$$\frac{\partial L}{\partial \lambda_k} = \sum_{x,y} \sum_{t=1}^{T} f_k(y_{t-1}, y_t, x_t) - \sum_{x,y} \sum_{t=1}^{T} \sum_{y,y'} f_k(y, y', x_t) p(y, y'|x) - \frac{\lambda_k}{\sigma^2} \tag{3}$$

The detail of learning the CRFs model can be referred to [19]. Then given a new word sequence x^* and a learned model $\Theta = \{\lambda_k\}_{k=1}^{K}$, the optimal label sequence y^* could be calculated by:

$$y^* = arg \max_{y \in \mathcal{Y}} p(y|x^*, \Theta) \tag{4}$$

where $\mathcal{Y}$ is the set of all possible label sequences for the given sentence x^*. We employ L-BFGS algorithm to learn the model and Viterbi algorithm to infer the optimal label sequence y^*.

3.4 Semi-supervised Learning Framework

Because the effort for labeling training data is extreme expensive, we propose a semi-supervised framework. We leverage the ideas of self training [12] and k nearest neighbors (k-NN). The intuition is that if an unlabeled sample is similar to a labeled sample in semantic space, the unlabeled sample is very probable to be successfully inferred by the model which is learned from all the current labeled data. Then the unlabeled sample is turned to a labeled one and can be added into the labeled dataset with model-inferred labels. A new model can be learned. We calculate the similarity by Cosine distance between two sentences. Sentence vector is denoted as:

$$VecSentence_i = \frac{1}{T} \sum_{t=1}^{T} VecWord_t \tag{5}$$

where $VecWord$ is learned by Word2Vec with default parameters. Algorithm 1 is the details of the semi-supervised version of training process.

The time complexity of Algorithm 1 is $O(NM^2) + \frac{M}{c}O(\text{TrainCRF})$ where N and M are the sizes of labeled set and unlabeled set respectively, and c is the number of unlabeled data which are selected to be inferred in each loop. The additional computing cost is deserved since human effort can be largely reduced, especially when N and M is not large.

Algorithm 1. k-NN Self Training for SSC-CRF

INPUT: labeled dataset $X_L = \{(x, y)\}$, unlabeled dataset $X_U = \{x\}$, number of candidates c

OUTPUT: model Θ

1:**repeat**

2:     $\Theta = \text{TrainCRF}(X_L)$

3:     $X_{c-nearest} = \varnothing$

4:     **for** $i=1$:c

5:         $x = \arg\min_{x \in X_U} \text{Cosine\_distance}(x, X_L)$

6:         $X_U = X_U - \{x\}$

7:         $X_{c-nearest} = X_{c-nearest} \cup \{x\}$

8:     $Y_{c-nearest} = \text{InferCRF}(X_{c-nearest}, \Theta)$

9:     $X_L = X_L \cup \{(X_{c-nearest}, Y_{c-nearest})\}$

10:**until** $X_U = \varnothing$

11:$\Theta = \text{TrainCRF}(X_L)$

12:**return** Θ

---

4   Experiment

In this section, firstly we evaluate the performance of supervised and semi-supervised framework for concept extraction. Then we conduct a task of identifying need of concept comprehension with forum content. Last we discuss the contribution of different features to the model.

4.1   Dataset Collection

We collect the corpus of an interdisciplinary course conducted in the fall of 2013 on Coursera. The course contains computer science, social science and economics. Textual content include video subtitles, PPTs, questions and forum contents (i.e. threads, posts and comments). Table 1 lists the statistics of the content. We invited the instructor and two TAs to help label the data. As seen in Table 1, the number of concepts in questions and PPTs are much smaller than that in subtitles. This implies subtitles may cover other kind of textual content. Based on our observation, during labeling the data, the instructor and TAs would spend much time on understanding each sentence. By statistic, labeling 3,000 sentences would spend about 8 hours per person (in average 10 seconds per sentence). So due to the complexity of labeling, we have not labeled concepts in forum content yet. However we conduct another task which also demonstrates the effectiveness of our model.

A preprocessing step of word segment for Chinese may be necessary. We adopt the Stanford Word Segmenter[3] [3]. All data are randomly shuffled before they are learned and validated. 5-cross-fold validation is adopted.

Label of Word. The label of a word is defined as three classes: *NO*, *ST* and *IN*. They respectively mean *not a concept, the beginning word of a concept* and *the middle word of a concept*. So the label variable is $Y \in \{NO, ST, IN\}$. The example sentence with labels is shown as Fig. 2.

---

[3] Stanford Chinese word segment:http://nlp.stanford.edu/software/segmenter.shtml.

Table 1. Corpus statistic of the course.

Source	# sentence	# word	# concept
Subtitles	3,036	69,437	402
PPTs	2,823	22,334	249
Questions	268	7,138	95
Threads (title and initial post)	213	12,759	-
Posts	704	28,095	-
Comments	691	27,803	-

> we will discuss two tools *graph theory* and *game theory*
> *NO NO   NO   NO   NO   ST   IN   NO   ST   IN*
> ......
> **English word sequence with labels**

Fig. 2. Examples of word sequences with labels. Underlined words are concepts.

4.2 Baselines

We propose several baselines to extract concepts for comparison with our approach. The preprocessing is identical for baselines as for our method.

- **Term Frequency (TF)**: Words are ranked by their term frequency. If a word is a concept, the instructor may say it repeatedly in lecture.
- **Bootstraping (BT)**: Instructors may have personal language styles to give talks. So we design the rule-based algorithm by giving several patterns containing true concepts. This method is actually course- and instructor-dependent.
- **Stanford Chinese NER (S-NER)**: This is an exiting tool developed for named entity recognition, whose model is already trained and we just use it to infer concepts in our educational datasets[4] [14].
- **Terminology Extraction (TermExtractor)**: This is an exiting tool for terminology extraction[5]. The well-trained model is also only used to infer concepts in our datasets.
- **Supervised Concept-CRF (SC-CRF)**: This is a method of supervised learning based conditional random fields with all features as defined before.
- **Semi-supervised Concept-CRF (SSC-CRF)**: This is the semi-supervised version for concept extraction. The parameter of c, number of candidates, is empirically set as 20.

We adopt three metrics, precision, recall and F1-value, to measure the results.

---

[4] Stanford Chinese Named Entity Recognizer (NER): http://nlp.stanford.edu/softw are/CRF-NER.shtml.

[5] Terminology Extraction by Translated Labs: http://labs.translated.net/terminology-extraction/.

4.3 Task of Concept Extraction

In this subsection, we use subtitles, PPTs and questions to evaluate the proposed models.

Table 2 shows the comparison of performance between baselines. We use 30% data of subtitles as training data for SC-CRF and SSC-CRF, and the rest are for evaluation. Especially for SSC-CRF, half of the training data are unlabeled. The statistic-based methods (TF@500 and TF@1000) are unreliable due to many stopwords may degrade the performance. The rule-base method (BT) is highly dependent on human experience, and the low precision means plenty of subsequent work for filtering the outputs is required. On the other hand, Stanford Chinese NER and TermExtractor do not perform well maybe because of two reasons: (1) named entity and terminology are different from concepts, (2) the models are not learned from our dataset. The semi-supervised CRF is comparable to the supervised version.

Table 2. Performance of baselines. SC-CRF and SSC-CRF use 30% data of subtitles for training. Half of the training data as unlabeled for SSC-CRF.

	Precision	Recall	F1
TF@500	0.402	0.500	0.446
TF@1000	0.600	0.746	0.665
BT	0.099	0.627	0.171
S-NER	0.131	0.080	0.099
TermExtractor	0.202	0.107	0.140
SC-CRF	**0.914**	**0.897**	**0.905**
SSC-CRF	0.889	0.825	0.856

Figure 3 manifests that the semi-supervised learning would be comparable to the supervised version, especially when less than 20% data are used for training. Half of training data is identically regarded as unlabeled by SSC-CRF. Note that the amount of labeled data when using 10% training data by SC-CRF are equivalent to that of using 20% training data by SSC-CRF, but SSC-CRF performs better than SC-CRF. This result means the semi-supervised framework can obtain satisfactory performance by only labeling a handful of data.

Now we evaluate the different model ability among various MOOC textual content. As shown in Table 3, the items in row are training dataset while those in column are testing dataset. This table can explain some common situations of educational settings. Subtitles can cover almost all the concepts. They are ideal to be regarded as the training data. PPTs is also decent to be as training data seeing from the precisions, but the recalls are low. Maybe due to usually in PDF format, PPTs may cause incomplete sentences when being converted to text. Questions could lead to lower recalls than PPTs, because not all concepts are present in questions as shown in Table 1. In summary, different kinds of MOOC textual content have different model ability, so they should be separately considered.

Feature Contribution. We analyze how the different kinds of features contribute to the model. The result is shown as Table 4. Dictionary Feature has a predominant influence on the final results, and Structure Feature is second important. Other features are also contributive but the difference is small. Even so, every kind of features contribute to the model positively.

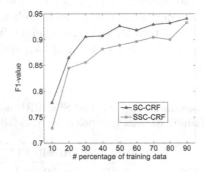

Fig. 3. Performance between supervised and semi-supervised models.

Table 3. Mutual learning between various content. The rows are training data and the columns are testing data.

	Subtitles			PPTs			Questions		
	P	R	F1	P	R	F1	P	R	F1
Subtitles	-	-	-	0.816	0.838	0.827	0.860	0.800	0.829
PPTs	0.868	0.764	0.813	-	-	-	0.857	0.685	0.761
Questions	0.846	0.349	0.494	0.722	0.360	0.480	-	-	-

4.4 Task of Concept Identification

Since we do not obtain the labels of forum content, we conduct a task of concept identification. The model is learned from subtitles, and then the model is used to infer concepts on forum content. We use 30% of subtitles to learn the semi-supervised model. Unlike directly identifying concepts, we use our model to identify the *need of concept comprehension*. The new task is actually a binary classification of forum threads, that is to identify whether a thread is about concept comprehension. Only threads title and the initial post are inferred by our model, instead of all the posts. The classification result is post-evaluated which means: as to each thread, that the assessed score is '1' means two situations:

- If no concept is identified and this thread is not about need of concept comprehension.
- If at least one concept is identified and the definition of identified concepts can at least partially answer the thread.

Other situations are assessed as '0'. The result is shown as Table 5. The accuracy is not bad. And the relatively high recall is useful. It can suggest which threads instructors should intervene. Our model applied in this task can not only identify whether a thread is about need of concept comprehension, but also identify which concept should be explained.

Table 4. Efficacy of features. 10% of data are used for training by SSC-CRF.

	Precision	Recall	F1
All	**0.780**	**0.775**	**0.777**
Without text style feature	0.768	0.776	0.772
Without structure feature	0.722	0.683	0.702
Without context feature	0.757	0.753	0.755
Without semantic feature	0.772	0.757	0.764
Without dictionary feature	0.689	0.235	0.350

Table 5. Result of identifying threads about need of concept comprehension by SSC-CRF.

Accuracy	Precision	Recall	F1
0.822	0.523	0.784	0.627

5 Discussion and Conclusion

Along with the coming of MOOCs, large-scale online educational resources are unprecedentedly produced. Instructors can provide videos, subtitles, lecture notes, questions and etc. While learners can generate forum content, Wiki, log of homework and etc. How to process these data from unstructured to structured is a challenging problem. In this paper, we explore the task of concept extraction on MOOC resources.

Concept extraction can benefit a lot of subsequential applications. Firstly, it is a kind of annotation for MOOC resources. The annotation can be used for studying machine learning methods for MOOC-related natural language processing tasks, such as information extraction, information retrieval and question answering. Secondly, concept extraction can pick up domain-specific or cross-domain knowledge points from complex text. This result can be further processed to build knowledge graph or concept map. With the graph (or map), instructors can better organize the course, and learners can plan their own learning paths more easily. Then by collecting the feedback from learners, the whole teaching and learning process can be a virtuous cycle.

Getting back to the topic of this paper, we are faced with two challenges: (1) MOOCs are cross-domain, (2) labeling training data is extremely expensive. So we propose a flexible framework, called MOOCon, based on semi-supervised machine learning with domain-agnostic features. Experiments demonstrate the efficacy of our framework. Using very a little labeled data can achieve decent performance. We find that various kinds of MOOC content, e.g. subtitles and PPTs, have different modeling ability for concept extraction. So they should be separately treated in future work. Our framework also can be applied to the task of concept identification on MOOC forum content.

In the future, methods of transfer learning and deep learning may be better for extracting cross-domain concepts. External resources of knowledge, e.g. Wikipedia, may be helpful. The relationship between concepts is deserved to be paid more attention for building a domain-specific or even cross-domain concept map.

Acknowledgments. This research is supported by NSFC with Grant No. 61532001 and No. 61472013, and MOE-RCOE with Grant No. 2016ZD201.

References

1. Anderson, A., Huttenlocher, D., Kleinberg, J., Leskovec, J.: Engaging with massive online courses. In: WWW 2014, pp. 687–698 (2014)
2. Bin, Y., Shichao, C.: Term extraction method based on mutual information with threshold interval. In: Zhang, J. (ed.) ICAIC 2011. CCIS, vol. 227, pp. 186–194. Springer, Heidelberg (2011). doi:10.1007/978-3-642-23226-8_25
3. Chang, P.C., Galley, M., Manning, C.: Optimizing Chinese word segmentation for machine translation performance. In: WMT 2008, pp. 224–232 (2008)
4. Collier, N., Nobata, C., Tsujii, J.: Automatic acquisition and classification of terminology using a tagged corpus in the molecular biology domain. Terminology **7**(2), 239–257 (2002)
5. Dong, X., Gabrilovich, E., Heitz, G., Horn, W., Lao, N., Murphy, K., Strohmann, T., Zhang, S.S.W.: Knowledge vault: a web-scale approach to probabilistic knowledge fusion. In: KDD 2014, pp. 601–610 (2014)
6. Frantzi, K., Ananiadou, S., Mima, H.: Automatic recognition of multi-word terms: the c-value/nc-value method. Int. J. Digit. Libr. **3**(2), 115–130 (2000)
7. Hasan, K.S., Ng, V.: Automatic keyphrase extraction: a survey of the state of the art. In: ACL 2014, pp. 1262–1273 (2014)
8. Huang, J., Dasgupta, A., Ghosh, A., Manning, J., Sanders, M.: Superposter behavior in MOOC forums. In: L@S 2014, Atlanta, GA, pp. 117–126, March 2014
9. Jiang, Z., Zhang, Y., Liu, C., Li, X.: Influence analysis by heterogeneous network in MOOC forums: what can we discover? In: EDM 2015, Madrid, Spain, pp. 242–249, June 2015
10. Justesona, J.S., Katza, S.M.: Technical terminology: some linguistic properties and an algorithm for identification in text. Nat. Lang. Eng. **1**(1), 9–27 (1995)

11. Lafferty, J.D., McCallum, A., Pereira, F.C.N.: Conditional random fields: probabilistic models for segmenting and labeling sequence data. In: ICML 2001, pp. 282–289 (2001)
12. Liu, A., Jun, G., Ghosh, J.: A self-training approach to cost sensitive uncertainty sampling. Mach. Learn. **76**(2–3), 257–270 (2009)
13. Mikolov, T., Chen, K., Corrado, G., Dean, J.: Efficient estimation of word representations in vector space. In: Workshop at ICLR 2013, pp. 1–12 (2013)
14. Nadeau, D., Sekine, S.: A survey of named entity recognition and classification. Lingvisticae Investig. **30**(1), 3–26 (2007)
15. Nojiri, S., Manning, C.D.: Software document terminology recognition. In: AAAI Spring Symposium, pp. 49–54 (2015)
16. Qin, Y., Zheng, D., Zhao, T., Zhang, M.: Chinese terminology extraction using EM-based transfer learning method. In: Gelbukh, A. (ed.) CICLing 2013. LNCS, vol. 7816, pp. 139–152. Springer, Heidelberg (2013). doi:10.1007/978-3-642-37247-6_12
17. Ratinov, L., Roth, D.: Design challenges and misconceptions in named entity recognition. In: CoNLL 2009, pp. 147–155 (2009)
18. Robertson, S., Zaragoza, H., Taylor, M.: Simple BM25 extension to multiple weighted fields. In: CIKM 2004, pp. 42–49 (2004)
19. Sutton, C., McCallum, A.: An introduction to conditional random fields. Mach. Learn. **4**(4), 267–373 (2011)
20. Toutanova, K., Klein, D., Manning, C., Singer, Y.: Feature-rich part-of-speech tagging with a cyclic dependency network. In: HLT-NAACL 2003, pp. 252–259 (2003)
21. Wang, X., Yang, D., Wen, M., Koedinger, K., Rosé, C.P.: Investigating how studentqs cognitive behavior in MOOC discussion forums affect learning gains. In: EDM 2015, Madrid, Spain, pp. 226–233, June 2015
22. Wen, M., Yang, D., Rose, C.: Sentiment analysis in MOOC discussion forums: what does it tell us? In: EDM 2014, pp. 130–137 (2014)

What Decides the Dropout in MOOCs?

Xiaohang Lu[1], Shengqing Wang[2(✉)], Junjie Huang[1], Wenguang Chen[1(✉)],
and Zengwang Yan[1]

[1] Department of Information Management, Peking University, Beijing 10086, China
chenwg@pku.edu.cn
[2] Teaching Development Center, Peking University, Beijing 10086, China
wangsq@pku.edu.cn

Abstract. Based on the datasets from the MOOCs of Peking University running on the Coursera platform, we extract 19 major features of tune in after analyzing the log structure. To begin with, we focus on the characteristics of start and dropout point of learners through the statistics of their start time and dropout time. Then we construct two models. First, several approaches of machine learning are used to build a sliding window model for predicting the dropout probabilities in a certain course. Second, SVM is used to build the model for predicting whether a student can get a score at the end of the course. For instructors and designers of MOOCs, dynamically tracking the records of the dropouts could be helpful to improve the course quality in order to reduce the dropout rate.

Keywords: MOOCs · Dropout rate · Sliding window model · Dropout prediction

1 Introduction

Big impacts and challenges on teaching methods, learning methods, student credit affairs management and so on in higher education are brought by MOOCs. MOOCs are also changing the lifelong professional training system gradually. In typical MOOCs platforms, not only could the learners access lecture videos, assignments and examinations, but they can also use collaborative learning approaches such as online forums [6]. A lot of universities and institutions at home and abroad are exploring and researching MOOCs in different ways. For example, the US Department of Education [2] has begun to focus on how to improve student performance using online learning systems. They use automated data mining and analysis techniques to monitor teaching quality, to test teaching improvement, to help teachers understand the performance of students. After that, teachers can help students adapt to teaching individualized. Therefore, the excavation and analysis of teaching data can lead to an important direction for the future development of education.

MOOCs, despite of being open, free and many other advantages, lacks of face to face interaction coupled with such high student teacher ratios [4], leading to

Z. Bao et al. (Eds.): DASFAA 2017 Workshops, LNCS 10179, pp. 316–327, 2017.
DOI: 10.1007/978-3-319-55705-2_25

new problems that traditional teaching hasn't encountered. The most serious problem is the surprisingly high dropout rate. According to a previous study, most MOOCs courses have a passing rate less than 13% [1]. Thus, the teachers and designers of MOOCs are deeply concerned about how to improve the MOOCs' passing rate and the course quality.

In this paper, a total of 32 MOOCs data of Peking University are collected from Coursera platform in Autumn 2013, Spring to Autumn 2014, as the basic datasets. To analyze the dropout problem of MOOCs learners, the characteristics of dropout time and start time are analyzed statistically. Besides, we propose a sliding window model to dynamically predict the overall dropout rate of the learners in the course and a model to predict whether a student can get a score at the end. It can help teachers improve the quality of the course, communicate with the potential dropout in time, provide timely help and feedback, and increase the passing rate.

2   Related Work

Many scholars at home and abroad have studied on when the learners of MOOCs will drop out of the course. There are two kinds of analysis data in the current research: the data of the forum and the data of the clickstream. Here, several typical dropout studies are analyzed and presented.

Amnueypornsakul et al. [1] used learners' clickstream data to predict whether or not a student would drop out of the course. Researchers formed a sequence of weekly learning behaviors for each learner. Then the researchers defined three learners: active, inactive, drop. The results showed that the accuracy rate was significantly improved when the inactive learners were excluded from the model construction, and when including inactive learners, the accuracy rate of modeling inactive learners as active learners was relative higher, but lower than the baseline.

Sinha et al. [12] leveraged combined data of video clickstream and forum to form the action sequence to seek traits that were predictive of decreasing engagement over time. The results showed that dropout behavior was more affected by learning behavior of recent weeks. And most of the dropout students started classes a few weeks after the beginning of the course. There were two possible explanations. One was that these dropout students have needs in specific information, and they ceased to attend classes after they met their needs. Or, the students who joined later had to give up due to the excessive material and work to keep up with the course.

Taylor et al. [13]used different machine learning approaches to predict the dropout students, including logistic regression, support vector machines, deep belief networks, decision trees, and hidden Markov models. The researchers defined dropout as a learner no longer submitting any assignments and tests, and screened 14 weeks' data of learning behavior as training and testing. Researchers divided learners into four categories: passive collaborator, wiki contributor, forum contributor, fully collaborator, and constructed the model for the four types of learners respectively. The researchers proposed a lead and lag prediction model, which predicted the rest $14-i$ weeks using the previous week's data when given a week i. Therefore, there were a total of 91 forecast tasks. The results showed that the prediction accuracy of the passive collaborator was the highest, and the accuracy of wiki contributor and fully collaborator was low due to the lack of data. However, the feature of editing wiki was better to reflect whether a learner will persist to the end of the course. Unless the amount of data is insufficient, the prediction accuracy of the various methods of building the model is similar. Moreover, for a given week forecast, the data of the most recent four weeks was more predictive. In terms of predictive features, the results showed that those who were familiar with MOOCs can present better predictive features.

In addition, some other relevant studies also provide some inspiring ideas. Kloft et al. [6] used click stream data and machine learning algorithms to predict dropout behavior. In the prediction process, a predictive test was performed for each feature. The results showed that the prediction accuracy is not high in the first eight weeks due to the insufficiency of data, followed by weekly prediction accuracy increasing, and suggested to involve the forum data in prediction. Sharkey et al. [11] described in detail the iterative process of using machine learning techniques to predict dropouts and derive predictive features as well as their relative weights. The results showed that the prediction accuracy of the model using machine learning algorithms was above average, and the predictive features, as people expected, showed whether students were enthusiastic to participate in the course. Another study by Yang et al. [14] showed that social factors do have an effect on dropout, and gave MOOCs designers inspiration to design social engagement activities that can enhance students' participation to prevent dropout behavior.

3   Data Cleaning and Learning Metrics

In this paper, a total of 32 sessions-based MOOC courses data were collected from Coursera platform in Autumn 2013, Spring to Autumn 2014, as sample datasets. Through the preliminary analysis and screening of the course data, we find that:

1. the three semester courses of bioinformatics are independent to each other and the contents of the three semesters are exactly the same;
2. the course of social survey and research method is divided into two consecutive courses each occupying one semester with associated contents.

These two modes are highly representative in the MOOCs course. Therefore, we mainly use the data of three semesters of bioinformatics, pkubioinfo-001(BIO01), pkubioinfo-002(BIO02), pkubioinfo-003(BIO03)and two semesters of social survey and research methods, methodologysocial-001(MS11), methodologysocial2-001(MS21), a total of five courses.

To understand the phenomenon of MOOCs dropout in an instant way, we compute the number of enrollment, the number of students whose score were greater than 0, the number of students passing the course, and the passing rate of each course (as shown in Table 1).

Table 1. Course information

Couse ID	Registers	Records	Score > 0	Score > 60	Passing rate
MS21	3566	3184	371	185	0.051879
MS11	7836	6051	6051	255	0.032542
BIO02	16714	15790	1268	510	0.030513
BIO01	18367	18367	1620	520	0.028312
BIO03	16958	16072	909	360	0.021229

3.1 Feature Extraction

Usually, every MOOCs course includes course videos, quiz, forums, and other learning modules. In order to acquire the accuracy of prediction, multiple learning module data were extracted. In the process of obtaining specific data, we extracted the key words from the URL in the log file to identify the learning modules used by the learners and to obtain the learning behavior data for a certain time.

Learners participate in the course by watching videos, taking part in forums, conducting quiz and so on. The participatory process has two important characteristics:

1. Course progress is based on weekly units. A learner can complete the week's learning tasks at any time within a week. In fact, the tracing of learning behavior can be discussed in the context of smaller granularity.
2. The learning behavior is one-direction-oriented mostly, bi-directional occasionally. Most of the learning behaviors, such as clicking and browsing, aim at absorbing knowledge. The interactive activities are relatively rare. This is also the reason why a new form of peer review is introduced to many current courses to enhance bi-directional participation.

Interestingly, disagreement among researchers lies in whether forum data can be introduced as features. Amnueypornsakul and other researchers believe that only 5–10% of the students would participate in the forum, implying that most learners do not have any forum behavior data. For this majority of learners, it is not appropriate to use forum data to make predictions, and therefore they decided not to use forum data. Yet, we carry out the correlation test between the participation of the forum and whether the learner gets score ≥ 60.

The correlation test formula is shown below:

$$R = \frac{SET1 \cap SET2}{SET2} \tag{1}$$

We define SET1 as the set of users who participate in forum at least once, and SET2 as the users who get a score $\geq$ 60. We find that SET1 and SET2 have high coincidence rate in each course, as shown in Table 2.

Table 2. The correlation test

CourseID	Forum-participate	Score > 60	Forum-participate Score > 60	Ratio
BIO01	2645	580	511	0.881034
BIO02	1425	508	395	0.777559
BIO03	1523	358	316	0.882682
MS11	1165	290	269	0.927586
MS21	326	203	153	0.753695

In the methodologysocial-001 course, the coincidence rate of SET1 and SET2 is 92.8%. That is to say, 92.8% of the learners who have had score $\geq$ 60 in this course were involved in the course forum. Therefore, this paper argues that the participatory behavior of the forums has a significant effect on the learners' adherence to learning, hence the data of the students' participation in the forum are added to the forecasting model.

The final feature list we extract is as shown in Table 3.

Table 3. Feature list

Clickstream		Assignment and test		Forum	
Field	Data type	Field	Data type	Field	Data type
page_view	Int	try_hw	Int	view_forum	Int
page_view_quiz	Int	try_quiz	Int	thread_forum	Int
page_view_forum	Int	try_lec	Int	post_thread	Int
page_view_lecture	Int			post_comments	Int
page_view_wiki	Int			Upvote	Int
video_view_times	Int			Downvote	Int
video_pause_times	Int			add_tag	Int
video_pause_speed	Float			del_tag	Int

3.2 Learning Cycle

Learning cycle refers to the learners' start and end time. In order to unify the standard, the start and end of the learning time needs a unified standard definition, and the duration needs to be divided in weeks. Because of the large number of courses and the difference of starting time, we define the starting time of each course as the time when the first video click data appears. To determine the end of the course and to maintain the accuracy of the prediction, we count the time of the last learning behavior of the learners in these five courses, and choose the time point when the first 80% students got a score.

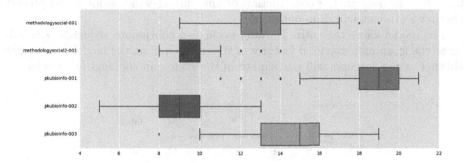

Fig. 1. The learning cycles of different courses

As shown in Fig. 1, among the learners of 2013 Bioinformatics 001 who got a score, the earliest learning behavior ended in the 9th week and the last 19th week. 80% learners ended the study in the 14th week. Accordingly, the 14th week would be regarded as the end time of 2013 bioinformatics 001. Other courses share the same principle.

3.3   The Starting Point and Dropout Point

We do a more in-depth statistical analysis to each persons start time and dropout time. We define whether or not a learner participates in the course in a given week as 1 or 0, then we get a sequence about whether the learner participates in the course every week. Then, we define the point when the number 1 first comes up as the starting point, which means when the learner begins the course. We define the week after the number 1 last comes up as the dropout point, from which the learner no longer comes up.

Then we examine the relationship between the starting point and the dropout point. We find:

1. The later a learner begins, the higher the rate of dropping out after learning a week.
2. The earlier a learner begins, the higher the rate of persisting to the end of the course.

This shows that the earlier the start time, the more likely to stick to the end. The later you start, the more likely to drop out of the course in the following week. Here we can refer to the explanation in [12]. One possibility is that people who start late are difficult to follow due to excessive material, and the other may be people who start late are more likely to seek specific information and no longer learn after acquisition. [12]

4   Modeling

4.1   The Model of Predicting Dropout

The Sliding Window Model. Based on the above analysis, the following discussion focuses on the construction of the sliding window model to predict whether a student will drop out or not.

The model views the entire learning cycle as a continuous sequence. According to the learning behavioral features of the weeks before, the model will predict whether or not learners will participate in the course in the next few weeks.

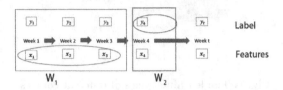

Fig. 2. The diagram of the sliding window model. The first window length is w_1, the length of the weeks before current week n. The second window length is w_2, the length of the weeks thereafter current week n [7].

As shown in Fig. 2, the model uses the learning behavioral features in the first window to predict the label of the second window. The definition of dropout here is that in the second window the learner does not have learning behavior. The model does not focus on individual learner dropping out, and mainly focuses on the overall situation of the learners' behavior. That is, in the current window, how many learners drop out.

As is mentioned above, most learners in MOOCs will not stick to the end of a course. Therefore, the definition of baseline in this paper is to predict that all learners will not appear in the course of the next week, and then compare the improvement of each model with respect to the baseline. This prediction actually product a large number of applications in reality, such as teachers will send a series of emails to encourage learners to follow during the course even though the learner keeps learning. This paper introduces the method of machine learning to improve this strategy. For the prediction model, we use Logistic Regression (LR), Support Vector Machine (SVM), MLP, LSTM.

Results. Experiments on the sliding window model were carried out in the five courses mentioned above. If the value of w_1 is too small, such as $w_1 = 1$, that is, data of one week is used to predict the current week, the model would be rough and simple. If too large, it would be redundant. To make a compromise, according to previous studies, we do the experiments on $w_1 = 3, w_2 = 1$, using a 5-fold cross validation. Take $w_1 = 3$ as an example, other results ($w_1 = 2, w_1 = 4$) work well. The results are shown below (Fig. 3)

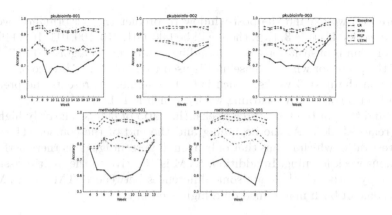

Fig. 3. Prediction accuracy of five courses ($w_1 = 3, w_2 = 1$)

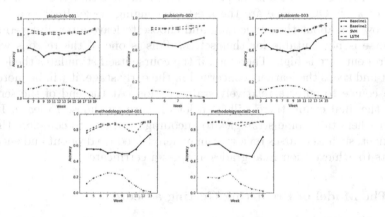

Fig. 4. Prediction accuracy of five courses ($w_1 = 3, w_2 = 3$)

The experiment result shows that:

1. The prediction accuracy of baseline is generally high. This is due to the large number of dropout every week with a dropout rate of 70% in general.
2. Drop-out rate reaches at a peak at the start and end of the course. There were many people who leave early for the course was not suitable for them.
3. In general, the machine learning methods have high prediction accuracy. In different machine learning methods, logistic regression represents the predictive effect in the basic case. In contrast to baseline, machine learning is better at identifying learners who can persist in learning. However, its prediction ability is limited due to the greater need for data.
4. LSTM and SVM have better prediction effect. Compared to multi-layer perceptron and logistic regression, these two methods, the result comes out of which could hardly affected by the data volume, have better prediction ability and can be more stable (Fig. 4).

Moreover, we expand the post-window to predict the performance over the next few weeks. If $w_2 = 3$, then there will be $000, 001, 010, 100, 011, 101, 110, 111$, a total of 8 kinds of situations, corresponding to different categories 0 to 7. In the case of the posterior window, baseline1 is used to predict the absence of learning behavior in the next 3 weeks, which is 000; baseline2 represents the predicted behavior for 3 consecutive learning weeks, which is 111.

We find that in the next three weeks, the ratio of 000 is relatively high, and 111 is relatively low. At the beginning and the end of the course, 111 reaches the lowest ratio, which means that only a small percentage of learners performed continuous week learning. In addition, SVM is effective to the multi-classifying. More to say, although there are some differences between LSTM and SVM, the results come of both maintained on a high accuracy.

Application. The sliding window model solves the problem of monitoring and forecasting the course at different stages, and finds out that the beginning and ending of a MOOC course are the most challenging stage. In the early stage, they know less about the curriculum and choose to leave after discovering that the course is not suitable for themselves. This is one of the reasons why the early dropout rate is high. Therefore, if the course itself stimulates the learners' interest and keeps the learners concerned in the early stage, it will be more likely for the course to enter a relatively stable period. At the end of the semester, due to the final exam, there will be a lot of learners choose to leave. In fact, if the teachers take some strategies to encourage learners to complete the final assessment, such as courses review, it may help to reduce dropout and encourage students to achieve their final grades and get a certificate.

4.2 The Model of Predicting Getting a Score

This model needs to find a course stage as early as possible from the beginning of the course to accurately predict whether the student can get a score. Here, it is defined as the course prediction point. It means after the feature extraction of the learning behavior data before the prediction point, we can judge whether a learner will get a score at the end of the course. This model uses SVM method and the extracted feature data mentioned above. Previous studies have shown that the proportion of learners who can get a score is low, at 5%, showing a positive and negative imbalance in the data. Therefore, AUC and ROC are used to evaluate the model. The classifier with larger AUC value is more effective. It is generally believed that when AUC is greater than 0.9, the classifier has better classification effect. The results are shown below (Figs. 5, 6, 7, 8, and 9).

The time when the AUC reaches 0.9 is inconstant in different courses. In fact, after the prediction point, the model tends to be stable, and the AUC value does not increase significantly. The results of 5-fold cross-validation also show that the model is relatively stable and efficient to predict whether a learner can get a score at the prediction point. The importance of the prediction point is reflected in two ways. On the one hand, the prediction point reflects the number of weeks

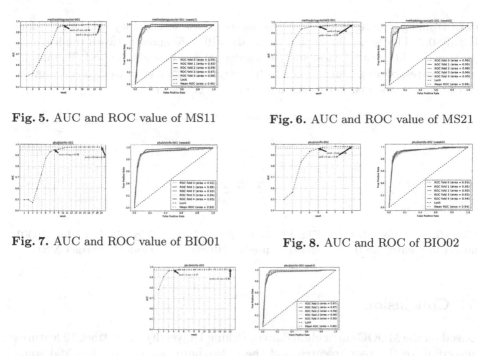

Fig. 5. AUC and ROC value of MS11

Fig. 6. AUC and ROC value of MS21

Fig. 7. AUC and ROC value of BIO01

Fig. 8. AUC and ROC of BIO02

Fig. 9. AUC and ROC of BIO03

of learning behavior data needed to classify the classifier. The results show that most of the prediction points are in the middle of the course, indicating that whether a learner can get a score has been determined in the middle of the course. On the other hand, it shows that the course can distinguish the different learners mainly through the first half of the learning stages. The classifier functions better in prediction since the middle of the course.

Further, we discuss whether the classifier can perform well in different courses (Fig. 10). In the model of serial courses (MS11 and MS21), the prediction effect is great with AUC = 0.96. This shows that in the serial courses, the classifier can be generalized on account of the similar learning pattern. But it still needs a certain length of learning week data to predict. In the case of using data of only 2 weeks, the AUC decreased to 0.8, which means the predictive ability decreases.

In the case of the same course (Fig. 11), data from bioinformatics-001 and bioinformatics-003 were used for model test; The AUC is 0.97, which means the prediction effect is great. It indicates that in the same course, the learning pattern is fixed. The classifier can easily identify whether a learner can get a score at the end of the course. Likewise, with a shorter length of data, the predictive power decreases.

In different courses (Fig. 12), with the use of a longer length of learning weeks' data, the AUC value is relatively high, indicating that even in the MOOCs learning, whether a learner can adhere to the medium term basically determines whether a learner can get a score at the end. While with short length of learning data training, the result of the performance is not satisfactory. The AUC value is only 0.7. The result shows that there are still differences between the learning patterns of different courses, and the model cannot be easily reused. The model of the corresponding course needs to be retrained. This is related to the content of the course, the instructor, the nature of the course, etc.

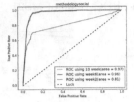

Fig. 10. ROC for MS11 model to predict MS21

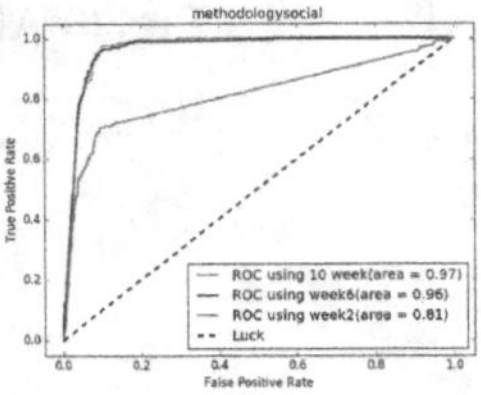

Fig. 11. ROC for BIO01 model to predict BIO03

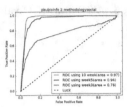

Fig. 12. ROC for BIO01 model to predict MS11

5   Conclusion

Based on the MOOC curriculum data of Peking University, we extract 19 features according to the past research and the data characteristic. We first did some basic calculation about the dropout students and summarized the characteristics of leaners' starting point and dropout point. We find that the MOOCs course dropout is high, and the learners who start later are more likely to drop out of school. Then, two predictive models are constructed. Model 1 focuses on the overall learner dropout of the course, while model 2 focuses on predicting whether a learner will get a score. The results of Model 1 show that the machine learning and sliding window model can predict the loss of students with high accuracy, and can help teachers track the curriculum, predict the loss of students, grasp the progress of the class. Model 2 results show that the classifier for predicting whether a learner can get a score achieve high accuracy, and whether a learner can adhere to the middle of the course basically determines their probability to get a score. Predictive results of both models can help designers and faculty adjust curriculums from quick feedback to reduce dropout rates.

In the future, we will focus on the similarities and differences of the dropout patterns among different courses and analyze the similarities and differences of learners' dropping motivations among different courses (even in different universities). Then we will do further experiments on a larger dataset, improve the model and give different classes different feedback advice to assist teachers to design their own courses better and to help learners more effective in learning, thus reduce the dropout rate.

Acknowledgments. This paper is supported by Peking University Education Foundation (2015ZD05) and National Natural Science Foundation of China (61472013).

References

1. Amnueypornsakul, B., Bhat, S., Chinprutthiwong, P.: Predicting attrition along the way: the UIUC model. In: Proceedings of the EMNLP 2014 Workshop on Analysis of Large Scale Social Interaction in MOOCs, pp. 55–59 (2014)
2. Bienkowski, M., Feng, M., Means, B.: Enhancing teaching and learning through educational data mining and learning analytics: an issue brief. US Department of Education, Office of Educational Technology, pp. 1–57 (2012)
3. Elouazizi, N.: Point-of-view mining and cognitive presence in MOOCs: a (computational) linguistics perspective. In: EMNLP 2014, p. 32 (2014)
4. Guo, P.J., Reinecke, K.: Demographic differences in how students navigate through MOOCs. In: Proceedings of the first ACM Conference on Learning@ Scale Conference, pp. 21–30. ACM (2014)
5. Jordan, K.: MOOC completion rates: the data (2013). http://www.katyjordan.com/MOOCproject.html. Accessed 27 Aug 2014
6. Kloft, M., Stiehler, F., Zheng, Z., Pinkwart, N.: Predicting MOOC dropout over weeks using machine learning methods. In: Proceedings of the EMNLP 2014 Workshop on Analysis of Large Scale Social Interaction in MOOCs, pp. 60–65 (2014)
7. Mi, F.: Machine learning models for some learning analytics issues in massive open online courses. Ph.D. thesis, The Hong Kong University of Science and Technology (2015)
8. Mining, T.E.D.: Enhancing teaching and learning through educational data mining and learning analytics: an issue brief. In: Proceedings of Conference on Advanced Technology for Education (2012)
9. Moon, S., Potdar, S., Martin, L.: Identifying student leaders from MOOC discussion forums through language influence. In: Proceedings of the 2014 Conference on Empirical Methods in Natural Language Processing (EMNLP), pp. 15–20 (2014)
10. Rosé, C.P., Siemens, G.: Shared task on prediction of dropout over time in massively open online courses. In: Proceedings of EMNLP, vol. 14, p. 39 (2014)
11. Sharkey, M., Sanders, R.: A process for predicting MOOC attrition. In: Proceedings of the EMNLP 2014 Workshop on Analysis of Large Scale Social Interaction in MOOCs. pp. 50–54 (2014)
12. Sinha, T., Jermann, P., Li, N., Dillenbourg, P.: Your click decides your fate: inferring information processing and attrition behavior from MOOC video clickstream interactions. arXiv preprint (2014). arXiv:1407.7131
13. Taylor, C., Veeramachaneni, K., O'Reilly, U.M.: Likely to stop? Predicting stopout in massive open online courses. arXiv preprint (2014). arXiv:1408.3382
14. Yang, D., Sinha, T., Adamson, D., Rosé, C.P.: Turn on, tune in, drop out: anticipating student dropouts in massive open online courses. In: Proceedings of the 2013 NIPS Data-Driven Education Workshop, vol. 11, p. 14 (2013)
15. Yang, D., Wen, M., Rose, C.: Towards identifying the resolvability of threads in MOOCs. In: EMNLP 2014, p. 21 (2014)

Exploring N-gram Features in Clickstream Data for MOOC Learning Achievement Prediction

Xiao Li[1(✉)], Ting Wang[2], and Huaimin Wang[3]

[1] Information Center, National University of Defense Technology, Changsha, China
xiaoli@nudt.edu.cn
[2] College of Computer,
National University of Defense Technology, Changsha, China
tingwang@nudt.edu.cn
[3] National Key Laboratory of Parallel and Distributed Processing,
National University of Defense Technology, Changsha, China
whm_w@163.com

Abstract. MOOC is an emerging online educational model in recent years. With the development of big data technology, a huge amount of learning behavior data can be mined by MOOC platforms. Mining learners' past clickstream data to predict their future learning achievement by machine learning technology has become a hot research topic recently. Previous methods only consider the static counting-based features and ignore the correlative, temporal and fragmented nature of MOOC learning behavior, and thus have the limitation in interpretability and prediction accuracy. In this paper, we explore the effectiveness of N-gram features in clickstream data and model the MOOC learning achievement prediction problem as a multiclass classification task which classifies learners into four achievement levels. With extensive experiments on four real-world MOOC datasets, we empirically demonstrate that our methods outperform the state-of-the-art methods significantly.

Keywords: MOOC · Educational data mining · Learning analytics · Machine learning

1 Introduction

MOOC (Massive Open Online Course) is an emerging online educational model in recent years. In a typical MOOC learning scenario, a learner watches freely delivered online videos with supplemented in-video quizzes (or assignments) and interacts with MOOC cohorts and instructors through online forums. Due to well-designed course materials, rich online interaction and personalized learning management, MOOC has gained increasing popularity across the world. Since 2012, millions of people have enrolled in the main stream MOOC platforms such as Coursera[1], Edx[2] and Udacity[3].

---

[1] https://www.coursera.org.
[2] https://www.edx.org.
[3] https://www.udacity.com.

© Springer International Publishing AG 2017
Z. Bao et al. (Eds.): DASFAA 2017 Workshops, LNCS 10179, pp. 328–339, 2017.
DOI: 10.1007/978-3-319-55705-2_26

In the era of big data, a huge amount of learning behavior data can be seamlessly recorded, analyzed and mined by MOOC platforms. This not only brings great challenges of educational big data management but also brings research opportunities into the community of educational data mining and learning analytics.

Recently, analyzing massive MOOC clickstream data and predicting learners' future learning achievement by machine learning technology has become a hot research topic in educational data mining and learning analytics research [12]. The MOOC learning achievement prediction problem is that given a learner's past clickstream dataset in a MOOC course, we would like to predict her or his academic achievement (e.g., certification, course grades, assignment marks) in the future.

Previous work can be summarized in three different views. From the view of what learning objectives they achieve, these work can be categorized into certification prediction [1–9, 13, 18, 19], assignment mark prediction [4, 17] and forum thread reading prediction [14, 15]. From the view of what learning algorithms they use, these work can be categorized into single statistic machine learning models [1–3, 5–9, 13, 14, 17–19], graphic models [4] and matrix factorization methods [15]. From the view of what learning features they preprocess, these work can be categorized into static counting-based features [1–4, 7, 13, 17–19], natural language features [6, 8, 9, 14, 15], behavior sequence features [5] and social network inspired features [15, 21].

However, there still exists some limitations in the state-of-the-art research of MOOC learning achievement prediction problem. A major drawback of previous work is the underestimation of the importance of preprocessing learning behavior features. Most of previous work still preprocess clickstream datasets into static counting-based features, such as frequency of video watching, forum thread reading and quiz submissions. Such learning features ignore the correlative, temporal and fragmented nature of MOOC learning behavior.

For example, consider two learners with different MOOC learning habits. They study a MOOC chapter which contains three videos each of which has one in-video quiz (i.e., the material sequence is "V_1-Q_1-V_2-Q_2-V_3-Q_3"). One learner consumes these course materials sequentially and thus generates an action sequence as "V-Q-V-Q-V-Q" while the other learner watches all the three videos first and then submit all quizzes later (i.e., the action sequence is "V-V-V-Q-Q-Q"). For this case, the static counting-based feature methods will represent the two learners identically (i.e., both learners watch three videos and submit three quizzes) while ignore their inherent learning behavior differences. Besides, when we use static counting-based features, it is unclear how the transition of learning actions and action motifs affect the prediction of learning achievement. Thus, the predictive models built based on static counting-based feature methods have the limitations in both prediction accuracy and interpretability for predicted results.

Recently, researchers have found that in educational data mining, data preprocessing should be considered with more attention than learning algorithms [10]. Indeed, learning features with powerful representation ability can not only improve the accuracy of learning algorithms but also help MOOC instructors understand the learned predictive models better. Thus, it is important to conduct research on the preprocessing and representation of learning features for MOOC achievement prediction problem.

In this paper, we study the problem of MOOC learning achievement prediction with a focus on the preprocessing methods of learning behavior features. We explore the effectiveness of N-gram features in clickstream data for MOOC learning achievement prediction problem. Data visualization and clustering methods are utilized to evaluate the representation ability of the proposed features over the traditional static counting-based features. Finally, we formalize the MOOC learning achievement prediction problem as a multiclass classification task which aims at learning a highly accurate classification model to classify learners into four achievement levels. With extensive experiments on four real-world MOOC datasets, we demonstrate that our methods are highly effective and outperform the state-of-the-art methods significantly.

The rest of this paper is organized as follows. Section 2 describes the statistics of datasets. Section 3 studies the N-gram learning behavior features and analyzes their representation ability. Section 4 conducts experiments and discusses results. The last section presents the conclusion and future work.

2   Dataset Description

In this paper, we use clickstream data from four MOOC courses offered by Mengke MOOC platform hosted at National University of Defense Technology. Similar to the main stream MOOC platforms, the content of each MOOC course in Mengke is also organized into sequential chapters opening week by week. Each chapter contains a list of short videos with embedded quizzes and supplemented references.

To help understand the MOOC learning achievement problem better, it is helpful to describe the grade and certification policy in Mengke. Specifically, in order to receive a certification in a MOOC course, a learner has to obtain a minimal course grade of 60 out of 100. The course grade is composed of two parts. One part is the progress mark (40%) which is calculated based on a learner's learning progress of video watching, performance on quiz attempt and active rate of forum discussion. The other one is the final online examination mark (60%). Both marks are automatically evaluated by Mengke. In addition, it should be noted that if the progress mark of a learner is lower than 30%, the learner will not be allowed to attend the final online examination[4].

For the four MOOC course datasets, Table 1 summarizes their course IDs, titles, opening time, number of enrollment, certification rate and number of logs. The average certification rate of the four courses is 11.6%. The margin between the highest certification rate and the lowest rate is about 20%. The lowest certification rate of C4 is due to the reason that this course contains too many software instruction videos and quizzes which are difficult to finish.

We present an example of clickstream data from a MOOC course in Table 2. It describes a typical MOOC learning session containing a session begin mark, a series of actions and a session end mark.

---

[4] The minimal progress marks for attending final examinations are usually set in the range from 20% to 30% in Mengke. They can be changed by instructors before courses are opened.

Table 1. Statistics of MOOC course datasets.

ID	Title	Opening time	Enrollment	Certification rate	Logs
C1	Network security	2013 fall	3,990	9.7%	389,681
C2	Electromagnetic fields and electromagnetic waves	2013 fall	3,334	23.4%	323,568
C3	Physical education	2014 fall	9,639	11.8%	290,829
C4	Multimedia design	2014 spring	15,933	1.5%	243,643

Table 2. MOOC clickstream data in a learning session.

Timestamp	Action
2016-12-16 12:18:21	Open the MOOC app (session begin)
2016-12-16 12:20:30	Watch a video
2016-12-16 12:23:13	Watch a video
2016-12-16 12:30:11	Watch a video
2016-12-16 12:35:08	Submit a quiz
2016-12-16 12:40:39	Reply a forum thread
2016-12-16 12:45:30	Watch a video
2016-12-16 12:50:32	Switch to another app (session end)

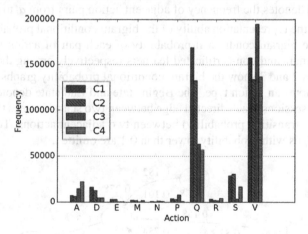

Fig. 1. Histogram of 10 action types in each course.

In this paper, we identify 10 types of major action types in the clickstream datasets. They are course navigation (A), reading discussion posts (D), replying posts (E), homepage setup (M), recording notes (N), posting discussion threads (P), submitting quizzes (Q), reading references (R), certification management (S) and watching videos (V). The histogram of the 10 action types is plotted in Fig. 1. We can see that the actions of watching videos (V) and submitting quizzes (Q) form two peaks compared to the other actions. This is because in the four courses learners have to finish a certain amount of videos and quizzes to obtain a minimal progress mark of 30% in order to be allowed to attend the final online examinations.

3   Exploring N-gram Features in Clickstream Data

In this section, we study two new learning behavior features for MOOC learning achievement prediction problem, namely the bigram conditional probability features and the N-gram action subsequence features.

3.1   Bigram Conditional Probability Features

For an action sequence $\left(a_1^{(i)}, a_2^{(i)}, \cdots, a_n^{(i)}\right) a_j \in A$ in a learning session S_i where A is the set of all the action types, we assume that each action is only dependent on its previous adjacent action. Thus, the action sequence has the Markov property. Based on this, we can estimate the bigram conditional probability between any pairs of action types. Specifically, the bigram conditional probability of action a_j given a_i can be estimated as

$$P(a_j|a_i) = N(a_i, a_j) / \sum_{a_k \in A} N(a_i, a_k) \tag{1}$$

where $N(a_i, a_j)$ denotes the frequency of adjacent action pairs from a_i to a_j.

To validate the representation ability of the bigram conditional probability features, we compute the bigram conditional probability of each pair of action types for both certificated learners and non-certificated learners respectively among the four MOOC courses. Figures 2 and 3 show the bigram conditional probability graphs. Each state in the Figures denotes an action type. The Begin state and End state denote the start and end of learning sessions. Each directed edge between two states denotes the conditional probability (i.e., transition probability) between two adjacent actions. To simplify our analysis, the edges with probability lower than 0.1 are omitted.

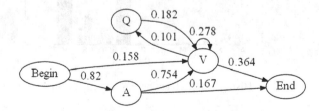

Fig. 2. Bigram conditional probability of action transition for non-certificated learners.

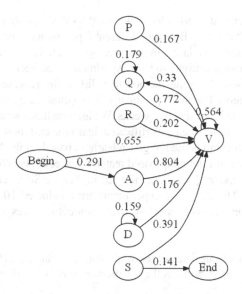

Fig. 3. Bigram conditional probability of action transition for certificated learners.

Comparing Figs. 2 and 3, we can find three differences between certificated and non-certificated learners. First, the certificated learners have more types of learning actions. Besides the common actions types (i.e., V and Q) appeared in both figures, the certificated learners have higher usage in posting discussion threads (P), reading references (R) and reading discussion posts (D). This means that the certificated learners are more likely to join discussion in online forums than the non-certificated learners. Second, the certificated learners have higher probability to rewatch videos (loops on V), resubmit quizzes (loops on Q) and switch their learning actions between V and Q. This indicates the higher active rate and stronger learning determination of certificated learners than non-certificated learners. Third, the certificated learners usually close their learning sessions after checking the certification status (S) while the non-certificated learners close sessions immediately after watching videos (V) and browsing course pages (A). This could be interpreted as evidence of different motivation in pursing certifications. From this analysis, we can see that the bigram conditional probability features can reveal important behavior difference between certificated and non-certificated learners. In static counting-based features, it may be difficult to reveal such differences of behavior regularity.

3.2   N-gram Action Subsequence Features

For an action sequence $(a_1^{(i)}, a_2^{(i)}, \cdots, a_n^{(i)})$ in a learning session S_i, we can also extract subsequence with length k and count their occurrence. For example, the 3-gram subsequence patterns derived in Table 2 include "Begin-V-V", "V-V-V", "V-V-Q", "V-Q-V", "Q-V-P" and "V-P-End". This is usually called N-gram method in data mining community [11]. The statistic distribution of N-gram subsequences can represent unique behavior motifs for different groups of MOOC learners [16]. We can interpret behavior motifs as special learning behavior patterns such as video watchers ("V-V-V"), quiz harvesters ("Q-Q-Q") and forum spammers ("P-P-P"). By using different k parameters, we can capture learners' studying habit in different granularity. In practice, we need to find optimal k parameter. We will discuss the parameter sensitivity in later experiments (see Sect. 4.3).

We use unsupervised clustering method to validate the representation ability of the N-gram action subsequence features. Specifically, in our clustering experiment, we use each learner's clickstream data in the first two weeks. We assume that learners in each course can be grouped into two clusters, e.g., certificated learners and non-certificated learners. K-means clustering algorithm is used to group learners based on the N-gram action subsequence features and the static counting-based features respectively. We measure the quality of clusters by Silhouette score [20]. Generally, the higher the Silhouette score is, the better the clusters quality is. The clustering experiments are conducted 10 times in each course with different initial centers. We report average Silhouette scores and t-test significance level in Table 3.

Table 3. Comparison of Silhouette score between the static counting-based features and the N-gram action subsequence features. *** means the significance level is less than 0.01 and – means no difference.

Course	Static	N-gram	t-test significance	Win/tie/lose
C1	0.779	0.804	***	Win
C2	0.730	0.776	***	Win
C3	0.836	0.817	***	Lose
C4	0.955	0.953	–	Tie

We can see that the N-gram action subsequence feature method wins in two cases, loses in one case and ties in one case over the static counting-based feature method. This demonstrates that in most cases the N-gram action subsequence feature method performs better than the static counting-based feature method in discriminating different MOOC learners.

4   Prediction Experiments and Discussion

In this section, we conduct prediction experiments to evaluate the generalization performance of learning algorithms based on the two proposed N-gram behavior features.

4.1   Experiment Configuration

Different from the previous works that view the MOOC learning achievement prediction problem as a binary classification task (e.g., classifying learners into certificated or non-certificated), in this paper, we model the prediction problem as a multiclass classification task. This is much more difficult than previous work as we are facing more serious class imbalance problem. Specifically, we set up four achievement levels based on the final course grades of learners, i.e., level A (90%–100%), level B (60%–89%), level C (30%–

59%) and level D (0–29%). It should be noted that we separate the failed level (i.e., grades lower than 60%) into level C and level D. Actually, the level C represents students who attend final examination but fail while the level D represents students not attending the final examination because their progress marks are lower than 30% (see grade policy in Sect. 2).

We develop four types of learning features in our experiments

- Base, the traditional static counting-based features.
- Bigram, the bigram conditional probability features.
- N-gram, the N-gram action subsequence features.
- Hybrid, combining the Bigram and the N-gram features.

We use the *one-vs-rest* strategy to build the multiclass classifier. In this strategy, a binary classifier is built to distinguish one class from the rest. For a multiclass classification task with k classes, k one-vs-rest binary classifies will be built. The final predicted class is estimated by the maximum a posteriori rule

$$\arg \max_{c \in \{A,B,C,D\}} P(c|x) \tag{2}$$

The basic binary classifier is implemented by logistic regression because of its good generalization ability and efficiency in training and predicting. It should be noted that our features are not dependent on specific classifiers. Other classic binary classifiers such as decision tree, SVM, neural network can also be used.

We use Macro F1-score to compare the prediction performance of the multi-class classifiers trained by the four different learning features. When conducting experiments in a course, we use 10-fold cross validation method to split the dataset into a training set (with nine folds) and a testing set (with one fold) and repeat experiments in 10 times. We report the average Macro F1-score in testing sets.

4.2   Experiment Results

In MOOCs, the knowledge and skills learned by learners is gradually improved as their learning progress advances. Thus, the learning component in MOOC platforms should output prediction dynamically and temporally. To implement this idea, we train our classifiers in the first t ($1 \le t \le 30$) days and make prediction for the future achievement of learners. This mimics the real-world online prediction scenarios where learners receive system prediction or recommendation as their learning progresses advance. We set N-gram parameter k as 3 in this experiment.

Figure 4 shows Macro F1-score curves along 30 learning days. First, comparing overall performance, we find that the performance curves of the Hybrid, N-gram and Bigram methods proposed in this paper dominate the curves of the Base method significantly along all days. This clearly demonstrates that the Bigram and N-gram features as well as their hybrid outperform the traditional static counting-based feature method significantly. Second, we analyze performance curves among the Bigram, N-gram and Hybrid methods. We find that although the N-gram and Bigram methods win each other in certain cases they could also lose or tie with others. However, when we combine the two methods together as

Hybrid, it performs the best among the three feature methods. This suggests that in practice we should combine the N-gram and Bigram methods together. Third, for the curves of hybrid method, after two weeks of rapid increasing, their growth rates become slow. This means that by using just two weeks of clickstream data from learners, the performance of the Hybrid method is already good enough. From the view of instructors, this could be very positive as they can foresee a learner's future achievement in an early learning period. Thus, proper guide and intervention can be projected into a learner's learning path in time based on the predicted academic achievement.

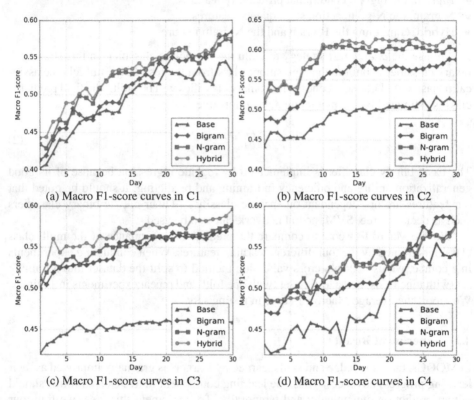

Fig. 4. Macro F1-score curves in different courses.

4.3 Parameter Sensitivity Discussion

In this subsection, we empirically study the impact of parameter k of N-gram feature method to the prediction performance. We use two weeks of clickstream data in this experiment as this setting is suggested in the previous experiments. We choose k parameter from the set $\{1, 2, 3, 4, 5, 6, 7, 8, 9, 10\}$ and conduct 10-fold cross validation experiments at each value of k. Besides the Macro F1-score metric, we also report the Macro Precision and Macro Recall in Fig. 5.

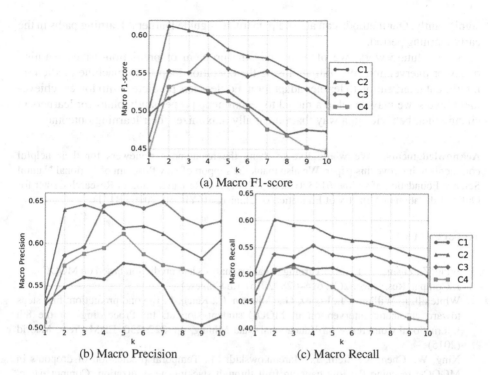

(a) Macro F1-score

(b) Macro Precision

(c) Macro Recall

Fig. 5. Macro F1-score, Macro Precision and Macro Recall at different k for N-gram.

We first analyze the performance curves of Macro F1-score in Fig. 5a. We find that almost all the curves increase sharply until $k = 3$ or $k = 4$. When the value of k is larger than 4, their growth trend is dramatically changed. The performance gets worse as the value of k becomes large. By analyzing the curves of Macro Precision and Macro Recall in Fig. 5b and c, we can find that when the value of k is larger than 4, both Macro Precision curves and Macro Recall curves decrease. This is the main reason why Macro F1-score drops so quickly when k is larger than 4. This study suggests that in practice we may choose k as 3 or 4 for the N-gram action subsequence feature method.

5   Conclusion and Future Work

In this paper, we study the problem of MOOC learning achievement prediction with a focus on the preprocessing method of learning behavior features. We explore the effectiveness of N-gram features in clickstream data for MOOC learning achievement prediction problem by studying the Bigram conditional probability feature method and the N-gram action subsequence feature method. We formalize the MOOC learning achievement prediction problem as a multiclass classification task which aims at learning a highly accurate classification model to classify learners into four achievement levels. With extensive experiments on four real-world MOOC datasets, we demonstrate that our methods are highly effective and outperform the state-of-the-art methods significantly. Our methods can also help instructors guide learners' learning paths in the early learning period.

In our future work, we plan to study the problem of prediction-driven learning behavior intervention. We aim at closing the operation loop from machine prediction, intelligent intervention to learner adaptation. For learners predicted with lower achievement levels, we may develop a model to recommend proper study plans for learners to change their behavior in a way that potentially maximizes their learning outcome.

Acknowledgments. We would like to thank the anonymous reviewers for their helpful comments to improve this paper. We also thank the support of Key Program of National Natural Science Foundation of China (61432020, 61532001) and Research Fund of Research Center for Online Education of Ministry of Education of China (2016YB150, 2016YB151).

References

1. Jiang, Z., Zhang, Y., Li, X.: Learning behavior analysis and prediction based on MOOC data. J. Comput. Res. Dev. **52**(3), 614–628 (2015). (in Chinese)
2. Whitehill, J., Williams, J., Lopez, G., Coleman, C., Reich, J.: Beyond prediction: first steps toward automatic intervention in MOOC student stopout. In: Proceedings of the 8th International Conference on Educational Data Mining, pp. 222–230. EDM Press, Madrid (2015)
3. Xing, W., Chen, X., Stein, J., Marcinkowskid, M.: Temporal predication of dropouts in MOOCs: reaching the low hanging fruit through stacking generalization. Comput. Hum. Behav. **58**, 119–129 (2016)
4. Qiu, J., Tang, J., Liu, T., Gong, J., Zhang, C., Zhang, Q., Xue, Y.: Modeling and predicting learning behavior in MOOCs. In: Proceedings of the 9th ACM International Conference on Web Search and Data Mining, pp. 93–102. ACM Press, San Francisco (2015)
5. Brooks, C., Thompson, C., Teasley, S.: A time series interaction analysis method for building predictive models of learners using log data. In: Proceedings of the 5th International Conference on Learning Analytics and Knowledge, pp. 126–135. ACM Press, Poughkeepsie (2015)
6. Robinson, C., Yeomans, M., Reich, J., Hulleman, C., Gehlbach, H.: Forecasting student achievement in MOOCs with natural language processing. In: Proceedings of the 6th International Conference on Learning Analytics and Knowledge, pp. 383–387. ACM Press, Edinburgh (2016)
7. He, J., Bailey, J., Rubinstein, I., Zhang, R.: Identifying at-risk students in massive open online courses. In: Proceedings of the 29th AAAI Conference on Artificial Intelligence, pp. 1749–1755. AAAI Press, Austin (2015)
8. Chaplot, D., Rhim, E., Kim, J.: Predicting student attrition in moocs using sentiment analysis and neural networks. In: Proceedings of AIED 2015 Fourth Workshop on Intelligent Support for Learning in Groups, pp. 7–12. AIED Press, Madrid (2015)
9. Crossley, S., Paquette, L., Dascalu, M., McNamara, D., Baker, R.: Combining click-stream data with NLP tools to better understand MOOC completion. In: Proceedings of the 6th International Conference on Learning Analytics and Knowledge, pp. 6–14. ACM, Edinburgh (2016)
10. Zhou, Q., Mou, C., Yang, D.: Research Progress on Educational Data Mining A Survey. J. Softw. **26**(11), 3026–3042 (2015). (in Chinese)

11. Xing, Z., Pei, J., Keogh, E.: A brief survey on sequence classification. SIGKDD Explor. **12**, 40–48 (2010)
12. Romero, C., Ventura, S.: Educational data mining: a review of the state of the art. IEEE Trans. Syst. Man Cybern. Part C Appl. Rev. **40**(6), 601–618 (2010)
13. Jiang, S., Williams, A., Schenke, K., Warschauer, M., Dowd, D.: Predicting MOOC performance with week 1 behavior. In: Proceedings of the 7th International Conference on Educational Data Mining, pp. 273–275. EDM Press, London (2014)
14. Kumar, M., Kan, M., Tan, B.: Learning instructor intervention from MOOC forums: early results and issues. In: Proceedings of the 8th International Conference on Educational Data Mining, pp. 218–225. EDM Press, Madrid (2015)
15. Yang, D., Piergallini, M., Howley, I., Rose, C.: Forum thread recommendation for massive open online courses. In: Proceedings of the 7th International Conference on Educational Data Mining, pp. 257–260. EDM Press, London (2014)
16. Davis, D., Chen, G., Hauff, C., Houben, G.: Gauging MOOC learners' adherence to the designed learning path. In: Proceedings of the 9th International Conference on Educational Data Mining, pp. 54–61. EDM Press, Raleigh (2016)
17. Ren, Z., Rangwala, H., Johri, A.: Predicting performance on MOOC assessments using multi-regression models. In: Proceedings of the 9th International Conference on Educational Data Mining, pp. 484–489. EDM Press, Raleigh (2016)
18. Kennedy, G., Coffrin, C., Barba, P., Corrin, L.: Predicting success how learners' prior knowledge, skills and activities predict MOOC performance. In: Proceedings of the 5th International Conference on Learning Analytics and Knowledge, pp. 136–140. ACM Press, Poughkeepsie (2015)
19. Sanchez-Santillan, M., Cerezo, R., Paule-Ruiz, M., Nuñez, J.: Predicting students' performance: incremental interaction classifiers. In: Proceedings of the Third ACM Conference on Learning @ Scale, pp. 217–220. ACM Press, Edinburgh (2016)
20. Peter, J.: Silhouettes: a graphical aid to the interpretation and validation of cluster analysis. Comput. Appl. Math. **20**, 53–65 (1987)
21. Tong, Y., She, J., Meng, R.: Bottleneck-aware arrangement over event-based social networks: the max-min approach. World Wide Web J. **19**(6), 1151–1177 (2016)

Predicting Student Examinee Rate in Massive Open Online Courses

Wei Lu[1,2], Tongtong Wang[1,2], Min Jiao[1,2], Xiaoying Zhang[1,2], Shan Wang[1,2], Xiaoyong Du[1,2], and Hong Chen[1,2(✉)]

[1] Key Laboratory of Data Engineering and Knowledge Engineering, MOE, Renmin University of China, Beijing, China
{lu-wei,wttrucer,shingle,xyzruc,swang,duyong,chong}@ruc.edu.cn
[2] School of Information, Renmin University of China, Beijing, China

Abstract. Over the past few years, massive open online courses (a.b.a MOOCs) has rapidly emerged and popularized as a new style of education paradigm. Despite various features and benefits offered by MOOCs, however, unlike traditional classroom-style education, students enrolled in MOOCs often show a wide variety of motivations, and only quite a small percentage of them participate in the final examinations. To figure out the underlying reasons, in this paper, we make two key contributions. First, we find that being an examinee for a learner is almost a necessary condition of earning a certificate and hence investigation of the examinee rate prediction is of great importance. Second, after conducting extensive investigation of participants' operation behaviours, we carefully select a set of features that are closely reflect participants' learning behaviours. We apply existing commonly used classifiers over three online courses, generously provided by China University MOOC platform, to evaluate the effectiveness of the used features. Based on our experiments, we find there does not exist a single classifier that is able to dominate others in all cases, and in many cases, SVN performs the best.

Keywords: MOOC · Machine learning methods · Examinee rate · Prediction

1 Introduction

MOOCs target at unlimited participation of learners and provide an open access to the courses via the web. Since first introduced in 2008, MOOCs have rapidly emerged as a popular online learning paradigm and a wide variety of MOOC providers, including Coursera [3], Udacity [6], edX [5], XuetangX [7], China University MOOC [2], offer high-quality courses from the world's best universities, institutions and enterprises to learners everywhere.

MOOCs are proposed as an important complement but rather than a replacement to the traditional classroom-style education. From the perspective of learners, MOOCs not only provide traditional course materials lectures notes, readings, question-and-answer drills, and quizzes, but also provide videos, and

Z. Bao et al. (Eds.): DASFAA 2017 Workshops, LNCS 10179, pp. 340–351, 2017.
DOI: 10.1007/978-3-319-55705-2_27

online/offiline discussion forum zones to support community interactions among students, lecturers, and teaching assistants [22,23,25–27]. More importantly, many MOOC providers offer certificates or statements of completion as long as students have successfully completed an online course, and in some cases, a verified certificate can help provide an important reference for job hunting and further education exploring.

Despite various benefits provided by MOOCs, as observed in the majority of current MOOC providers [8,11,18,28], one of the critical issues related to MOOCs is their low completion rate. That is, only quite a small percentage of learners finally earn the certificates. A recent research conducted over 29 online courses in Coursera shows that the averaged completion rate for the majority of MOOCs is less than 7% [1]. The reason of incurred low completion rate is that, unlike traditional classroom-style education, students enrolled in MOOCs often show a wide variety of motivations, and not all the students are well motivated to earn certificates. In this paper, we find that, although the final examination score only occupies 20% of total mark, being a examinee (participate in the final examination) is almost a necessary condition for a student to earn a certificate. Hence, to raise the completion rate for the learners of earning the certificates and help students improve the learning performance, it is necessary for us to analyze behaviours of the examinees by raising the examinee rate.

Predicting the examinee rate by analyzing the learners' behaviours is challenging. First, the motivations of students engaging in an online course are quite diverse. A student who does not engage in the final examination could either indicate poor learning attitudes, or give up the attendance of the final examination but with positive learning attitudes, or forget the attendance of the final examination due to some unknown reason, or mean that he or she already knows the knowledge point well and is not well motivated to earned a certificate. Due to the above reasons, the data for examinee rate predication is quite noisy. Second, MOOC providers maintain various fact table data and behavioural data, but only a very small fraction of them might be relevant to examinee rate predication. Consequently, the input data of predicting the examinee rate could be sparse. Third, there exist a wide spectrum of machine learning techniques, including naive Bayes classifier, decision tree, logistic regression [13], support vector machine [10], etc., among which selecting the best one that could accurately predict whether a student is an examinee is required to study.

Collaborating with China Higher Education Press[1], we open up an online course named as "Introduction to Database Systems"[2] in the China University MOOC platform and attract more than 20,000 students to engage in our course. As a parter, we are provided by the platform with students' behaviour data of three online courses. As the privacy concerns, we omit their specific names. In this paper, we take these online courses as the object of study, and explore the analysis over the examinees' behaviour data. Our contributions are listed below:

– We find that being an examinee for a learner is almost a necessary condition of earning a certificate and hence, we investigate the problem of examinee

---

[1] http://www.hep.edu.cn/.
[2] http://www.icourse163.org/course/RUC-488001#/info.

rate prediction, i.e., verifying whether a student will participate in the final examination of the course. To the best of our knowledge, we are the first group to investigate the examinee rate prediction in this research filed. Our research is of great importance to improve the learning performance and provide references for the teaching effectiveness.

– Collaborating with and generously supported by China University MOOC platform, we conduct extensive investigation of participants' operation behaviours, and carefully select a set of features that are closely reflect participants' learning behaviours. We apply existing commonly used classifiers over three online courses to evaluate the effectiveness of the used features. The experimental results show that there does not exist a single classifier dominating others in all cases, and in many cases, SVN performs the best.

The rest of the paper is organized as follows. Section 2 discusses the related work. Section 3 presents the feature selection. Section 4 briefly introduces the machine learning methods used as the comparison. Section 5 reports the experimental results and Sect. 6 concludes the paper.

2   Related Work

MOOCs have open up a new era of education and attracted a great deal of research interest in MOOC data analysis. Some top conferences launch special workshops or competitions for MOOC data analysis, such as the first data driven education workshop together with the twenty-seventh annual conference on neural information processing systems (NIPS) [4] and the SIGKDD Cup 2015 Workshop for MOOC dropout prediction [8]. For these works, many of them focus on collecting the statistics of the online courses and designing more effective assessment methodologies. For example, Nesterko et al. collect geographic data of analyzing 18 online courses in HarvardX [17]. As an indispensable link of the teach-learn-assess cycle in education, the assessment of students' performance is of great importance. For this reason, peer-assessment methods have been proposed in MOOCs to evaluate the works (especially for subjective questions) of students engaged in the courses [15, 20]. Recently, there is an increasing research interest in MOOC dropout prediction. A great deal of them focus on the prediction using contextual information like posts or click-stream data. Yang et al. try to model students' participation patterns by analyzing their posting behaviour in the discussion forum zones. This approach is restrictive in real applications because posts are just partial features of students' behaviour data. Similar work is proposed in [19] based on students' behaviours in discussion forum. Kloft, et al. target to model the relation between the dropout and the students' most active time. Kim et al. aim to model students' participation patterns by analyzing the students' video click-stream activities, such as skipping, zooming, pausing, playing. Another set of these research work aim to investigate more effective machine learning methods to model the dropout prediction. By extracting the features, [9,14] apply SVM, [24] applies logistic regression, and [21] applies decision tree to the extracted features to predict the active participation of a student. Besides using a single method, hybrid machine learning methods are used to predict the dropout [16]. Mi et al. take the drop prediction as the sequence labeling problem and hence apply the recurrent neural network (RNN) modeling method [12]. Although a large number of prior research has been conducted, the difference between our work and the state-of-the-art approaches is two-fold. First, we investigate the examinee rate prediction problem while they focus on the dropout prediction problem or others. Second, as we collaborate with the MOOC provider and they provide us with fairly complete student-behaviour data, in this way, we are able to capture more features of contextual information.

3   Feature Selection

China University MOOC platform provides us with three data sets, each of which involves in an individual online course. Without loss of generality, we refer to above three data sets as course A, course B, course C. We collect the statistics of used online courses which are shown in Table 1. We omit the description of these statistics as they are self-explained.

3.1   Preliminary Investigation

For ease of illustration, we list the definitions and symbols below for a student s in Table 2 that are used throughout the remainder of the paper.

We first study the necessity for a student to participate in the final examination in order to earn a certificate. We evaluate $P(E_1|E_2)$ and $P(E_2|E_1)$, which are formalized below.

$$P(E_1|E_2) = \frac{\text{Number of examinees with certificates}}{\text{Number of students with certificates}} \tag{1}$$

$$P(E_2|E_1) = \frac{\text{Number of examinees with certificates}}{\text{Number of examinees}} \tag{2}$$

Table 1. Statistics of three online courses

Attributes	Course A	Course B	Course C
Number of enrolled students	71,753	22,145	28,413
Number of students with certificates	371	176	7,790
Number of examinees	625	239	9,211
Number of students engaged in examination view	1,880	1,530	15,780
Number of students with quiz scores	2,855	753	7,413
Number of students engaged in quiz view	3,328	2,684	14,825
Number of students engaged in forum view	1,715	1,223	3,995
Number of students with posts in forum	1,074	304	1,530
Number of students engaged in page views	9,478	6,997	8,214

Table 2. Symbols and definitions

Event	Definition
E_1	Event that s participates the final examination
E_2	Event that s completes the course A (B or C) successfully with a certificate
E_3	Event that s views the examination page of the course A (B or C)
E_4	Event that s views the quiz page of the course A (B or C)
E_5	Event that s has quiz score of course A (B or C)
E_6	Event that s views a discussion forum page of the course A (B or C)
E_7	Event that s has a post in discussion forum zones of the course A (B or C)
E_8	Event that s views a video page of the course A (B or C)

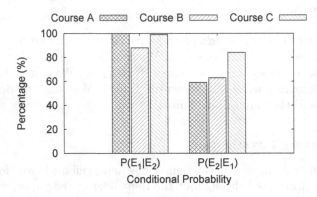

Fig. 1. Interrelationship between examinees and the students with certificates.

Figure 1 shows $P(E_1|E_2)$ and $P(E_2|E_1)$ for all used online courses. On one hand, as we can see from the values of $P(E_1|E_2)$, every student with the certificate almost participates in the final examination, i.e., although the final examination only occupies 20% of total mark, it almost becomes a necessary condition to earn the certificate. On the other hand, more than 59% of examinees in the end earn the certificates. Apparently, study of examinee rate prediction to improve the completion rate of in MOOCs is of great importance.

We then study $P(E_1|E_3)$, the conditional probability of E_1 given E_3 and present the formula below. We omit $P(E_3|E_1)$ that always equals to 1 since an examinee always view the examination page.

$$P(E_1|E_3) = \frac{\text{Number of examinees}}{\text{Number of students with examination page view}} \quad (3)$$

Figure 2 plots values of $P(E_1|E_3)$ for three used online courses. An interesting observation is that for Course C, $P(E_1|E_3)$ is 72%, i.e., 72% of the students who view the examination pages will participate in the final examination while that for Course A and Course B is just 22% and 15%. The reason of resulting in low $P(E_1|E_3)$, as we guess, is that the examination is too difficult to complete for students or there are too many questions in Course A and Course B.

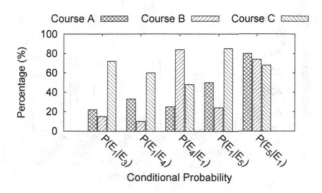

Fig. 2. Interrelationship between examinees and examination page view, quiz page view, quiz scores.

This observation could help the lecturer design the difficulty of the examination properly.

Similarly, we study $P(E_1|E_4)$, $P(E_1|E_5)$, $P(E_1|E_6)$, $P(E_1|E_7)$, $P(E_1|E_8)$ and $P(E_4|E_1)$, $P(E_5|E_1)$, $P(E_6|E_1)$, $P(E_7|E_1)$, $P(E_8|E_1)$, which are formalized from Eq. 4 to Eq. 13.

$$P(E_1|E_4) = \frac{\text{Number of examinees with quiz page view}}{\text{Number of students with quiz page view}} \qquad (4)$$

$$P(E_4|E_1) = \frac{\text{Number of examinees with quiz page view}}{\text{Number of examinees}} \qquad (5)$$

$$P(E_1|E_5) = \frac{\text{Number of examinees with quiz scores}}{\text{Number of students with quiz scores}} \qquad (6)$$

$$P(E_5|E_1) = \frac{\text{Number of examinees with quiz scores}}{\text{Number of examinees}} \qquad (7)$$

$$P(E_1|E_6) = \frac{\text{Number of examinees with forum page view}}{\text{Number of students with forum page view}} \qquad (8)$$

$$P(E_6|E_1) = \frac{\text{Number of examinees with forum page view}}{\text{Number of examinees}} \qquad (9)$$

$$P(E_1|E_7) = \frac{\text{Number of examinees with posts}}{\text{Number of students that have posts}} \qquad (10)$$

$$P(E_7|E_1) = \frac{\text{Number of examinees with posts}}{\text{Number of examinees}} \qquad (11)$$

$$P(E_1|E_8) = \frac{\text{Number of examinees with video page view}}{\text{Number of students with video page view}} \qquad (12)$$

$$P(E_8|E_1) = \frac{\text{Number of examinees with video page view}}{\text{Number of examinees}} \qquad (13)$$

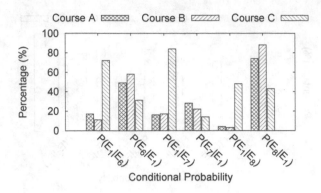

Fig. 3. Interrelationship between examinees and forum page view, number of posts, video page view.

Values of the above conditional probabilities engaged in three used online courses are plotted in Figs. 2 and 3. Interestingly, at most 60% of students (from Course C) who view the quiz pages finally participate in the final examination, and only up to 48% of examinees in Course A and C view the quiz pages. Although quiz scores show much higher correlated with the examinee likelihood, the conditional probability for Course A and Course B is still low. Similar observation of interrelationship is found between examinees and forum page view, number of posts, video page view. We omit the details in order to avoid repeatedly elaboration. To summarize, using individual features to predict whether a student is an examinee is far from satisfactory. Although examination view shows a fairly strong correlation with examinee, on one hand, it still leads to a large number of false positives; on the other hand, getting the behaviour of examination view is somehow difficult and one of our objectives in the paper is to motivate students, who do not view examination pages, to participate in the final examination. Therefore, we study of the problem of applying existing prediction models on a set of carefully selected features to predict the examinee rate.

3.2   Features of the Prediction Model

Following the above analysis, we use the features that are closely related to the examinee rate prediction in MOOCs and list them with details in Table 3. Our feature selection is conducted from three perspectives. First, from the perspective of diverse MOOC contents, our selected features (Feature 4, 5, 8–12) cover participants' behaviours over various course materials, including video, online discussion forum, quizzes, examinations. Second, feature 1, 2, 3, 6, 9 reflect the engagement of participants' behaviours from the perspective of learning attitude. Third, feature 12 captures the daily class performance from the perspective of learning ability.

Table 3. Used features of the prediction model

ID	Features	Description
1	Number of sessions	Session is a unit of measurement in web analytics, capturing either a user's actions within a particular time period, or a user's actions in completing a particular task. Session helps capture the time period from logging in to logging out the learning platform. A larger number of sessions reflects a higher engagement of user behaviours
2	Number of requests	Total number of requests including forum views, examination views, video views, etc.
3	Number of active days	We refer to a day on which the user has a request record as an active day
4	Number of video views	Total number of accessing the video page
5	Number of video views per session	Averaged number of accessing the video page per session
6	The time gap between video view and corresponding chapter release	Typically, online courses in MOOC platforms periodically release the chapters. A student can access the chapters at any time after they are released. We collect the averaged gap between the time of viewing the video page and the release time of the video. Apparently, a smaller time gap reflects a higher engagement and more active attitude of user behaviours
7	Number of forum views	Total number of accessing the pages in the course discussion forum zone
8	Number of posts	Total number of posts published in discussion forum zone
9	Number of examination view	Total number of accessing the examination page
10	Number of quiz view	Total number of accessing the quiz pages
11	Quiz score	The score of the quiz

4   The Prediction Models

We define each course with n students engaged and each student is characterized as p features. Therefore, each student is viewed as p-dimensional vector, which is formalized below:

$$X = \{x_1, x_2, \ldots, x_p\} \in \mathbb{R}^{n \times p}$$

The examinee prediction of X is:

$$Y = f(X) \in \mathbb{R}^n, y \in \{0, 1\}$$

where $y = 1$ represents the attendance of the final examination while $y = 0$ represents the absence of the final explanation.

Naive Bayes Classifier. Naive Bayes is a conditional probability model. In our case, naive Bayes classifier essentially assigns the possibility $P(y|x_1, x_2, \ldots, x_p)$

of a student $\{x_1, x_2, \ldots, x_p\}$ to be a class y ($\in \{0, 1\}$). Consider $P(y|x_1, x_2, \ldots,$ $x_p) = \frac{P(y, x_1, x_2, \ldots, x_p)}{P(x_1, x_2, \ldots, x_p)}$, where the numerator is equivalent to the joint probability model of X and class y. Together with the assumption in naive Bayes classifier that features of objects are pairwise independent, a Bayes classifier, is the function listed below:

$$\begin{cases} 1, & p(y = 1) \bullet \prod_{i=1}^{n} P(x_i|y = 1) > p(y = 0) \bullet \prod_{i=1}^{n} P(x_i|y = 0) \\ 0, & \text{otherwise;} \end{cases} \quad (14)$$

Decision Tree. Decision Tree is commonly used as a predictive model, in which each leaf represents a class label associated with a probability, and each internal node is labeled with a certain input feature. Given a student $X = x_1, x_2, \ldots, x_p$, it is able to predict the class label of a y based on several input variables, i.e., $x_1, x_2, \ldots, x_p$ by traversing the decision tree.

Binomial Logistic Regression. The binomial logistic model is used to estimate the probability of a binary classification based on one or more independent features. Logistic regression is based on logistic function with parameter W and b for input X as follows:

$$f(X) = \frac{e^{W \bullet X + b}}{1 + e^{W \bullet X + b}}$$

SVM. Supported vector machine (a.b.a. SVM) is a non-probabilistic binary linear classifier. SVM targets to separate the positive and negative samples using a hyperplane that maximizes the margin of distances from the nearest training-data point of positive and negative objects to the hyperplane. Generally, the SVM classifier targets to solve the following optimization problem to find the above hyperplane: minimize $\|w\|$ subject to

$$y(w \bullet X - b) \geq 1$$

All the above prediction models are provided by a popular machine learning toolkit Scikit-Learn[3]. All parameters of used prediction models are set by default.

5 Experiments

The input three data sets are divided into two parts, training set and test set, and the ratio of the size of training set to that of test set is 8:2. We use the training set to learn the parameters of each classifiers, and the test set to evaluate the performance of the classifiers, respectively. The metrics of evaluating the performance of classifiers are listed as follows:

- **Accuracy ($\mathcal{A}$).** The accuracy is defined as the probability of correctly verifying whether a participant is an examinee or not an examinee. Formally, it is quantified as:

$$\mathcal{A} = P(f(X) = y|y) \quad (15)$$

---

[3] http://scikit-learn.org/stable/.

Table 4. Student examinee rate prediction

Course	Classifier	Accuracy	Precision	Recall	F-score
A	Bayes	0.903	0.28	<u>0.72</u>	0.41
	DecisionTree	<u>0.973</u>	0.79	0.57	0.66
	Logistic Regression	0.972	0.78	0.55	0.64
	SVM	0.969	<u>0.83</u>	0.72	0.74
B	Bayes	0.958	0.31	0.64	0.41
	DecisionTree	0.99	0.83	<u>0.78</u>	<u>0.81</u>
	Logistic Regression	0.98	0.83	0.66	0.74
	SVM	<u>0.991</u>	<u>0.87</u>	0.74	0.8
C	Bayes	0.67	0.72	0.56	0.63
	DecisionTree	<u>0.852</u>	0.82	<u>0.9</u>	<u>0.86</u>
	Logistic Regression	0.803	0.84	0.73	0.78
	SVM	0.824	<u>0.87</u>	0.76	0.81

- **Precision ($\mathcal{P}$).** The precision is defined as the probability of correctly verifying whether a participant is an examinee. Formally, it is quantified as:

$$\mathcal{P} = P(f(X) = 1 | y = 1) \tag{16}$$

- **Recall ($\mathcal{R}$).** The recall is defined as the fraction of examinees that are retrieved. Formally, it is quantified as:

$$\mathcal{R} = \frac{|\text{examinees} \cap \text{retrieved examinees}|}{\text{number of examinees}} \tag{17}$$

- **F-score.** F-score is defined as the harmonic mean of precision and recall, shown below:

$$\text{F-score} = 2 \bullet \frac{\mathcal{P} \bullet \mathcal{R}}{\mathcal{P} + \mathcal{R}} \tag{18}$$

The experimental results are shown in Table 4. Generally speaking, the accuracy of the classifiers over three data sets is quite high (up to 99%) since the majority of the participants are not examinees. Except naive Bayes, the other three classifiers show very similar accuracy for each course. The precision of applying Naive Bayes classifier is much lower than that of the other classifiers, and SVM achieves the best precision. Decision tree takes the best F-scores over Course B and Course C while naive Bayes and SVM take the best recall over Course A. Based on the performance in precision and recall, SVM takes the best F-score over Course A and decision tree[4] takes the best F-score over Course B and Course C. To summarize, we find, (1) there does not exist a single prediction model that can outperform others in all cases, and (2) in terms of precision, SVM takes the best, and (3) in terms of recall and F-score, SVM and decision tree take the best.

---

[4] The performance of SVM is fairly close to that of decision tree in Course B and Course C.

6   Conclusion

In this paper, we find that being an examinee for a learner is almost a necessary condition of earning a certificate and hence, the problem of studying student examinee rate prediction is of great importance to improve the learning performance and provide references for the teaching effectiveness. By conducting extensive investigation of participants' operation behaviours over three online courses, we carefully select a set of features that cover participants' learning contents, learning attitudes and learning abilities. We apply existing commonly used prediction models, and report our findings in the experiment evaluation.

Although we use a wide spectrum of features to do the prediction, the recall and F-score are not as high as we imagined. The reason is that the granularity of the features is still too coarse. For example, the behaviours of participants that watch videos could be quite different. A good participant plays a video with many forward jumps to make a better understanding of the content while a passive participant may just close the video immediately after he plays the video. Our current features cannot differentiate these two guys. To address this issue, investigation of selecting features with finer-granularity will be our future work. Besides, deep learning has become a research hot spot recently and study of applying deep learning to improve the performance of the prediction could be another our future work.

Acknowledgements. Hong Chen is the corresponding author of this paper. The work in this paper was in part supported by the Fundamental Research Funds for the Central Universities, and the Research Funds of Renmin University of China under No. 297615121721, No. 297616331721, No. 2015-ms007, and No. 15XNLF09. Our experimental environment is in part supported by the National Virtual Experimental Teaching Center on Big Data Aided Comprehensive Training for Liberal Arts and Social Science, Renmin University of China.

References

1. Academic & university news — times higher education (the). https://www.times highereducation.com/news/mooc-completion-rates-below-7/2003710.article/
2. China university MOOC. http://www.icourse163.org/
3. Coursera. https://www.coursera.org
4. Data driven education workshop. https://nips.cc/Conferences/2013/
5. edX. https://www.edx.org/
6. Udacity. https://cn.udacity.com/
7. XuetangX. http://www.xuetangx.com/
8. KDD Cup 2015, MOOC dropout prediction. https://biendata.com/competition/kddcup2015/
9. Amnueypornsakul, B., Bhat, S., Chinprutthiwong, P.: Predicting attrition along the way: the UIUC model. In: Empirical Methods in Natural Language Processing Workshop on Modeling Large Scale Social Interaction in Massively Open Online Courses (2014)

10. Cristianini, N., Shawe-Taylor, J.: An Introduction to Support Vector Machines and Other Kernel-Based Learning Methods. Cambridge University Press, Cambridge (2010)
11. The Free Encyclopedia, Massive open online course. https://en.wikipedia.org/w/index.php?title=Massive_open_online_course&oldid=694372484/
12. Fei, M., Yeung, D.: Temporal models for predicting student dropout in massive open online courses. In: IEEE International Conference on Data Mining Workshop, pp. 256–263 (2015)
13. Freedman, D. (ed.): Statistical Models Theory and Practice. Cambridge University Press, Cambridge (2009)
14. Kloft, M., Stiehler, F., Zheng, Z., Pinkwart, N.: Predicting MOOC dropout over weeks using machine learning methods. In: Empirical Methods in Natural Language Processing Workshop on Modeling Large Scale Social Interaction in Massively Open Online Courses (2014)
15. Luaces, O., Díez, J., Alonso-Betanzos, A., Lora, A.T., Bahamonde, A.: A factorization approach to evaluate open-response assignments in moocs using preference learning on peer assessments. Knowl.-Based Syst. **85**, 322–328 (2015)
16. Manhães, L.M.B., da Cruz, S.M.S., Zimbrão, G.: WAVE: an architecture for predicting dropout in undergraduate courses using EDM. In: Symposium on Applied Computing, pp. 243–247 (2014)
17. Nesterko, S.O., Seaton, D.T., Reich, J., McIntyre, J., Han, Q., Chuang, I.L., Ho, A.D.: Due dates in MOOCs: does stricter mean better? In: First (2014) ACM Conference on Learning @ Scale, L@S 2014, Atlanta, GA, USA, 4–5 March 2014, pp. 193–194 (2014)
18. Qiu, J., Tang, J., Liu, T.X., Gong, J., Zhang, C., Zhang, Q., Xue, Y.: Modeling and predicting learning behavior in MOOCs. In: Proceedings of the Ninth ACM International Conference on Web Search and Data Mining, pp. 93–102 (2016)
19. Ramesh, A., Goldwasser, D., Huang, B., Daumé III., H., Getoor, L.: Learning latent engagement patterns of students in online courses. In: Proceedings of the Twenty-Eighth AAAI Conference on Artificial Intelligence, 27–31 July 2014, Québec City, Québec, Canada, pp. 1272–1278 (2014)
20. Shah, N.B., Bradley, J., Parekh, A., Wainwright, M.J., Ramchandran, K.: A case for ordinal peer-evaluation in MOOCs. In: Neural Information Processing Systems (NIPS): Workshop on Data Driven Education (2013)
21. Sharkey, M., Sanders, R.: A process for predicting MOOC attrition (2014)
22. She, J., Tong, Y., Chen, L.: Utility-aware social event-participant planning. In: SIGMOD 2015, pp. 1629–1643 (2015)
23. She, J., Tong, Y., Chen, L., Cao, C.C.: Conflict-aware event-participant arrangement and its variant for online setting. IEEE Trans. Knowl. Data Eng. **28**(9), 2281–2295 (2016)
24. Taylor, C., Veeramachaneni, K., O'Reilly, U.: Likely to stop? Predicting stopout in massive open online courses. CoRR, abs/1408.3382 (2014)
25. Tong, Y., She, J., Ding, B., Chen, L., Wo, T., Xu, K.: Online minimum matching in real-time spatial data: experiments and analysis. PVLDB **9**(12), 1053–1064 (2016)
26. Tong, Y., She, J., Ding, B., Wang, L., Chen, L.: Online mobile micro-task allocation in spatial crowdsourcing. In: ICDE 2016, pp. 49–60 (2016)
27. Tong, Y., She, J., Meng, R.: Bottleneck-aware arrangement over event-based social networks: the max-min approach. World Wide Web **19**(6), 1151–1177 (2016)
28. Zhuoxuan, J., Yan, Z., Xiaoming, L.: Learning behavior analysis and prediction based on MOOC data. J. Comput. Res. Dev. **52**(3), 614–628 (2015)

Task Assignment of Peer Grading in MOOCs

Yong Han[✉], Wenjun Wu, and Yanjun Pu

State Key Laboratory of Software Development Environment, School of Computer Science, Beihang University, Beijing, China
{hanyong,wwj,puyanjun}@nlsde.buaa.edu.cn

Abstract. In a massive online course with hundreds of thousands of students, it is unfeasible to provide an accurate and fast evaluation for each submission. Currently the researchers have proposed the algorithms called peer grading for the richly-structured assignments. These algorithms can deliver fairly accurate evaluations through aggregation of peer grading results, but not improve the effectiveness of allocating submissions. Allocating submissions to peers is an important step before the process of peer grading. In this paper, being inspired from the Longest Processing Time (LPT) algorithm that is often used in the parallel system, we propose a Modified Longest Processing Time (MLPT), which can improve the allocation efficiency. The dataset used in this paper consists of two parts, one part is collected from our MOOCs platform, and the other one is manually generated as the simulation dataset. We have shown the experimental results to validate the effectiveness of MLPT based on the two type datasets.

Keywords: Task assignment · Crowdsourcing · LPT · Peer grading

1 Introduction

Currently MOOCs learning is becoming a popular way to acquire knowledge besides the traditional classrooms. A typical MOOC course contains videos, exercises, online test, peer grading and forums. Such an online course often has thousands of active learners in every study session [9]. When students submit their homework to the course, there are a lot of submissions of each assignment for course staffs to grade. Given the scale of the submissions, if the staffs have to examine all the submissions, they will be overwhelmed by the grading workload so that they can't have sufficient time and energy to focus on other things that will be helpful for students.

Researchers have proposed the methods of peer grading which are similar to crowd sourcing. In the process of peer grading, students are asked to grade their submissions. Therefore, the students are both submitters and the graders, which allows us to exploit this unique feature to solve this peer grading problem. In general, there are two basic problems in peer grading: (1) one is how to assign the submissions to the graders considering the difference of knowledge level between graders, (2) the other is how to aggregate to peer grades to generate a final fair score for each student. In this paper we mainly focus on the first problem.

Z. Bao et al. (Eds.): DASFAA 2017 Workshops, LNCS 10179, pp. 352–363, 2017.
DOI: 10.1007/978-3-319-55705-2_28

In most MOOCs platforms, the systems adopt a simple random assignment approach. Their peer grading servers record the submitted time of an incoming submission and store it in queue. Then the distribution mechanism chooses an idle grader randomly and assign it to the grader without considering the grader's knowledge level. Such a random assignment approach often results in biased grading scores for students because the assignment plan may put graders with a high level of knowledge mastery in the same group to evaluate the same homework assignment. Previous research papers have shown that the reliability of every grader is related to his knowledge skills. The students with good knowledge mastery tend to deliver reliable and accurate grading scores. Therefore, in the uneven distribution plan, some submissions receive very accurate evaluations while the others receive with very inaccurate evaluations. The problem can be avoided if the peer grading server makes grading task allocation with the awareness of the knowledge level of each participating student. The task allocation algorithm should be to distribute students with different knowledge levels evenly to grading groups so that the difference among the average knowledge level in each group could be minimized.

Considering the unique characteristic of peer grading, we choose the method Longest Process Time (LPT) [5] as a heuristic algorithm for grading task assignment. The algorithm is used in parallel systems to deal with task scheduling. It is a static task allocation strategy, so it requires all the tasks are assigned before the processors begin to deal with the tasks. Assigning tasks in peer grading has its unique characteristics compared to other task scheduling problems. A homework submission in the submission set may be assigned to its author if the distribution mechanism is not restricted and this is not allowed in practice. Furthermore, one submission should be graded by several graders and conversely one grader should also grade the same number of submissions.

Based on the algorithm LPT, we propose our algorithm Modified LPT (MLPT) to solve the allocating submissions in peer grading. Most researchers on peer grading focus on developing statistical models to infer accurate scores based on peer grading results. Few work consider the impact of the peer grading task allocation. Our main contribution is the introduction of student knowledge mastery into the design of task allocation algorithm. By developing the student performance evaluation model, our peer grading system can accurately estimate the student knowledge level. And based on the, we redefine the LPT algorithm to adapt the peer grading process to the distribution of student knowledge level.

The rest of the paper is organized as follows: Sect. 2 compares our work with other related efforts. Section 3 presents the overview of our peer grading framework and describes the MLPT algorithm in detail. Sections 4 and 5 present our experimental datasets and analysis results. Discussion and Future work is given in Sect. 6.

2   Related Work

In this paper, before allocating tasks, we detailed analysis the student online performance of learning for accurately estimating the knowledge level of student. We have referred to the thought task assignment in Crowdsourcing and proposed an effective method by compare the previous models. We first review some existing researches for allocating tasks. Allocating tasks is a long historical issue and appears in many fields such as distributing conference papers or in the parallel systems and so on. It is also one of the core problems of the crowdsourcing.

In the mode of Crowdsourcing, a task is decomposed into multiple small tasks [8] which is assigned to massive online workers. Because of the existing of human error and fraudulent workers, the confident of the collection results which typically contains a lot of noise is usually very low [2, 3]. In order to ensure the reliability of the results, the measure of redundant allocation is typically taken, namely, assigning the same task to multiple workers, and aggregating the multiple results to generate a final one. Here the task allocation is to find a balance between the reliability of the results and the redundancy of the task allocation.

Ul Hassan and Curry [15] proposed a task allocation solution based on the human ability. This method first models the abilities of a worker according to historical data. Similarly, it models the types of workers' abilities that the task required. Then through adaptation, the algorithm allocates the task to the most suitable workers thus obtaining the optimal results with the minimal redundancy. Finally, they adjust the worker model by comparing the completing results to the aggregating result.

Jung and Lease [16] had solved the problem of data sparsity by decomposing matrix based on the research of Umair and Edward, increasing the universality of the method. In addition to model the abilities of the workers based on the historical data, other researchers also considered the demographic factors, education background, and especially the social information to determine the level of the workers' abilities which can allocate the task to the worker accurately.

Furthermore, because the most Crowdsourcing tasks are paid tasks, the capital budget is constraint [4], Karger et al. [17] inspired by the algorithm of propagation of confidence and the algorithm of low rank matrix approximation designed the minimum cost allocation strategy. The results in the strategy make the cost of the spending the least if it meets certain prerequisites of reliability.

Yan et al. [18] proposed an online adaptive task assignment algorithm by ensuring the confidence of the results. The algorithm allocates the subsequent redundancy and the workers according to the required of the task by analyzing the completion of the task time to time and the confidence of the results to make the final results highly reliability.

Piech et al. [6] exploited the confidence estimated to determine what point in the grading process is confident about each submission's score. They simulated grading taking place in rounds. For each round they ran model using the corresponding subset of grades and counted the number of submissions for which they were over 90% confident that they predicted grades were within 10pp of the students' true grade.

Besides, in the area of spatial crowdsourcing, tasks with spatiotemporal constraints are allocated [19–23].

In our work, there is no need to classify the students according to the types. The primary information about students is their historical learning data, thus we exploit some models compute the student knowledge level based on this data. By extracting student features, we using the method of logistic regression to fit the features and this part accounts for the proportion of our work. We also define a hyper-parameter as the redundancy value, which requires us to constantly experiment.

We have done experiments over different redundancy values and different number of students. The results can be seen in Sect. 5. Our another attempt is using the examination scores as the knowledge level, but it does not generate a good effect, the reasons may be that the score is a monotonous factor (Fig. 1).

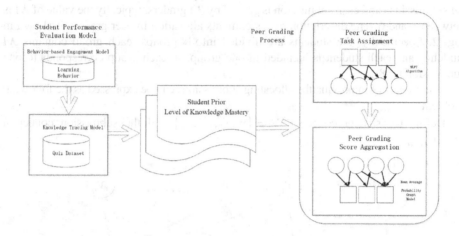

Fig. 1. Peer grading framework.

3 Methods

In this section, we present the overview of the entire peer grading framework and introduce the design of grading task assignment in detail. Figure 2 illustrates the basic framework of our peer grading process. It consists of three major components: the student performance evaluation model, the peer grading task allocator and the score aggregation model. The student performance evaluation model is responsible for computing each student's knowledge level by analyzing each student's behavior in the watching course videos and the performance exercises after the videos. The outcome from the model presents the prior knowledge level of students for the peer grading process to determine the arrangement of peer grading groups. The peer grading process includes two steps: grading task allocation and score aggregation model. In this paper, we introduce the MPLT algorithm to perform task allocation in order to minimize the difference among the students in each grading group. Based on the task assignment plan, students can evaluate other peers' submissions and give scores using their own judgement. As each submission in general receives multiple scores from its graders, the score aggregation model needs to infer the accurate and unbiased score for the submission from the scores recommended by peer graders.

At the beginning of the section, we describe the problem of assignment allocation in peer grading. And then we present the algorithm of LPT that is used in the field of task scheduling, and how we can extend the idea LPT for the purpose of peer grading and introduce the new algorithm MLPT. At the end of the section, we briefly discuss our research work on student performance evaluation model.

3.1   The Problem of Submission Allocation in Peer Grading

In the process of peer grading, each student plays two roles: a grader and a submitter. Suppose that there are $\mathbb{N}$ students participating peer grading, and there are also $\mathbb{N}$ submissions from these students. For convenience, we suppose that each student needs to grade $\mathcal{M}$ submissions and each submission is graded by $\mathcal{M}$ graders, typically the value of $\mathcal{M}$ is between 3 and 6. The description of assignments allocation in peer grading is shown in Fig. 2. Specifically, the students are divided into $\mathbb{N}$ groups: each group contains $\mathcal{M}$ students, and each student is included in one group, so each group corresponds to one submission [1, 8].

The process of improving the allocating submissions can be expressed as the following problem:

Define the student collection $V = \{v_1, v_2, \ldots, v_n\}$ and the submission collection $U = \{u_1, u_2, \ldots, u_n\}$.

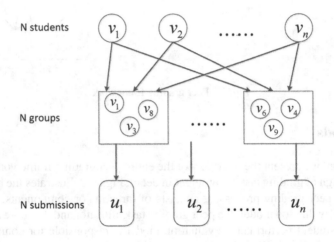

Fig. 2. The process of task assignment in peer grading.

The degree of a student v_i mastering the knowledge is denoted by $w = w(v_i)$ where $w(v_i)$ lies in the interval (0, 1) and the parameter i falls in the range (0, n) if student v_i submits the submission u_i. $w(v_i)$ that is generated by the student evaluation model, is regarded as the prior level of student knowledge mastery.

The problem is that given the parameter m how to divide the student collection $G_1, G_2, \ldots, G_n$, and make it satisfy the following conditions:

1. The grader collection $G_i = \{v_{i1}, v_{i2}, \ldots, v_{im}\}$ only contains m students.
2. For any student v_i, there are m groups including her.
3. $v_i \in G_i$, it means that all of the students grade the submission u_i, so the submitter v_i could be contained in the grading group G_i.

4. The variance of the students' comprehensive knowledge level must be the minimal, namely $\min\{var(w(G_1), w(G_2), \ldots, w(G_n))\}$, for the purpose that the grading level of each group is as close as possible in where $w(G_i) = \sum_{v_j \in G_i} w(v_j)$.

3.2 The Algorithm of LPT

The algorithm LPT is a basic method to solve the task scheduling problem in parallel computing system. The tasks scheduling problem in a parallel system can be described as the optimization process of minimizing the difference of processing duration among all the machines. Given the tasks $J = \{j_1, j_2, \ldots, j_n\}$, each of which has a processing time $t(j_i)$. Define $T = \{t(j_1), t(j_2), \ldots, t(j_n)\}$. Assume the tasks are allocated to the machines $M = \{m_1, m_2, \ldots, m_p\}$ with the same model. We assume that the tasks are independent and their arrivals occur at the same time point (Table 1).

Table 1. The process of the algorithm of LPT.

Algorithm: the LPT algorithm of task scheduling
1. Sort the tasks in the descend order by the costing time, make the result as $t(J_1) \geq t(J_2) \geq \cdots \geq t(J_n)$.
2. for $i = 1, 2 \ldots, p$:   Assign the load L_i of machine M_i as 0, namely $L_i \leftarrow 0$.   Assign the task collection $J(i)$ of the machine M_i as empty, namely $J(i) \leftarrow \phi$.
3. for $k = 1, 2 \ldots, n$:   Select the current machine with the smallest load, recorded as M_i.   Add the current task J_k to the task collection of the machine $J(i)$, namely $J(i) \leftarrow J(i) \cup J_k$.   Update the load of machine M_i, namely $L_i \leftarrow L_i + t(J_k)$.

The task scheduling has been proved to be a NP-complete problem [7, 14], namely the complexity of finding the optimal solution increases as the number of tasks and the number of machines in the exponential way. To find the acceptable solution in practice, it is necessary to employ the heuristic algorithm to compute an approximate solution.

The LPT algorithm adopts the greedy strategy for exploring approximate solutions according to the static task assignment policy. At the initial moment, all of the tasks are allocated to the machines, and during the executions of the tasks, the task schedule must remain until the end. The LPT algorithm arranges the pending tasks in a descending order and always assigns the tasks to the machines as soon as they become available. The detail process of the algorithm is described as follows.

The process of the algorithm LPT is simple and clear, and the time complexity can reach $O(n \log n)$ if the step of finding the earliest free machine adopts the minimum heap algorithm [10, 12]. Furthermore, the algorithm also has the optimal approximate result.

It can be proved that if $F^*(J)$ denotes the optimal costing time when assigning the task collection J to machine collection M, F(J) denotes the costing time adopting the algorithm LPT, so we can obtain:

$$\frac{F(J)}{F^*(J)} \leq \frac{4}{3} - \frac{1}{3m}$$

From the approximation relation we can draw that LPT is an efficient algorithm. The detailed process of proving the approximation relation can be found in references [9].

3.3   The Algorithm of MLPT

Similarly, there are many similarities between assigning submissions and task scheduling in parallel systems. We introduce a MLPT algorithm by redefining the process of LPT towards the needs of grading task assignment. Because for the problem of assigning submissions, assigning submissions to the graders can be viewed in another angle as assigning the graders to the submissions, so the student collection $V = \{v_1, v_2, \ldots, v_n\}$ is considered as the task collection $J = \{j_1, j_2, \ldots, j_n\}$ and the knowledge level $w(v_i)$ of each student is the processing time $t(j_i)$, then the submission collection $U = \{u_1, u_2, \ldots, u_n\}$ is considered as the machine collection $M = \{m_1, m_2, \ldots, m_p\}$. According to these above assumptions, assigning the graders to submissions in peer grading is equivalent to allocating the tasks to machines in the scheduling process, the MLPT algorithm can ensure the sum of the knowledge levels of the graders in each group is close to any of the others (Table 2).

Table 2. The algorithm of MLPT.

Algorithm: the algorithm of MLPT allocating submissions
1.   Sort the graders by the level of knowledge in descending order, namely $w(v_1) \geq w(v_2) \geq \cdots \geq w(v_n)$
2.   For i = 1, 2, ..., n:
The submission collection U(i) graded by grader v_i is assigned as empty, Namely, U(i) ← φ.
3.   For j = 1, 2, ..., n:
The comprehensive grading level L_j corresponding to the submission u_j is assigned to 0, namely L_j ← 0.
The student group V(j) corresponding to the submission u_j is assigned to 0, namely V(j) ← 0.
4.   For i = 1, 2, ..., n:
For k = 1, 2, ..., m:
Select the submission with the smallest comprehensive grading level from submission collection $\{u_j \| j \neq i\}$, recorded as u_j.
Insert the current grader v_i to the submissions collection u_j, namely V(j) ← V(j) ∪ v_i.
Update the comprehensive grading level L_j of the submission u_j, namely L_j ← $L_j + w(v_i)$.
Insert the submission u_j to the collection U(i) of graded by grader v_i, namely U(i) ← U(i) ∪ u_j.

Moreover, there are two conditions in assigning submissions compared to task scheduling. One condition is that the redundant grading parameter m guarantees that the number of different submissions for each grader is the same [13]. The other is that the grader could not evaluate his own submission.

3.4 Student Performance Evaluation Model

We establish a two-stage model to assess student mastery level of each knowledge skill, in other words, to estimate the probability that a student has learned the specific knowledge of a chapter. The first stage is the behavior-based student engagement model, which first analyze student video-watching activities within a chapter and extract interpretive quantities to predict the probability that a student has mastered the knowledge of that certain chapter. This model extract eight features including the total video playing time, the number of video played, the number of pauses, the number of sessions, the number of requests, the number of rewinds, the number of slow play rate setting, the number of fast forwards. The logistic regression method is used to fit these features and predict the engagement level of every student.

The second stage is the knowledge tracing model to infer the level of student knowledge mastery. The knowledge tracing model, popularized in Intelligent Tutoring System (ITS), leverages the quiz response sequence of every student to assess their learning outcomes. Our work adopts the knowledge tracing model and ameliorates it by combining the prediction results obtained in the behavior-based student engagement model. Essentially, it is based on a 2-state dynamic Bayesian network where student responses are the observed variable and student knowledge is the latent one. The model takes student response sequences and uses them to estimate the student's level of knowledge. The initial state of the model can be trained by the results obtained in the behavior-based student engagement model.

4 Datasets

The dataset using in our paper is collective from the course of computer network experiment abbreviated as CNE) of our MOOCs platform of Beihang Xuetang (http://mooc.buaa.edu.cn). The MOOC course CNE includes three assignments for the Sophomore undergraduate who had not studied this course before and we only use assignment 1 and assignments 2. The course required that the students had to upload the submissions before the end of each week.

Specially, the session course was open in the year of 2015 and active for a 11-week semester. The course contains 10 chapters where there are several video units per chapter and a final examination. After each video unit, there are multiple choice problems as quiz for students. The number of the problems ranges from 5 to 13 in each unit. Also, the course contains 4 open-ended design problems that demand the students to design and implement the experimental programs independently. The score of these open-ended problems are generated through the peer grading mechanism in our MOOC platform. There are approximately 600 active students each semester since the course CNE went online and roughly 500 students to take part in the final exams. The number of the records collected by the system is 2 million, which logs every user interaction with the courseware including scanning and clicking, video interaction, posting questions and comments in the course forum.

After removing the incomplete items in the original dataset, we choose 210 students including 210 items, and in our setting each student grade 4 submissions and each submission must be graded by 4 graders. In order to prove the advantage of the MLPT algorithm, we also produced two simulation datasets imitating the real data. In the simulation datasets, the number of items is 200 and we valued the redundancy peer grading number as 3, 4 and 5 in our experiments. The major difference between the real data and the simulation data is that the prior is generated from the normal distribution and evenly distribution, respectively.

5   Experimental Results

This section mainly verifies the algorithm of MLPT from both the real dataset and simulation dataset. For each dataset, we compute the variance of all the groups using the sum of each group as one item and we have run 10 times for the computing an average value. In both Tables 3 and 4, the simulation dataset 1 is generated from an evenly distribution and the simulation dataset 2 is generated from a random normal assignment distribution.

Table 3. The comparison between random assignments and MLPT based on the simulation datasets.

	Random assignment	MLPT
Simulation dataset 1	0.381	0.009
Simulation dataset 2	0.266	0.017

Table 4. The comparison between random assignments and MLPT based on the real datasets.

	Random assignment	MLPT
Assignment 1	0.355	0.053
Assignment 2	0.376	0.046

Table 3 shows the results from the simulation datasets, the dataset 1 and dataset 2, which are generated from the evenly distribution and normal distribution, respectively. The MLPT algorithm reduces the variances of the knowledge master level among the groups from 0.381 to the 0.009 on the simulation dataset 1 and from 0.266 to 0.017 on the simulation dataset 2.

Similarly, Table 4 is the results generated from the real datasets. The value of the variance is 0.053 after using MLPT in Assignment 1, and the optimization percentage is 85.0%. The optimization percentage is 87.7% and the MLPT value is decreased from 0.376 down to 0.046 in Assignment 2. Note that the MLPT optimization results on the simulation datasets seem better than the results on the real dataset. The reason may be that the prior inferred from the performance of quiz doesn't completely associated with the actual knowledge level of students in the open-ended problems.

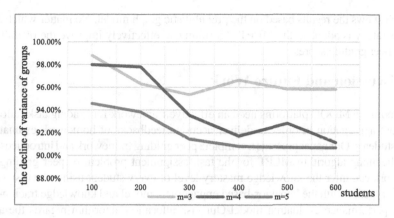

Fig. 3. The degree of declining of variance from MLPT algorithm to the random algorithm.

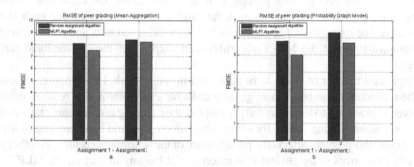

Fig. 4. The effectiveness of task assignment on RMSE of peer grading.

Through simulation, we can explore different settings such as the various numbers of students and the different values of the parameter m to compare the performance between the MLPT algorithm and the random algorithm. In the simulation dataset, we have verified a series of datasets with the number of students varying from 100 to 600 and the value of parameter m ranging from 3 from 5. And in each simulation, we compute the decline in the group variance caused by the MLPT algorithm over the Random algorithm. Figure 3 plots the results generated from the Random and MLPT algorithms. It demonstrates that the MLPT algorithm can reduce the variance of all the groups significantly in comparison with the Random algorithm, especially when there are not many students and the parameter m is relatively small. Figure 4 shows the comparison results generated from the MLPT and Random algorithm based on the real datasets in the process of peer grading. The vertical axis indicates the Root Mean Square Error (RMSE), which is used to compute the accurate of grading scores. In the research of Mi and Yeung [11], researchers proposed new score aggregation methods based on probabilistic graph models to infer accurate scores of student submissions. These methods outperform the method using the mean aggregation strategy. It is necessary to investigate whether the MLPT performance has any impact on different aggregation methods. The result in Fig. 4(a) shows the difference of RMSE between the MLPT algorithm and the RMSE algorithm based on the mean aggregation strategy and

Fig. 4(b) shows the results based on the probabilistic graph model. No matter which aggregation strategy is adopted, the MLPT algorithm can effectively improve the overall accuracy of peer grading scores.

6   Discussion and Future Work

Peer grading in MOOC platforms need an effective framework for grading tasks allocation and score aggregation in order to deliver accurate feedbacks of homework evaluation to online students. Our paper presents a complete peer grading framework and introduces a new task scheduling algorithm MLPT for the task assignment problem of peer grading. This framework can infer the knowledge mastery level of every student using a two-stage data analysis pipeline with the behavior-based engagement model and knowledge tracing-based student performance evaluation model Our task allocation algorithm regards the assessment outcomes of each student in the quizzes of the MOOC courses as the priors and utilizes the priors to build peer grading groups. It aims at minimizing the capability difference among grading groups and generating a fair and accurate evaluation for each student submission. The experimental results from both simulation datasets and real course datasets demonstrates that the MLPT algorithm can outperform the conventional random strategy.

There are a number of issues to be addressed in future work. In this paper, we assume that the redundancy number of peer grading tasks for every submission remains constant. However, in practice, students may fail to complete their grading tasks in time, thus resulting in less redundant grading results for some submissions. It would be interesting to investigate whether this problem affects the performance of the schedule algorithm. Another issue is whether the errors in the student evaluation model has any impact on the MLPT algorithm. Given the probability nature of the knowledge tracing model, it is unavoidable that it may give a biased estimation of knowledge levels of some students when there are noises in the data of user interactions and question answering. One of the feasible solutions is to have staffs to cross-examine some peer grading results and correct the potential errors.

Acknowledgments. This work was supported in part by grant from State Key Laboratory of Software Development Environment (Funding No. SKLSDE-2015ZX-03) and NSFC (Grant No. 61532004).

References

1. Fonteles, A.S., Bouveret, S., Gensel, J.: Heuristics for task recommendation in spatiotemporal crowdsourcing systems. In: Proceedings of the 13th International Conference on Advances in Mobile Computing and Multimedia, pp. 1–5. ACM (2015)
2. Cheng, J., Teevan, J., Bernstein, M.S.: Measuring crowdsourcing effort with error-time curves. In: Proceedings of the 33rd Annual ACM Conference on Human Factors in Computing Systems, pp. 1365–1374. ACM (2015)
3. Qiu, C., Squicciarini, A.C., Carminati, B., et al.: CrowdSelect: increasing accuracy of crowdsourcing tasks through behavior prediction and user selection. In: Proceedings of the 25th ACM International on Conference on Information and Knowledge Management, pp. 539–548. ACM (2016)

4. Howe, J.: The rise of crowdsourcing. Wired Mag. **14**(6), 1–4 (2006)
5. Wiki for Multiprocessor Scheduling Information. https://en.wikipedia.org/wiki/Multi processor_scheduling
6. Piech, C., Huang, J., Chen, Z., et al.: Tuned models of peer assessment in MOOCs. arXiv preprint arXiv:1307.2579 (2013)
7. Coffman, Jr. E.G., Sethi, R.: A generalized bound on LPT sequencing. In: Proceedings of the 1976 ACM SIGMETRICS Conference on Computer Performance Modeling Measurement and Evaluation, pp. 306–310. ACM (1976)
8. Alfarrarjeh, A., Emrich, T., Shahabi, C.: Scalable spatial crowdsourcing: a study of distributed algorithms. In: 2015 16th IEEE International Conference on Mobile Data Management, vol. 1, pp. 134–144. IEEE (2015)
9. Baneres, D., Caballé, S., Clarisó, R.: Towards a learning analytics support for intelligent tutoring systems on MOOC platforms. In: 2016 10th International Conference on Complex, Intelligent, and Software Intensive Systems (CISIS), pp. 103–110. IEEE (2016)
10. Gonzalez, T., Ibarra, O.H., Sahni, S.: Bounds for LPT schedules on uniform processors. SIAM J. Comput. **6**(1), 155–166 (1977)
11. Mi, F., Yeung, D.Y.: Probabilistic graphical models for boosting cardinal and ordinal peer grading in MOOCs. In: AAAI, pp. 454–460 (2015)
12. Massabò, I., Paletta, G., Ruiz-Torres, A.J.: A note on longest processing time algorithms for the two uniform parallel machine makespan minimization problem. J. Sched. **19**(2), 207–211 (2016)
13. Gardner, K., Zbarsky, S., Harchol-Balter, M., et al.: The power of d choices for redundancy. ACM SIGMETRICS Perform. Eval. Rev. **44**(1), 409–410 (2016)
14. Feier, M.C., Lemnaru, C., Potolea, R.: Solving NP-complete problems on the CUDA architecture using genetic algorithms. In: International Symposium on Parallel and Distributed Computing, ISPDC 2011, Cluj-Napoca, Romania, pp. 278–281. DBLP, July 2011
15. Ul, Hassan U., Curry, E.: Efficient task assignment for spatial crowdsourcing. Expert Syst. Appl. Int. J. **58**(C), 36–56 (2016)
16. Jung, H.J., Lease, M.: Crowdsourced task routing via matrix factorization. Eprint Arxiv (2013)
17. Karger, D.R., Oh, S., Shah, D.: Budget-optimal crowdsourcing using low-rank matrix approximations. In: 2011 49th Annual Allerton Conference on Communication, Control, and Computing (Allerton), pp. 284–291. IEEE (2011)
18. Yan, Y., Fung, G.M., Rosales, R., et al.: Active learning from crowds. In: Proceedings of the 28th International Conference on Machine Learning (ICML 2011), pp. 1161–1168 (2011)
19. Tong, Y., She, J., Ding, B., et al.: Online minimum matching in real-time spatial data: experiments and analysis. Proc. VLDB Endow. (PVLDB) **9**(12), 1053–1064 (2016)
20. Tong, Y., She, J., Ding, B., et al.: Online mobile micro-task allocation in spatial crowdsourcing. In: Proceedings of the 32nd International Conference on Data Engineering (ICDE 2016), pp. 49–60 (2016)
21. Tong, Y., She, J., Meng, R.: Bottleneck-aware arrangement over event-based social networks: the max-min approach. World Wide Web J. **19**(6), 1151–1177 (2016)
22. She, J., Tong, Y., Chen, L., et al.: Conflict-aware event-participant arrangement and its variant for online setting. IEEE Trans. Knowl. Data Eng. (TKDE) **28**(9), 2281–2295 (2016)
23. She, J., Tong, Y., Chen, L., et al.: Conflict-aware event-participant arrangement. In: Proceedings of the 31st International Conference on Data Engineering (ICDE 2015), pp. 735–746 (2015)

Predicting Honors Student Performance Using RBFNN and PCA Method

Moke Xu[1(✉)], Yu Liang[2], and Wenjun Wu[2]

[1] Shenyuan Honors College, Beihang University, Beijing, China
08753@buaa.edu.cn
[2] State Key Laboratory of Software Development Environment,
Department of Computer Science and Engineering, Beihang University, Beijing, China
{liangyu,wwj}@nlsde.buaa.edu.cn

Abstract. This paper proposes a predictive model based on Principle Component Analysis (PCA) combining with radical basis function Neutral Network (RBFNN) to accurately predict performance of honors student through the analysis of personalized characteristics. This model consists of two phases: PCA is firstly adopted to apply dimension reduction to the honors student dataset; extracted principle features are then employed as the input of RBF Neutral Network so as to build a three-layer RFF Neutral Network predictive model. Compared with other Neutral Network models, the PCA-RBF predictive model demonstrates a faster convergence speed, a higher predictive accuracy and stronger generation ability. Moreover, this model enables honors programmer administrators to identify those honor students at early stage of risk, and allow their academic advisors to provide appropriate advising in a timely manner.

Keywords: Data mining · Predictive model · PCA · RBFNN

1 Introduction

Honors Education has been proposed for server decades in many famous public universities as a new education model that intends to cultivate more talented students. It emphasizes a personalized education paradigm by providing top students with distinct competence the best faculty and research resources. Those students enrolled honors education programs are often called as Honors Student [1]. Given the limited resource of honors education programs in public universities, it is necessary to introduce a performance-based scholarship to only keep the qualified students in the honor program based on their academic performance in studying honors courses. Thus it becomes a challenge to predicate performance of honors students by analyzing their personalized characteristics. Meanwhile, this method may be of considerable usefulness in identifying students at early stage of risk, especially in very large classes, and allow the instructor to provide appropriate advising in a timely manner [2], which is of great practical significance.

The rest of the paper is organized as follows. Section 2 presents a summary of related work. Section 3 describes the predicative model for analyzing student performance. Section 4 presents data analysis and the experimental results. Finally, Sect. 5 concludes the paper.

© Springer International Publishing AG 2017
Z. Bao et al. (Eds.): DASFAA 2017 Workshops, LNCS 10179, pp. 364–375, 2017.
DOI: 10.1007/978-3-319-55705-2_29

2 Related Work

Prediction of student scores is an important research topic in the field of educational data mining. Many researchers have proposed predicative models based on a variety of machine learning techniques.

Elbadrawy et al. [3], proposed a predictive model based on regression-based and matrix factorization–based methods, which can predict student performance using personalized analytics. *Al-Radaideh et al.* [4], introduced a decision tree model to predict the final scores in studying a C++ programming courses, in which the referred decision tree prediction is considered as the better prediction method than other models.

Dekker et al. [5], presented a case study to evaluate multiple dropout prediction models based on to multiple classification techniques. After analyzing several data sets larger than 500, it comes to the conclusion that the simpler classifier (J48, CART) shows better predication performance in comparison with other algorithms (e.g. Bayes Net or JRip). *Tanner and Toivonen* [6] introduces the k-Nearest Neighbor (KNN) based prediction method and achieves more accurate result. Other than that, early technique test is proved to be a strong prediction factor of final scores and other courses that based on the particular technique. *Kotsiantis et al.* [7] compares six classification algorithms of predicting dropouts of students, among which Naive Bayes and Neutral Network are justified as the best algorithms.

Multiple dimensional features of every honors student's behavior pose a great challenge for conventional methods [8]. With the progress of modern data mining technique, especially rapid development of artificial Neutral Network, we propose a novel prediction method based on learning behavior of honors students by employing Neutral Network. With the fine generation ability and fast learning convergence speed [9, 10], Radical Basis Function Neural Networks (RBFNN) has been widely applied in multiple fields [11–14], which is able to approximate any non-linear functions and handle regularity that is hard to analysis within system. To obtain a better prediction of performance and more accurate model with better effectiveness, this paper approaches Principal Component Analysis (PCA), adopting dimension reduction to various figures in study behavior to eliminate the interaction between each data while minimizing the lost of data and using fewer general indictors to replace the redundant indictors [15]. Comprehensive figures with the biggest influence are adopted as the input variable of RBFNN so that predictive model of performance of honors student based on PCA-RBF network is formed to resolve low accuracy of traditional methods.

Besides, to predict scores of students, the social network relationship among students is also an import factor [16–19].

3 Predicative Model for Analyzing Student Performance

3.1 Factor Analysis of Student Behavior

For honors students, the performance (Y_1) is related to their mean score of honors courses (H) and mean score of other courses (O). H_{max} donates the maximum value of the mean score of honors courses among all the honors student, O_{max} represents the maximum value of the mean score of other courses, the performance of honors student is then calculated as Eq. (1):

$$Y_1 = 0.6 \times \frac{H}{H_{max}} + 0.4 \times \frac{O}{O_{max}} \tag{1}$$

For honors students, their ranking range of performance is of great value to be a reference rather than merely specific figures. We can classify the ranking range of honors student according to practical situations into 5 categories (Y_2): those ranking at the top 20% is Category 1 referred to as excellent student; those 20–40% is Category 2 as good student; those 40–60% is Category 3 as average student; those 60–80% is Category 4 as backward student; those at the last 20% is Category 5 as rudimentary student.

The performance of honors student is influenced by multiple personalized characteristics [3]. A personalized assessment indicators (X) of honors students is ultimately established covering 10 assessment factors that describe the main features of honors students in Table 1. By applying the behavioral factors of the students, we are able to develop a prediction model on both performance and ranking ranges of honors students.

Table 1. The performance factors of the honors students

Symbol	Meaning	Symbol	Meaning
X_1	College entrance score	X_6	Consumption level
X_2	Middle-term test score	X_7	Moral performance
X_3	Social work performance	X_8	Poverty student range
X_4	Book circulation	X_9	Extra competition points
X_5	Online time in one academic year	X_{10}	Athletic performance

3.2 Overview of PCA Algorithm and RBF Network

Principal Component Analysis (PCA) Algorithm: Based on traditional mathematics statistic and analysis methods, PCA transforms various figures of study behaviors to independent assessment parameters. The dimension reduction process could reach the removal of the overlapping influence of information among different figures, which is illustrated in details as follows:

Assume there are m sample data and each involves n figures, thus the original sample could be stated in matrix as:

$$X = \left(x_{ij} \right)_{m \times n} \tag{2}$$

(a) Approach standardization to raw sample M. The process is necessary as the influence of dimension among figures which will shade light on the final analysis is needed to be eliminated, which often appears in academician sample with various dimension and units. After standardization, figures in raw data will be in the same order of magnitudes which will be more suitable for comprehensive comparison and assessment [20]. The zero-mean normalization method is applied in this paper to standardize both mean and standard deviation of raw data. Data being processed follows Gaussian distribution, of which mean value equals 0 and standard deviation is 1, and the transformed function is showed as Eq. (3):

$$X^* = \frac{X - \mu}{\sigma} \tag{3}$$

Among which, μ and σ are mean value and standard deviation of all sample data [21].

(b) Calculation of co-variance matrix R:

$$R = \left(r_{ij}\right)_{n \times n} = X'X \tag{4}$$

R refers to intimacy level of correlation among different assessment figures.

(c) Calculate eigenvalue and eigenvector of R. The eigenvalue of R is resolved as $\lambda_1, \lambda_2, \cdots, \lambda_n$ and after being arranged from small to big, the eigenvector is obtained as $\alpha_1, \alpha_2, \cdots, \alpha_n$.

(d) Calculate the cumulative and pick the principle component. The calculation method of proportion and cumulative of the kth principle component are illustrated respectively in Eqs. (5) and (6). Normally, if cumulative reaches 85%, replace the primary n figures with the first k principle components, and in turn most of information of raw sample matrix is kept.

$$G_k = \frac{\lambda_k}{\sum_{l=1}^{n} \lambda_l}, \, k < n \tag{5}$$

$$G_{l\_k} = \frac{\lambda_{h=1}^{k} \lambda_h}{\sum_{l=1}^{n} \lambda_l}, \, k < n \tag{6}$$

(e) Calculate scoring coefficients of each principle component. The calculation equation is shown in Eq. (7), in which scoring coefficients of each principle component needed come up to establish new training sample set and testing sample set so that the dimension of data is reduced.

$$S = X_{m \times n}\left(x_{ij}\right) \cdot \alpha_{n \times k}\left(\alpha_1, \alpha_2, \cdots, \alpha_k\right), \, k < n \tag{7}$$

Principle of RBF Neutral Network: The RBF Neutral Network is combined by input layer, hidden layer and output layer. Its topological structure is stated in Fig. 1. There are n neutral cells in the input layer, k neutral cells in the hidden layer and p neutral cells in the output layer [22]. Weighted sum of a basis function set is adopted in the RBF Neutral Network to reach the approximation of function, which has high local approximation ability without local minimum point and suitable for non-linear systematic prediction at high latitude [9, 10].

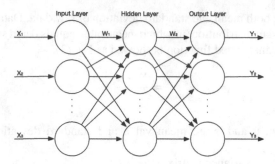

Fig. 1. RBF neutral network construction

The most commonly adopted radical basis function is Gaussian function [23, 24], as shown in Eq. (8). In this paper, the personalized characteristic-related assessment feature (X) represents m-dimensional input vector, b_i is the center of the basis function and σ_i indicates the distribution density of basis function. m is the number of neutral cells in output layer, $||X - b_i||$ is the distance between vectors, and when it becomes larger, rad(X) is more tend to 0, which means that centers closer to X is activated.

$$rad(X) = \exp\left(-\frac{||X - b_i||^2}{2 \cdot \sigma_i^2}\right), i = 1, 2, \cdots, m \tag{8}$$

The input of output layer is the weighted sum output by neutral cells in each hidden layer. The output equation is illustrated in Eq. (9), in which p is the number of neutral cells in output layer.

$$Y = \sum_{i=1}^{m} W_{ik} \times rad(X), k = 1, 2, \cdots, p \tag{9}$$

3.3   Overview of the PCA-RBF Network Predictive Model

This paper collects the personalized assessment indicators of performance of honors students as the experimental dataset.

Step (A) PCA analysis: Analyzing principle components in the dataset through PCA algorithm to extract reconstituted sample set so as to eliminate minor features.
Step (B) RBF-model training: Partition the new dataset whose dimension is reduced in Step A into a training dataset and a testing dataset. Applying RBF network training process in testing set, in which the best parameters is obtained by repeated fitting, the ultimate RBF network is the trained PCA-RBF network performance predictive model.
Step (C) Model Verification: Fit the training dataset into the RBF network model in (b). Comparing it with true values as well as both prediction from RBF and BP Neutral Network model so that the assessment of PCA-RBF network performance predictive model could be approached.

4   Data Analysis

4.1   Dataset

Personalized characteristic data of 205 honors students in one academic year in certain university is employed as the dataset in this paper. Considering education level among different areas varies, college entrance score of honors student is the ratio of their actual score and excellent value in that area; numbers 0–2 refer to respectively non-poor student, middling-poor student and sever-poor student; numbers 0–3 correspond to first/second/third prizes and students without any prizes will be scored as 0; rest figures are directly derived from student management system. All data stated above are all real.

This paper has applied anonymization technique to the data set to protect the privacy of students, in which student number and name are innominate. Table 2 lists the raw data of personalized assessment indicators of honors students.

Table 2. Raw data of personalized assessment indicators of honors students

No.	X_1	X_2	X_3	X_4	X_5	X_6
1	1.255	74.8	80	0	4072	5663.18
2	1.246	66.5	86	0	3372	4869.35
3	1.318	88.0	82	0	3627	5140.90
...	...	...	...	...	...	...
203	1.413	52.5	81	36	4994	5431.24
204	1.388	91.5	90	18	4158	9497.14
205	1.320	68.8	83	9	2723	6110.78
No.	X7	X8	X9	X10	Y1	Y2
1	86.50	0	0	93.50	0.898	2
2	87.00	0	1	90.50	0.929	1
3	87.50	0	0	97.50	0.960	1
...	...	...	...	...	...	...
203	85.00	1	0	90.00	0.799	4
204	88.00	0	0	93.00	0.884	2
205	91.00	0	0	87.50	0.882	2

In the following process of model fitting, the first 80% (165 data items) are used as training data set and the last 20% (40 items) as testing data set.

4.2   Principle Component Analysis

First, the algorithm standardizes each data item in the honors student dataset by Z-score method. Table 3 illustrates the outcome of the standardization.

Table 3. Standardized data in the honors student dataset

No.	X_1	X_2	X_3	X_4	X_5	X_6
1	−0.171	0.045	−1.441	−0.801	1.331	−0.732
2	−0.296	−0.473	0.184	−0.801	0.290	−1.243
3	0.684	0.864	−0.900	−0.801	0.670	−1.068
...	...	...	...	...	...	...
203	1.981	−1.344	−1.171	2.000	2.703	−0.882
204	1.635	1.082	1.267	0.599	1.460	1.733
205	0.712	−0.328	−0.629	−0.101	−0.676	−0.445
No.	X7	X8	X9	X10	Y1	Y2
1	−0.453	−0.364	−0.220	1.275	0.898	2
2	−0.249	−0.364	2.288	0.667	0.929	1
3	−0.046	−0.364	−0.220	2.085	0.960	1
...	...	...	...	...	...	...
203	−1.063	1.831	−0.220	0.566	1.981	4
204	0.158	−0.364	−0.220	1.174	1.635	2
205	1.379	−0.364	−0.220	0.060	0.712	2

Second, the algorithm calculates the co-variance coefficient matrix of the raw dataset. We can see that remarkable difference exists in co-variance coefficient, which indicates overlapping among each feature. If we directly input these raw features in a RBF network, the convergence speed of the network model will be definitely influenced, which results in the reduction of the prediction accuracy. Thus, it is necessary to apply a dimension reduction to the raw dataset.

Table 4. Co-variance coefficient matrix

	X_1	X_2	X_3	X_4	X_5	X_6	X_7	X_8	X_9	X_{10}
X_1	1.00	−0.13	−0.03	−0.04	0.07	0.07	−0.07	0.12	0.00	0.02
X_2	−0.13	1.00	0.12	0.18	−0.21	−0.14	0.02	−0.09	−0.02	0.10
X_3	−0.03	0.12	1.00	0.06	−0.17	0.09	0.07	−0.12	0.09	0.08
X_4	−0.04	0.18	0.06	1.00	−0.17	−0.03	0.12	0.17	0.00	0.00
X_5	0.07	−0.21	−0.17	−0.17	1.00	−0.14	0.00	0.13	0.01	−0.06
X_6	0.07	−0.14	0.09	−0.03	−0.14	1.00	−0.08	−0.17	−0.04	0.02
X_7	−0.07	0.02	0.07	0.12	0.00	−0.08	1.00	−0.01	0.12	0.01
X_8	0.12	−0.09	−0.12	0.17	0.13	−0.17	−0.01	1.00	0.02	−0.03
X_9	0.00	−0.02	0.09	0.00	0.01	−0.04	0.12	0.02	1.00	0.06
X_{10}	0.02	0.10	0.08	0.00	−0.06	0.02	0.01	−0.03	0.06	1.00

The algorithm calculates eigenvalue, eigenvector and proportion of co-variance coefficient matrix and further extracts the principle component. Apply data from Table 4 to Eqs. (4) and (5) to obtain the proportion and cumulative of each principle component, as displayed in Fig. 2. The first principle component has the highest proportion which contains information involved in 15.63% of raw variables. Cumulative of the first 8 principle components reaches 86.96% that larger than 85%, consequently, original 10 figures are replaced to achieve dimension reduction according to principle component decision law.

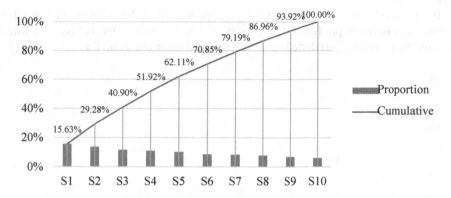

Fig. 2. Proportion and cumulative of principle component

The algorithm extracts eigenvector of principle component. The first 8 principle components extracted from 205×10 dimension being standardized is employed to establish the reconstructed sample set (205×8 dimension) of PCA-RBFNN predictive model, the results are illustrated in Table 5.

Table 5. Standardized data

No.	S_1	S_2	S_3	S_4	S_5	S_6	S_7	S_8	Y_1	Y_2
1	1.31	0.15	0.41	−1.01	1.75	0.83	0.40	−0.18	0.90	2.00
2	0.17	0.08	2.19	−0.07	0.98	−1.13	0.94	−0.13	0.93	1.00
3	0.32	0.22	0.49	−0.36	2.62	0.91	−0.35	−0.66	0.96	1.00
4	−1.91	−0.55	−0.91	1.54	−0.27	−0.53	−0.94	0.20	0.85	3.00
5	0.61	−2.13	−0.38	0.34	1.13	−0.58	0.39	1.83	0.76	5.00
...	...	...	...				...	...	...	...
201	−0.88	−0.06	0.41	−1.59	−0.87	0.04	−0.14	−0.07	−1.51	1.00
202	1.45	−1.67	0.12	−0.72	−0.80	0.40	1.59	1.76	−1.74	5.00
203	2.76	2.41	−0.67	2.00	0.98	0.47	0.03	0.94	1.98	4.00
204	−0.50	−0.96	0.04	1.74	0.85	0.63	−1.21	0.53	1.64	2.00
205	0.00	0.25	0.43	0.17	−0.33	0.94	−0.67	−1.17	0.71	2.00

4.3 Model Training

In this paper, we develop a 3-layer RBF neural network model. There are 8 neural cells in the input layer in the backward RBFNN since 8 principle components are extracted. Targeting at the performance and ranking range of honors students, the number of neural cells in output layer is 2. The optimal value of neutral cells in the hidden layer is determined to be 100 after the repeated experiments. Consequently, the topological structure of the PCA-RBFNN model is finally defined as 8-100-2. Targeted mean square error is 0.0001, expansion rate of radical basis function is 1, as well as that Gaussian function is adopted to be the radical basis function of cells in the hidden layer of the RBF network.

Employ 40 items in testing data set in pre-determined model to get prediction value. Comparison between predicted performance and real ones is stated in Fig. 3, in which absolute value error of predicted score is 0.057 and error rate is 6.78%.

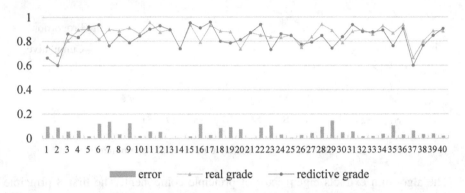

Fig. 3. Fitting of PCA-RBFNN performance prediction model

For ranking range of scores: value below 1.5 is defined as student in Category 1; 1.5–2.5 as student in Category 2; 2.5–3.5 as Category 3; 3.5–4.5 as Category 4, and value above 4.5 as Category 5. Comparison between predicted ranking range and real ones is illustrated in Fig. 4.

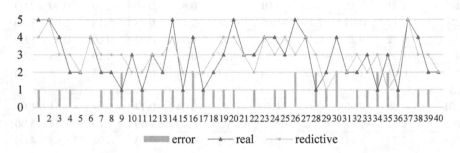

Fig. 4. Fitting of ranking category of PCA-RBFNN performance prediction model

As listed in Table 6, for students in training set, fitting result of 61.21% of student is entirely the same with their real category, while there are 92.12% of them have difference between fitting and real category within 1. For training set, 25% of students have 0 difference, and most of students (82.50%) possess differences within 1, which is a desirable outcome.

Table 6. Fitting of ranking percentage of PCA-RBFNN performance prediction model

	Differences among ranking range of predicted performance	No. of students	Ratio
Training data set (165)	0	101	61.21%
	1	51	30.91%
	2	13	7.88%
Testing data set (40)	0	10	25.00%
	1	23	57.50%
	2	7	17.50%

4.4 Model Verification

In order to state the effectiveness of PCA-RBFNN predictive model, this paper has also established standard RBF Neutral Network model (RBFNN) and BP Neutral Network (BPNN) model constructed by resilient BP algorithm to compare. Adopt standardized data in Table 2 directly into both referred models and apply the first 80% of data (165 items) as training set while the rest 20% (40 items) as testing set. After repeated experiments, with highest prediction accuracy and fastest convergence speed, the final topology structure is determined: 10-100-1 for standard RBFNN and 10-5-1 for BPNN. The targeted accuracy of 3 model is set as: 0.0001 of targeted mean square error, 20000 of maximum training times of BPNN, and 0.1 of learning rate.

Employ generalized ability comparison test to all PCA-RBFNN predictive model, single RBF Neutral Network model and BP Neutral Network model. Contrast between prediction and real data of the three models is shown in Fig. 3. Mean error rate of PCA-RBFNN predictive model, single RBF Neutral Network model and BP Neutral Network model are respectively 6.78%, 10.30% and 11.76%, and maximum error are 16.53%, 26.52% and 20.85% (Fig. 5).

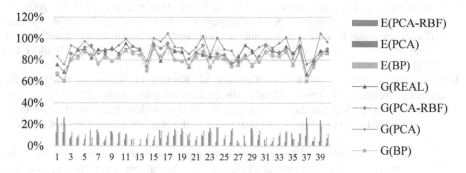

Fig. 5. Fitting of PCA-RBFNN performance prediction model

Model As concluded from the comparison, PCA-RBFNN model performs better on approximation ability than single RBFNN and BPNN. Moreover, simulation experiment has stated that PCA-RBFNN possess the fastest convergence speed among all three predictive models. The reason exists in the simplest topology structure of PCA-RBFNN model and less complexity of model than the other two, which ultimately determines the referred faster convergence speed.

5   Conclusions

Based on PCA, this paper proposed a method mapping primary multi-dimensional vectors to low dimension space and form low-dimension vectors mutually independent so that interference between each vector is reduced to the biggest extent. Consequently, the structure of RBF Neutral Network is simplified, in which the number of neurons in input layer and hidden layer is cut down, moreover, the calculation of network is reduced so that the convergence rate and prediction accuracy are improved. The simulation result has stated that the built PCA-RBFNN predictive model is better than single RBF Neutral Network model and BP Neutral Network model. With faster convergence speed and better operable features in practical, the proposed model is reliable in the prediction of performance and ranking range.

Among honors students, the result of performance and ranking range prediction could be performed to identify students at early stage of risk and those with huge potential to be top talents. It could be further referred as guidance to allow instructors to provide appropriate advising, as well as for management departments to make more scientific decisions.

Acknowledgement. This work was supported in part by grant from State Key Laboratory of Software Development Environment (Funding No. SKLSDE-2017ZX-03) and NSFC (Grant No. 61532004).

References

1. Achterberg, C.: What is an honors student? J. Natl. Coll. Honor. Counc. **6**, 75–81 (2005)
2. Minaei-Bidgoli, B., Kashy, D.A., Kortmeyer, G., Punch, W.F.: Predicting student performance: an application of data mining methods with an educational web-based system. In: The Proceedings of the 33rd ASEE/IEEE Frontiers In Education Conference, Boulder, CO (2003)
3. Elbadrawy, A., Polyzou, A., Ren, Z., Sweeney, M., Karypis, G., Rangwala, H.: Predicting student performance using personalized analytics. Computer **49**, 61–69 (2016)
4. Al-Radaideh, Q., Al-Shawakfa, E., Al-Najjar, M.: Mining student data using decision trees. In: International Arab Conference on Information Technology, Yarmouk University, Jordan (2006)
5. Dekker, G., Pechenizkiy, M., Vleeshouwers, J.: Predicting students drop out: a case study. In: Proceedings of the 2nd International Conference on Educational Data Mining, Cordoba, Spain, pp. 41–50 (2009)
6. Tanner, T., Toivonen, H.: Predicting and preventing student failure – using the k-nearest neighbour method to predict student performance in an online course environment. Int. J. Learn. Technol. **5**, 356–377 (2010)
7. Kotsiantis, S.B., Pierrakeas, C.J., Pintelas, P.E.: Preventing student dropout in distance learning using machine learning techniques. Appl. Artif. Intell. Int. J. **18**, 411–426 (2004)

8. David, A.H., Daniel, R.J.: Mean squared error of estimation or prediction under a general linear model. J. Am. Stat. Assoc. **87**, 724–731 (1992)
9. Yilmaz, I., Kaynar, O.: Multiple regression, ANN (RBF, MLP) and ANFIS models for prediction of swell potential of clayey soils. Expert Syst. Appl. **38**, 5958–5966 (2011)
10. Han, H.G., Chen, Q.L., Qiao, J.F.: An efficient self-organizing RBF neural network for water quality prediction. Neural Netw. Off. J. Int. Neural Netw. Soc. **24**, 717–725 (2011)
11. Zhao, Z.Q., Huang, D.S., Sun, B.Y.: Human face recognition based on multi-features using neural networks committee. Pattern Recogn. Lett. **25**, 1351–1358 (2004)
12. Li, M., Tian, J., Chen, F.: Improving multiclass pattern recognition with a co-evolutionary RBFNN. Pattern Recogn. Lett. **29**, 392–406 (2008)
13. Deng, J.: Structural reliability analysis for implicit performance function using radical basis function network. Int. J. Solids Struct. **43**, 3255–3291 (2006)
14. Zhang, A., Zhang, L.: RBF neural networks for the prediction of building interference effects. Comput. Struct. **82**, 2333–2339 (2004)
15. Yang, J., Zhang, D., Frangi, A.F., Yang, J.Y.: Two-dimensional PCA: a new approach to appearance-based face representation and recognition. IEEE Trans. Pattern Anal. Mach. Intell. **26**, 131–137 (2004)
16. Tong, Y., She, J., Meng, R.: Bottleneck-aware arrangement over event-based social networks: the max-min approach. World Wide Web J. **19**(6), 1151–1177 (2016)
17. She, J., Tong, Y., Chen, L., et al.: Conflict-aware event-participant arrangement and its variant for online setting. IEEE Trans. Knowl. Data Eng. (TKDE) **28**(9), 2281–2295 (2016)
18. She, J., Tong, Y., Chen, L.: Utility-aware event-participant planning. In: Proceedings of the 34th ACM SIGMOD International Conference on Management of Data (SIGMOD 2015), pp. 1629–1643 (2015)
19. She, J., Tong, Y., Chen, L., et al: Conflict-aware event-participant arrangement. In: Proceedings of the 31st International Conference on Data Engineering (ICDE 2015), pp. 735–746 (2015)
20. Shanker, M., Hu, M.Y., Hung, M.S.: Effect of data standardization on neural network training. Omega **24**, 385–397 (1996)
21. Montminy, D.P., Baldwin, R.O., Temple, M.A., Laspe, E.D.: Improving cross-device attacks using zero-mean unit-variance normalization. J. Cryptogr. Eng. **3**, 99–110 (2013)
22. Issariyakul, T., Hossain, E.: Introduction to Network Simulator NS2. Springer US, New York (2009)
23. Chen, S., Cowan, C.F.N., Grant, P.M.: Orthogonal least squares learning algorithm for radicalbasis function networks. IEEE Trans. Neural Netw. **2**, 302–309 (1991)
24. Wang, M., Yang, S., Wu, S., Luo, F.: A RBFNN approach for DoA estimation of ultra wideband antenna array. Neurocomputing **71**, 631–640 (2008)

DKG: An Expanded Knowledge Base for Online Course

Haimeng Duan[✉], Yuanhao Zheng, Lei Shi, Changhong Jin, Hongwei Zeng, and Jun Liu

SPKLSTN Lab, Department of Computer Science, Xi'an Jiaotong University, Xi'an, China
duanhaimeng@gmail.com, yuanhaozheng521@gmail.com, xjtushilei@foxmail.com, jinchanghonguk@gmail.com, zhw1025@gmail.com, liukeen@mail.xjtu.edu.cn

Abstract. Recent years have witnessed a proliferation of large-scale online education platforms. However, the learning materials provided by online courses are still finite. In this paper, to expand the learning materials on MOOC platforms, we construct an expanded knowledge base named DKG. DKG combines priori knowledge from concept map with extended textual fragments collected from web sources. For the sake of DKG's quality, we also propose a supervised method with four novel features to evaluate the quality of textual fragments. Finally, we conduct experiments on four online courses. The results show that our method can find good textual fragments efficiently and expand learning materials successfully.

1 Introduction

Recently, several large-scale Massive Open Online Courses (MOOC) platforms such as Coursera[1], xuetangx.com[2] have been built. High quality learning resources from top universities around the world make online education widely popular. However, massive courses do not mean real convenient learning. Actually, most MOOC institutions merely provide lecture videos, syllabuses and other original learning materials. Furthermore, as the main exchange platform, the discussion forum is not perfect and cannot resolve learners' difficulties in time [1]. As a result, online learners usually need to search for information on the Internet manually when they have some doubts [2].

Meanwhile, community networks [3] such as Quora[3] and Stackoverflow[4] are major sources but not educational data sources for online learners to acquire knowledge now for they provide sufficient information for learners. This motivates us to construct a novel knowledge base combining knowledge extracted from online courses with relevant texts collected from community networks. This

---

[1] www.coursera.org/.
[2] www.xuetangx.com/.
[3] https://www.quora.com/.
[4] http://stackoverflow.com/.

© Springer International Publishing AG 2017
Z. Bao et al. (Eds.): DASFAA 2017 Workshops, LNCS 10179, pp. 376–386, 2017.
DOI: 10.1007/978-3-319-55705-2_30

expanded knowledge base can provide better services for online learners since it offers abundant knowledge of relevance directly.

In this paper, for a given course on MOOC platform, we first obtain priori knowledge by constructing a concept map, and then collect relevant textual fragments from community websites as expanded knowledge. To ensure the quality of the collected textual fragments, we evaluate them by employing a supervised method with four novel features. Finally, both of the concept map and fragments consist of DKG (Domain Knowledge Graph). The contributions of this paper are as follows:

- We construct a novel knowledge base named DKG. DKG is different from previous work in education area because it combines structured priori course information with unstructured expanded knowledge from community networks.
- We apply a supervised learning method with four features to evaluate the quality and relevance of textual fragments collected from community networks. Experiments on four real datasets have been conducted to test the effectiveness of our method. The results show our method works well.

For the rest of this paper, Sect. 2 will introduce related work. Section 3 gives the definitions and overview of our problem. The details of our construction model is presented in Sect. 4. Section 5 discusses the experimental results. Section 6 presents the conclusion and future work.

2 Related Work

For we need to construct a structured concept map as priori knowledge with textual fragments of high quality from community networks as expanded knowledge, we have focused on two main researches.

A concept map is a graphical tool which shows concepts and their kinds of relationships [4]. Researches about concept map are mainly divided into extracting key concepts and identifying relations between concepts. Such as Navigli et al. [5] used ontology learning to extract concepts of a specific domain, Liu et al. [2] and Ruiz-Sánchez et al. [6] used knowledge acquisition from text taxonomic to find relationships between concepts. Additionally, Wang et al. [7] presented a framework for constructing a concept map from textbooks which jointly optimizes concept extraction and prerequisite relations identification. Association-rule mining [8] and text mining [9] techniques are often used to complete these related functions.

Quality evaluation of textual fragments from community networks also attracts many researchers attention. Studies are presented to evaluate the quality of questions, answers, blogs and comments respectively. Agichtein et al. [11] proposed a method to find high-quality answers by using question-answer relationships. Weimer et al. [12] utilized a classification framework to find high-quality content in social media. Massoudi et al. [13] evaluated the quality of posts on online discussions in the software domain with five classes of features. Other works such as [13, 14] studied how to find high-quality blogs and comments of blog.

3   Approach Overview

3.1   Problem Definition

Before introducing our whole framework to construct a DKG, we first introduce some important definitions and take the course "Data Structure" as an example to illustrate.

Definition 1 (Concept Map). A concept map CM is represented as a set of triples in the form of $\{(c_1, c_2, r)|c_1, c_2 \in C, r \in R\}$ where $C = \{c_i\}_n$ is a set of n concepts, such as "Binary tree", "Shortest path", "Hashtable". $R = \{0, 1\}$ is the prerequisite relationship. $r = 1$ means c_1 is the prerequisite of c_2 and $r = 0$ means c_1 and c_2 have no prerequisite relation.

Definition 2 (Textual Fragment). For a concept $c_i \in C$, it has a set of fragments $F_i = \{f_{i1}, f_{i2}, ..f_{ij}, ...\}$. The j-th fragment of it, f_{ij}, may be a blog or a question and answer (Q&A) pair. If f_{ij} is a blog, it consists of the title and content of the blog which represented as b_{ij}. If f_{ij} is a Q&A pair, it consists of a problem and the top-5 answers ranked by votes which represented as $qap_{ij} = \{q_{ij}, a_{j1}, a_{j2}, ...a_{j5}\}$.

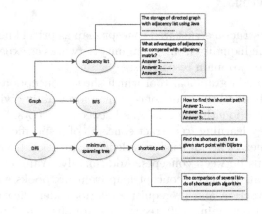

Fig. 1. Part of the DKG of the course "Data Structure".

Now we formalize the problem of constructing DKG. Given a course on a MOOC platform, the output will be a knowledge graph named DKG represented as a set of triples in the form of $\{(c_1, r, c_2, F_2)|c_1, c_2 \in C, r \in R, F_2 \in FRQ\}$, where $C = \{c_i\}_n$ is a set of n concepts and $R \subseteq C \times C$ represents the prerequisite relations. C and R are obtained from a priori concept map CM constructed from learning materials on MOOC platform. $FRQ = \{F_i\}_n$ is a collection of all the textual fragments sets, each fragments set $F_i = \{f_{i1}, f_{i2}, \ldots f_{ij}, \ldots\}$ corresponding to a concept $c_i \in C$. $f_{ij} \in F_i$ represents the j-th fragment and is obtained from community networks as the expanded knowledge to c_i. f_{ij} is considered to have high quality and relevance. f_{ij} can be represented as b_{ij} or qap_{ij} according to whether it is a blog or a Q&A pair in specific scene.

Figure 1 shows an example of a partial DKG of the course "Data Structure", where each circle node is a concept in "Data Structure", such as "BFS", "shortest path", a direct link indicates prerequisite dependency between the two concepts (from prerequisite to subsequence), fragments may include Q&A pair to describe "How to find the shortest path?" and blog about "Find the shortest path for a given start point with Dijkstra", and all the assembled fragments have high quality and relevant to the concepts.

3.2 Workflow

A DKG can be generated by employing a two-module framework. The construction process of DKG is shown in Fig. 2.

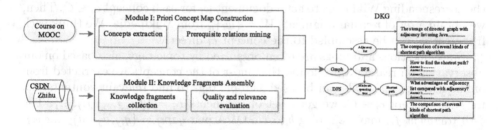

Fig. 2. The construction process of DKG.

Module I constructs a prior concept map CM through extracting key concepts and mining prerequisite relations. This step will produce the set of concepts C and the set of prerequisite relations R.

Module II collects the set of knowledge fragments $F_i = \{f_{i1}, f_{i2}, ..f_{ij}, ...\}$ from community networks for each concept $c_i \in C$ extracted in module I, and then assembles the good textual fragments to the concept map. This module also trained a classifier offline to evaluate the relevance and quality of the collected textual fragments by employing a supervised method with four novel features.

4 DKG Construction

4.1 Priori Concept Map Construction

Learning materials of online courses often provide a comprehensive list of domain concepts. They consist of major educational resources from universities and almost contain all the contents of the textbooks. Since Wang [7] proposed a jointly framework to extract key concepts and mining prerequisite relations from textbooks at the same time, we use their method directly and focus on online materials of a specific course. After this step we acquire a concept map CM, that is a set of n concepts $C = \{c_i\}_n$ and the prerequisite relations $R \subseteq C \times C$.

4.2   Knowledge Fragments Assembly

This part presents how to assemble knowledge fragments on concept map. The first step is collecting fragments from community networks, the second step is assessing the quality and relevance of the collected fragments to assemble the good ones. Here we have a basic heuristic behind our model based on common sense.

Heuristic 1. Textual fragment for a certain term in Wikipedia is authoritative and high-quality that can be assembled to the CM directly.

Knowledge Fragments Collection: According to **Heuristic 1**, we select Wikipedia as important reference for assessing quality. More concretely, we crawl the corresponding Wiki page to get a document w_i for each concept $c_i \in C$. Then, we obtain a set of textual fragments $W = \{w_i\}_n$ from Wikipedia. We think each fragment w_i can be assembled to the concept c_i directly.

Since questions and blogs are often organized by topic tags, also based on our observation on real data, we found that most of the concepts we extracted from online learning materials can be used as topic tags to find relevant contents. So, for each concept $c_i \in C$, we get a set of fragments $F_i = \{f_{i1}, f_{i2}, ..f_{ij}, ...\}$. The j-th fragment f_{ij} may be a blog b_{ij} or a Q&A pair $qap_{ij} = \{q_{ij}, a_{j1}, a_{j2}, ...a_{j5}\}$. So we can also represent $F_i = \{b_{i1}, b_{i2}, ...b_{ik}, qap_{i1}, qap_{i2}, ...qap_{im}\}$.

Except the textual contents of a Q&A pair or a blog, we also obtain the following information about community and users of each Q&A pair and blog. We list the information in Table 1.

Quality and Relevance Evaluation: In this sub-section, we introduce the details of our classification method to check whether a fragment is of high quality. In particular, for a Q&A pair $qap_{ij} = \{q_{ij}, a_{j1}, a_{j2}, ...a_{j5}\}$ and a blog b_{ij}, we evaluate its quality with four novel features. One of the features called *Text Similarity* evaluates the relevance of the fragment by contrasting it with corresponding Wiki document. Other features use statistical features, community features, and user features respectively named *Information Sore, Popularity Degree, User Authority*.

1. ***Information Score* (IS).** This feature is used to measure the amount of information contained in a fragment f_{ij}. We consider two kinds of words contribute to the amount of information of a fragment, that is more stop-word means more useless and more concept words means more useful. We represent the frequency of all the stop-words as $fstop$, the frequency of all

Table 1. Community and user information for quality evaluation.

Community information	
For a Q& A pair $qap_{ij} = \{q_{ij}, a_{j1}, a_{j2}, ...a_{j5}\}$	
interestN	Number of votes mark question q_{ij} as interesting
answerN	Total number of answers under question q_{ij}
timespan	The time span between q_{ij} and its latest answer
good1N	Total number of votes of the top-5 answers $a_{j1}, a_{j2}, ...a_{j5}$
comment1N	Total number of comments under this Q& A pair
For a blog b_{ij}	
good2N	Number of votes mark this blog as good
viewN	Number of views visited by users
comment2N	Number of comments under this blog
User information	
For users of a Q&A pair/blog	
submitN	Number of answers/blogs submitted by this users
fansN	Number of followers of this user
idolN	Number of following of this user

the concept words as $fconcept$. So the information score can be calculated by using this formula $IS(f_i) = \frac{1+e^{-fstop}}{1+e^{-fconcept}}$.

2. **Text Similarity (TS).** According to **Heuristic 1**, we calculate the similarity of a fragment f and the corresponding concept's Wiki document. Here we use LDA [15] to represent the document as word frequency vector. Then fragment f_i and corresponding Wiki document w_i can be represented as vector $v(f_i)$ and vector $v(w_i)$. The Text Similarity can be calculated by using cosine function $TS(f_i) = \frac{v(f_i) \cdot v(w_i)}{\|v(f_i)\| \times \|v(w_i)\|}$.

3. **Popularity Degree (PD).** We can see that for a Q&A pair qap_{ij}, if its question has more answers ($answerN$, represent as p_1), more votes ($interestingN$, represent as p_2), its answers have more votes ($good1N$, represent as p_3), it has more comments ($comment1N$, represent as p_4) and longer time span ($timespan$, represent as p_5), it is more likely to be popular and valuable. Here we use an empirical parameter max stands for the maximum of the sum of above features which set 2,000,000. Then the popularity degree for this Q&A pair can be calculated as $PD(qap_{ij}) = \frac{\sum_{k=1}^{5} p_k}{max}$. Similarly, more votes ($good2N$, represent as p_1), comments ($commen2t$, represent as p_2), views ($viewN$, represent as p_3) also means more popular and valuable for a blog b_{ij}, and its popular degree can be represented as $PD(b_{ij}) = \frac{\sum_{k=1}^{3} p_k}{max}$.

4. **User Authority (UA).** Features about users' popularity and authority can also be a standard measure of the quality of a fragment. For a blog b_{ij}, the user means the author of this blog, the number of fans of b_{ij}'s author represented as $fans(b_{ij})$, the number of idols of b_{ij}'s author represented as $idol(b_{ij})$. So

$UA(b_{ij}) = \frac{fans(b_{ij})}{fans(b_{ij})+idol(b_{ij})}$. For a Q&A pair $qap_{ij} = \{q_{ij}, a_{j1}, a_{j2}, ...a_{j5}\}$, the user means the "average user" of the five authors of the top-5 answers, so $UA(qap_{ij}) = \frac{\sum_{k=1}^{5} \frac{fans(a_{jk})}{fans(a_{jk})+idos(a_{jk})}}{5}$.

5   Experiments

In this section we first introduce our experiments on constructing concept maps and collecting fragments for four courses including "Data Structure", "C Language", "Advanced Math" and "Geometry". Then we introduce our experiments on training a classifier to determine whether a fragment is of high quality.

Table 2. Experimental data statistics.

Course	Concept	Prerequisite relations	Q&A pairs	blogs	Total fragments	Good fragments
Data Structure	94	282	800	542	1342	510
C Language	109	278	584	487	1071	356
Advanced Math	86	224	322	209	531	232
Geometry	42	112	156	134	290	117

5.1   The Datasets

We established a platform to construct four concept maps using the method above. The statistical results of four datasets for online course "Data Structure", "C Language", "Advanced Math" and "Geometry" are shown in Table 2, the second and third column represent the number of concepts and prerequisite relations respectively, that is the size of C and R. After this step, we exactly get four CMs. Since the method to construct concept map has been proved to be effective, we do not evaluate it here.

Then, in module II, for each concept c_i we got in module I, we collected Q&A pairs from Zhihu[5] and blogs from CSDN[6] by using a Java crawler. The popular Chinese Word Segmentation tool NLPIR[7] was used to split the fragments. The original Q&A pairs and blogs number are shown in the fourth and fifth column respectively. The following experiments evaluated the quality of the textual fragments collected.

---

[5] https://www.zhihu.com/.
[6] http://www.csdn.net/.
[7] http://ictclas.nlpir.org/.

Table 3. Community and user information for quality evaluation.

	LR	DT	SVM	RF	Naïve Bayes
Precison	0.729	0.756	0.896	0.852	0.827
Recall	0.794	0.861	0.925	0.929	0.678
F1-Measure	0.760	0.805	0.910	0.889	0.745

5.2 Quality Evaluation of Knowledge Fragments

We applied a supervised method to evaluate whether a fragment is a good one. The popular java data mining toolkit Weka[8] was used to train our learning model in this module. We selected 100 concepts randomly from the four datasets we constructed in last step, and then submitted them to the search bar of Zhihu and CSDN respectively. We got 632 Q&A pairs and 461 blogs. There were totally 1093 fragments and average 10 fragments for each concept. For labeling whether a fragment is a good one, we hired nine undergraduate students in their junior or senior year from the computer science department. The students were asked to annotate the collected fragments by using their own knowledge background. We annotated examples and offered guidelines to them. At last we got 408 positive fragments and 685 negative fragments. We selected 800 fragments from positive and negative examples equally to train model.

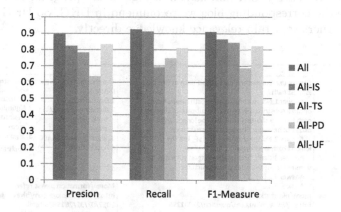

Fig. 3. Quality evaluation results with features removed.

We then used different learning algorithms to train various models and compare their performance under these models. 10-fold cross validation was performed during model training. Here, we compared Logistic Regression (LR), SVM, Decision Tree (DT), and Random Forest (RF), Naïve Bayes, the results are shown in Table 3. The learned SVM model achieved the best F1-Measure.

---

[8] http://www.cs.waikato.ac.nz/ml/weka/.

Finally, we studied the contributions of different features. Here, we chose five groups of feature combinations, namely all features (All), all features except *Information Score*, *Text Similarity*, *Popularity Degree*, *User features* (All-IS, All-TS, All-PD, All-UF) respectively. The same training data was used to learn five different SVM models. The result was shown in Fig. 3. The model using all features performs best. The F1-Measure scores of other models decrease to a certain extent, which indicates all features have some positive impacts to boost the performance of quality evaluation. Moreover, when removing *Popularity Degree* from the feature set, the learned model has the lowest performance. This means that the *Popularity Degree feature* is a key factor to assess the quality of a fragment.

After getting the trained SVM classifier, we evaluated all the fragments collected. The number of good fragments of each course is shown in the last column in Table 2. Good fragments for each course make up two-fifths of all fragments we collected, which is consistent with the ratio of good fragments in our annotation data. This means our classifier can find good fragments efficiently. Meanwhile, each concept gets 4–7 fragments, for the concept "Shortest path" a real Q&A pairs and a real blog included by our constructed DKG of the course "Data Structure" are listed below in Fig. 4. With this DKG, online learners don't need to post on the discuss forum and wait for a long time when they want to know "Whether the method for shortest path can be used to find the longest path", the relevance Q&A pair will give him a satisfactory answer. And if he wants to fulfill his own Dijkstra algorithm, he can also get a full program in C language by reference the corresponding blog we recommend in DKG. So DKG is helpful for online learners to attain relevance knowledge directly.

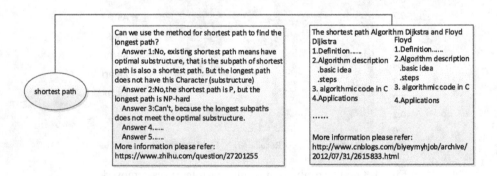

Fig. 4. Real Q&A pairs and blog included in DKG.

6   Conclusion

To offer a comprehensive and convenient learning style for online learners, we proposed a novel knowledge base called DKG in this paper. Considering that online learners usually need to get help from community networks when they have difficulties in learning, DKG combines priori knowledge from concept map constructed from online materials with extended knowledge collected from community networks. We also proposed a supervised method with four novel features to evaluate the quality of textual fragments collected from community networks. Results of experiments on four online courses showed our method can find good textual fragments efficiently and expand learning materials successfully. The potential direction for future research is focusing on a more fine-grained approach to obtain reliable facts from fragments.

Acknowledgments. We would like to thank the anonymous reviewers for their great efforts in improving the quality of the paper. The work was supported in part by the National Science Foundation of China under Grant Nos. 61672419, 61532004, 61532015, the National Key Research and Development Program of China under Grant No. 2016YFB1000903, the MOE Research Program for Online Education under Grant No. 2016YB166, the Fundamental Research Funds for the Central Universities.

References

1. Vigentini, L., Clayphan, A.: Exploring the function of discussion forums in moocs: comparing data mining and graph-based approaches. In: Proceedings of the Second International Workshop on Graph-Based Educational Data Mining (GEDM 2015). CEUR-WS (2015)
2. Liu, J., Jiang, L., Wu, Z., Zheng, Q., Qian, Y.: Mining learning-dependency between knowledge units from text. VLDB J. **20**(3), 335–345 (2011)
3. Wellman, B.: Networks in the Global Village. JSTOR (1999)
4. Novak, J.D., Cañas, A.J.: The theory underlying concept maps and how to construct and use them (2008)
5. Navigli, R., Velardi, P., Gangemi, A.: Ontology learning and its application to automated terminology translation. IEEE Intell. Syst. **18**(1), 22–31 (2003)
6. Ruiz-Sánchez, J.M., Valencia-Garcıa, R., Fernández-Breis, J.T., Martınez-Béjar, R., Compton, P.: An approach for incremental knowledge acquisition from text. Expert Syst. Appl. **25**(1), 77–86 (2003)
7. Wang, S., Ororbia, A., Wu, Z., Williams, K., Liang, C., Pursel, B., Giles, C.L.: Using prerequisites to extract concept maps from textbooks. In: Proceedings of the 25th ACM International on Conference on Information and Knowledge Management, pp. 317–326. ACM (2016)
8. Tseng, S.-S., Sue, P.-C., Su, J.-M., Weng, J.-F., Tsai, W.-N.: A new approach for constructing the concept map. Comput. Educ. **49**(3), 691–707 (2007)
9. Chen, N.-S., Wei, C.-W., Chen, H.-J., et al.: Mining e-learning domain concept map from academic articles. Comput. Educ. **50**(3), 1009–1021 (2008)
10. Wang, X.-J., Tu, X., Feng, D., Zhang, L.: Ranking community answers by modeling question-answer relationships via analogical reasoning. In: Proceedings of the 32nd International ACM SIGIR Conference on Research and Development in Information Retrieval, pp. 179–186. ACM (2009)
11. Agichtein, E., Castillo, C., Donato, D., Gionis, A., Mishne, G.: Finding high-quality content in social media. In: Proceedings of the 2008 International Conference on Web Search and Data Mining, pp. 183–194. ACM (2008)

12. Weimer, M., Gurevych, I., Mühlhäuser, M.: Automatically assessing the post quality in online discussions on software. In: Proceedings of the 45th Annual Meeting of the ACL on Interactive Poster and Demonstration Sessions, pp. 125–128. Association for Computational Linguistics (2007)
13. Massoudi, K., Tsagkias, M., Rijke, M., Weerkamp, W.: Incorporating query expansion and quality indicators in searching microblog posts. In: Clough, P., Foley, C., Gurrin, C., Jones, G.J.F., Kraaij, W., Lee, H., Mudoch, V. (eds.) ECIR 2011. LNCS, vol. 6611, pp. 362–367. Springer, Heidelberg (2011). doi:10.1007/978-3-642-20161-5_36
14. FitzGerald, N., Carenini, G., Murray, G., Joty, S.: Exploiting conversational features to detect high-quality blog comments. In: Butz, C., Lingras, P. (eds.) AI 2011. LNCS (LNAI), vol. 6657, pp. 122–127. Springer, Heidelberg (2011). doi:10.1007/978-3-642-21043-3_15
15. Blei, D.M., Ng, A.Y., Jordan, M.I.: Latent dirichlet allocation. J. Mach. Learn. Res. 3(Jan), 993–1022 (2003)

Towards Economic Models for MOOC Pricing Strategy Design

Yongzheng Jia[1(✉)], Zhengyang Song[1], Xiaolan Bai[2], and Wei Xu[1]

[1] Institute of Interdisciplinary Information Sciences, Tsinghua University,
Beijing, China
jiayz13@mails.tsinghua.edu.cn
[2] Faculty of Education, The University of Hong Kong, Pok Fu Lam, China

Abstract. MOOCs have brought unprecedented opportunities of making high-quality courses accessible to everybody. However, from the business point of view, MOOCs are often challenged for lacking of sustainable business models, and academic research for marketing strategies of MOOCs is also a blind spot currently. In this work, we try to formulate the business models and pricing strategies in a structured and scientific way. Based on both theoretical research and real marketing data analysis from a MOOC platform, we present the insights of the pricing strategies for existing MOOC markets. We focus on the pricing strategies for verified certificates in the B2C markets, and also give ideas of modeling the course sub-licensing services in B2B markets.

1 Introduction

Going on eight years since *massive open online courses* (MOOCs) first entered the scene, MOOCs go from the cameras at the back of college classrooms to new forms of online education ecosystems in the global industry. MOOCs bring a revolution to the worldwide higher education for growing opportunities in the verticals such as online education, lifelong learning, professional training, by offering freely accessible college education to everyone. However, MOOCs have also been criticized heavily by academics and industries for operational sustainability, low completion rate, unprofessional teaching methods, as well as the corporatization of higher education.

MOOC is an ecosystem involving efforts from many parties. The *MOOC platforms* are the core of the ecosystem. Every MOOC platform is a market place where *MOOC producers* (usually universities) delivers their MOOCs to the *users*. The users in the MOOC ecosystem consist of both *Internet users* and *institutional users*. The platforms also offer the users with various value-added *MOOC services* to increase their profitability. There are both for-profit platforms (e.g. Coursera and Udacity) and non-profit platforms (e.g. edX and FUN), and the format of the *MOOC services* varies from platform to platform.

One critical issue to discuss is how does the MOOC ecosystem stay financially stable. In the beginning, both public and private sector fundings flood into these MOOC platforms so they can focus on adding contents and expanding the market

© Springer International Publishing AG 2017
Z. Bao et al. (Eds.): DASFAA 2017 Workshops, LNCS 10179, pp. 387–398, 2017.
DOI: 10.1007/978-3-319-55705-2_31

share. However, after a few years, people not only want MOOC platforms to be financially independent but also provide financial incentives to the MOOC providers. Some MOOC platforms (e.g. xuetangX) are still in the primary stage of marketing, and struggling to make money and show their investors that they can be sustainable, or at least reach the breakeven point in the future.

Unfortunately, there is few academic research formally analyzing the business models for MOOCs in scientific and structured ways. This research is based on our research and the first-hand operation experience in industry. We focus on the pricing strategies of MOOC services in this paper, since the pricing strategy design is a key component of the MOOC business models, and also a common interest of the online education community, networking community, as well as researchers from marketing and economics.

There are two existing business models of MOOC platforms with totally distinct pricing strategy, the *business-to-customer* (B2C) model and *business-to-business* (B2B) model. Unlike the peer-education platforms (e.g. Skillshare and Udemy), there is usually no consumer-to-consumer model (C2C) as MOOCs are usually provided by organizations instead of individuals. While by definition, the MOOCs are free and open-to-all, the MOOC platforms sell value-added *MOOC services* for profit, and it is a common model in Internet services called the *freemium strategy*.

In the B2C markets, MOOC platforms make money by selling services to the *Internet users*, such as *verified certificates*, *specializations* and *online degrees*. In the B2B markets, MOOC services can be used as the form of *Small Private Online Courses* (i.e. SPOCs). The traditional in-classroom higher education can be transferred into a blended mode by using the high-quality MOOC content, and MOOC platforms can make money from *sub-licensing* the content to institutional users, and providing *bundled education services* including content customization, teaching assistant services, SaaS services, and technical supports.

In this work, we discuss our initial attempts to construct reasonable mathematical formulations for the pricing models based on our marketing experience. Due to the space reason, we focus on the pricing model for *verified certificates* in B2C markets. We summarize the key ideas of this paper as following:

First, We focus on the theoretical framework for the pricing strategies for the *verified certificates* in B2C markets. We first propose the basic model of certificate pricing, to solve the case of pricing the certificates for a single MOOC on one MOOC platform without competition. We analyze the profit maximization pricing strategies for MOOC platform and the market equilibrium with maximum social welfare. We further review the user-platform interaction of certificate pricing by using the *Stackelberg game*. Then we analyze the case of considering the purchasing power of users with budget constraints. In this model, we formulate a utility maximization framework to depict the consumers' buying behavior, and we can use the pricing strategies to increase per-user revenue. Finally, we summarize the model of bundled course services and their business initiatives.

Second, We analyze the sales data of certificates from 1236 real MOOCs based on our model. By using data-driven marketing approaches, we get some business and education insights of the users' behavior. We first present an overview of the marketing performance of selling certificates, and further analyze the users' behavior in different cases: the users' *willingness to pay* (i.e. WTP) for best-selling MOOCs and MOOCs with highest payment rate, as well as the users' behavior when a MOOC is repeatedly offered.

Third, We further present the future directions of the work on designing the pricing strategies. There are some other factors which may affect the B2C market in real settings and we should consider them in industry. We also propose the ideas of modeling the B2B sub-licensing markets, and the hybrid model of B2B2C (business-to-business-to-customer) with *cross-platform MOOC exchange and internationalization*. We will try to solve these problems in our future work.

The rest of the paper is organized as follows. We review related work in Sect. 2. We present the pricing strategy for B2C markets in Sect. 3 and analyze the real marketing data in Sect. 4. Finally, we present the directions of our future work in Sect. 5 and conclude the paper in Sect. 6.

2 Related Work

To the best of our knowledge, our work is the first to study the business model and marketing strategy of MOOCs with both theoretical models and data-driven analysis. There are many discussions on the business model and sustainability of MOOCs from the industry and the media. For instance, [1] shows the latest experience of finding niche and business model for MOOC in 2016, and [2] summarizes the business innovations and landscape changes for MOOC in 2016. Existing academic work on MOOC business model is based on case studies, surveys, and other social science methodologies. [3] presents the ideas of involving adaptive learning into the business model design of MOOCs. [4] shows the ideas of designing sustainable MOOC business models by carefully reviewing them for both US and Europe-based MOOC aggregators.

From a theoretical perspective, literatures in multiple fields give us ideas on analyzing pricing strategies by applying economics, optimization theory, and game theory. [5] presents a generalized *Smart Data Pricing* scheme of pricing the network applications to increase efficiency and cope with increased network congestion. [6,7] show the flat-rate pricing scheme for data traffic to obtain higher profits with game theory. To analyze the user behavior by economics, [8] reviews the methodologies of WTP estimation in marketing science, and [9] is a classical structural model of demand estimation in economics.

3 B2C Business Model

The fundamental B2C model of MOOC platforms is to make money from the Internet users with a *freemium* strategy: The basic materials of MOOCs are open and free to all users, and the MOOC platforms also offer fee-based *online value added services* to the users. The basic strategy is to grow the user base of the platform, and then try to cultivate the users' payment habits with online marketing strategies. In this section, we focus on the pricing models of the *verified certificates*, which is the fundamental B2C value-added services. We mainly discuss two models of certificate pricing: One pricing strategy is to maximize the total profit from each MOOC, and the other is to increase the profit from each paying user.

3.1   The Basic Model of Certificate Pricing

We observe the basic market structure of the verified certificates by modeling the following straightforward scenario. We consider that a MOOC (labeled as $\mathcal{M}$) is released on only one MOOC platform. Furthermore, we ignore the competitive relationship across different MOOCs (e.g. MOOC $\mathcal{M}$ may be a machine learning MOOC on Coursera, and there is another similar machine learning MOOC on edX). To simplify the market structure, we treat the MOOC producer (i.e. content providers) of $\mathcal{M}$ and the MOOC platform as a single entity[1] (i.e. the seller in the market).

Therefore the market for $\mathcal{M}$ incorporates a single seller (i.e. the MOOC platforms) and a set of users (i.e. learners on the Internet). Suppose MOOC $\mathcal{M}$ offers the verified certificates with price p for the paying users. We first consider users' decisions of whether to buy the verified certificate of MOOC $\mathcal{M}$. In modeling the user behavior, we suppose that each user acts so as to maximize her net benefit (i.e. consumer surplus), denoted by $U_j(x_j, p)$ for each customer $j \in \{1, 2, \cdots, J\}$ and $x_j \in \{0, 1\}$. The function U_j is the net benefit to a consumer from the utility received in buying the verified certificate or just audit the course. We further use $\bar{V}_j$ to denote the utility to user j of just auditing MOOC $\mathcal{M}$ or earning a free certificate, and use V_j to denote the utility to user j of taking the course and earning a verified certificate (i.e. the *willingness to pay* for the verified certificate). Then we formulate $U_j(x_j, p)$ as following:

$$U_j(0, p) = \bar{V}_j, \quad U_j(1, p) = V_j - p, \quad \forall j \in \{1, 2, \cdots, J\} \tag{1}$$

We then denote the decision of user j under price p as $x_j^*(p)$ to maximize her net benefit, which is also the *demand functions* for user j, and we can calculate $x^*(p)$ as follows:

$$x_j^*(p) = \begin{cases} 1 & \text{if } U_j(1, p) > U_j(0, p) \\ 0 & \text{otherwise} \end{cases} \quad \forall j \in \{1, 2, \cdots, J\} \tag{2}$$

By adding up the demand function of $x_j^*(p)$ for all the users $j \in \{1, 2, \cdots, J\}$, we obtain the *aggregate demand function* of $D(p) = \sum_{j=1}^{J} x_j^*(p)$ to capture the total demand of MOOC $\mathcal{M}$ in the market.

---

[1] The MOOC producers and the platforms build the collaboration based on agreements with revenue sharing terms. So in a single-seller market, we can treat the MOOC producer and the platform as a unity, without considering their internal interest exchanges.

Then we consider the strategy of the MOOC platform. We first observe the cost of the verified certificates. Due to the *economies of scales* [10] for Internet services, the MOOC services has huge fixed cost (including high production cost for a MOOC) but very low marginal cost. To offer verified certificate to one more paying user, the platform will only incur traffic cost, verification cost (i.e. identify the authentic user) and cost for some other value-added services. The marginal cost is small and fixed, we denote it as $\bar{c}$. Therefore the price p should satisfy $p > \bar{c}$. If we can figure out the aggregate demand function of MOOC $\mathcal{M}$, then we can get a profit maximization pricing strategy from the following theorem:

Theorem 1. *For the basic model of verified certificate, the best (i.e. profit maximization) pricing strategy for MOOC $\mathcal{M}$ is:*

$$\bar{p} = argmax_p[D(p) \cdot (p - \bar{c})] \tag{3}$$

Where $\bar{p}$ is the platform's best pricing strategy for MOOC $\mathcal{M}$.

We then analyze the market equilibrium. As the marginal cost $\bar{c}$ is a constant, the market equilibrium occurs at $p = \bar{c}$. One often-analyzed property of this equilibrium is the *social welfare*, which is the sum of the producer surplus and consumer surplus in the market. We use $\mathcal{SW}(p)$ to denote the social welfare at price p, and we have:

$$\mathcal{SW}(p) = \sum_{j \in [J]} U_j(1, p) + \sum_{j \in [J]} x_j^*(p) \cdot (p - \bar{c}) \tag{4}$$

$\mathcal{SW}(p)$ will get its maximum value at the *market equilibrium price* when $p = \bar{c}$ in a perfectly competitive market. We can see that when the MOOC market is highly competitive, the net profit of the platform may diminish. However, MOOC market is at least an oligopoly market. There are limited number of MOOC producers and MOOC platforms in the market. For high-quality MOOC content, the number of competitive players is even smaller. Therefore in the MOOC market, users act as price-takers and have weak bargaining power against the platforms.

The above reasoning, in which a platform chooses a price to offer subject to users' behavior as a function of the price chosen, is an example of a *game* between users and platform. In such a game, several players interact with each other, and each player acts to maximize her utility, which may be influenced by other players' decisions. For instance, in this basic model, we only consider the interaction between the users and MOOC platform for the pricing strategy. We can also consider the interaction between MOOC platforms. This idea leads us to think about some basic principles of game theory in relation to the pricing strategies of the verified certificates.

3.2 A Game-Theoretic View

To demonstrate some of the basics of game theory, we again consider the example above. The user-platform interaction is an example of a *Stackelberg game* [11] in which one player, the "leader" (i.e. platform), makes a decision (i.e., the platform sets a price p for course $\mathcal{M}$'s verified certificate). The remaining players, or "followers" (i.e. users), then make their decisions based on the leaders actions. In the basic model of MOOC certificate pricing, it reflects that the price p will influence the users' decision of whether to buy the certificate (i.e. the demand function $x_j^*(p)$). Stackelberg games often arise in user-platform interactions of the network economy, and we can use the method of *backwards induction* [12] to analyze the Stackelberg games: First, we computes the followers' actions as a function of the leaders decision (in our example, we compute the function $x_j^*(p)$ for user $j \in [J]$). The followers' actions are also called the best response to the leader. Then in the second step, the leader (i.e. platform) takes these actions into account and makes its decision to best respond the users.

Due to the space reason, we omit the analysis of Stackelberg game in details. From the ideas of Stackelberg game, we can see that the best response to the followers of the platform depends on the distribution of the demand function $x_j^*(p)$, and furthermore, the WTP distribution of the verified certificates and the utility gained from auditing the course. In practice, we can also use the ideas from the *backwards induction* to develop experiments to estimate the WTP of the users: The platform can dynamically change the price for certificates (e.g. make a discount) to figure out the WTP distribution at each price level.

3.3 Taking Multiple Courses with Budget Constraints

From the above basic model for certificate pricing, we consider a more complicated model in which each user buys multiple course certificates with budget constraints. Consider a user $j \in [J]$, she plans to take a number of courses on one MOOC platform with a fixed budget constraint B_j during a certain period (e.g. a month or a semester). Due to time limitation, she can take at most K_j MOOCs. There is a total of M courses in the market. Her WTP for the verified certificate of course $m \in [M]$ is $V_{j,m}$ and her utility of audit course m is $\bar{V}_{j,m}$. The price for course m's certificate is p_m. Similar to the basic model, we use $x_{j,m}(p_m)$ to denote whether user j will pay for the certificate of course m under price p_m. Then the user j's behavior can be formulated by solving an integer programming as follows:

$$Z : \text{maximize} \quad \sum_{m \in [M]} x_{j,m}(p_m) \cdot (V_{j,m} - p_m) \tag{5}$$

s.t.

$$x_{j,m}(p_m) \cdot (V_{j,m} - \bar{V}_{j,m} - p_m) \geq 0, \quad \forall j \in [J]; \tag{6a}$$

$$\sum_{m \in [M]} x_{j,m}(p_m) \cdot p_m \leq B_j, \quad \forall j \in [J]; \tag{6b}$$

$$\sum_{m \in [M]} x_{j,m}(p_m) \le K_j, \quad \forall j \in [J]; \tag{6c}$$

$$x_{j,m}(p_m) \in \{0,1\}, \quad \forall m \in [M], j \in [J]. \tag{6d}$$

The objective function (5) is the total net benefit gained of user j. Constraint (6a) guarantees that user j is better-off from buying the certificate, and (6d) is the budget constraint. The user j's strategy is a vector of $x_{j,m}(p_m)$ of whether to buy the certificate for course m or not.

The user j's strategy is the optimal solution of (5), then we try to calculate the best pricing strategy for the platform. Note that when p_m is different for each course, the problem is hard to solve and we will omit the discussions here. However, when p_m is the same for each course $m \in [M]$. We can estimate the demand function of user j as follows:

Theorem 2. *If p_m is same for each course (i.e. $p_m = p, \forall m \in [M]$), the platform tries to maximize the total number of certificates bought from user j. The demand function of user j can be calculated by a function of p, $\{V_{j,m}\}_{m \in [M]}$, $\bar{V}_{j,m}$, K_j and B_j, such that:*

$$D_j(p) = \sum_{m \in [M]} x_{j,m}(p_m) = \mathcal{F}_j\Big(p, B_j, K_j, \{V_{j,m}\}_{m \in [M]}, \{\bar{V}_{j,m}\}_{m \in [M]}\Big) \tag{7}$$

and the aggregate demand function is $D(p) = \sum_{j \in [J]} D_j(p)$.

If we can estimate the WTP for each course of user j and the user's budget distribution, we can use the methodology from Theorem 1 to design the platform's pricing strategy for profit maximization. From this model of analyzing user's purchasing behavior with budget constraints, we get the solution of maximizing the total profit from each paying user.

3.4 Bundled Course Services

Offering non-free verified certificate is the initial and basic business model for the B2C markets. For Coursera, the certificate services expand quickly of featuring a flat-rate price of $49 for all certificates. Revenue estimates suggest that Certificates generated between $8 and $12 million for Coursera in 2014 [13].

In practice, the MOOC platforms always put some of their courses together and add more value-added services to form the *bundled course services*, such as the *Specializations* on Coursera (or the *XSeries* on edX), the *Online Micro Masters* on edX (or the *Nano-degrees* on Udacity), the *Advanced Placement* (i.e. AP) courses and so forth. Offering the *bundled course services* is a strategy to improve the quality of value-added service and provide opportunities to charge the users with higher prices. It is a direction of our future work to model the bundled course services and show their performance.

4 Data-Driven Analysis

In this section, we present the results of applying our models to analyze the real sales data from a MOOC platform with millions of users and give some insights using data-driven marketing methods.

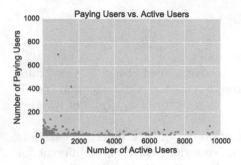

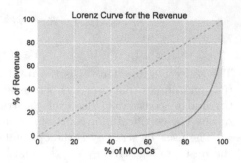

Fig. 1. Relationship between paying users and active users

Fig. 2. Lorenz Curve for the revenue

We use the real sales data of the platform during 2016 for certificate services. The platform offers three kinds of certificate: (1) the *free electronic certificates* that users can download the PDF when they successfully pass the MOOC; (2) the *paper certificates* with counter-forgery marks but no signature tracks from the instructors, and the unit price is 100RMB (about US$14.5), including shipping and handling. (3) the verified certificate with counter-forgery marks, signature tracks and the authentication to the users, and the unit price is 300RMB (about US$43).

There is a total of 1236 MOOCs available on the platform during 2016, and some MOOCs offer several times in different semesters, and we treat them different MOOCs. By the end of December, there are 1140 MOOCs already closed. We use the sales data of the closed courses in the analysis. Due to confidentiality reasons, we only show some basic statistical results.

4.1 Differentiation and Inequality of Revenue Generation

Figure 1 shows the relationship between the active users and the paying users. The *active users* are the users who have learning activities such as doing the homework, and the paying users are the users who buy either the paper certificate or the verified certificate. We can see that most courses are located at the bottom-left corner with a small number of active users and paying users. There is no direct relationship between the number of active users and paying users, and many factors of the courses such as difficulties, popularities, and practicability, may affect the relationship.

We also use the *Lorenz Curve* [14] to show the inequality of the revenue generation for different MOOCs. Figure 2 shows the Lorenz Curve for the revenue of the MOOCs on the platform. We can calculate that the *Gini coefficient* for the certificate market is 0.838, and the top 15% of the most profitable MOOCs create more than 80% of the total revenue for B2C services.

Based on our model in Sect. 3, we use the following method to infer the WTP of the users from the sales data: Recall the definition V_j as the WTP of user j for the certificate. If user j complete the course, we assume $V_j > 0$; if user j buys a

Table 1. WTP distributions for the best-selling MOOCs.

Subject category	Completion rate	$WTP > 0$	$100 \leq WTP < 300$	$WTP \geq 300$
Accounting	2.9%	870	315	381
Marketing	1.3%	362	73	69
Entrepreneurship	1.2%	385	48	63
Accounting	1.6%	110	24	48

Table 2. WTP distributions of offering a MOOC multiple times.

Semester	Completion rate	$WTP > 0$	$100 \leq WTP < 300$	$WTP \geq 300$
Fall 2015	2.9%	870	315	381
Spring 2016	1.3%	566	184	236
Summer 2016	1.8%	257	73	99

paper certificate, then $100 \leq V_j < 300$; if user j buys a verified certificate, then $V_j \geq 300$. From the sales data, we get the number of users with their WTPs located in the three intervals of $(0, 100)$, $[100, 300)$ and $[300, \infty)$. Now we show some insights of analyzing the data based on our model (Tables 1 and 2).

4.2 The Best-Selling MOOCs

We observe the data from the best-selling courses and estimate the user's WTP. The top four best-selling MOOCs are all economic and management courses: two accounting courses, one marketing course, and one entrepreneurship course.

We find that the best-selling economic courses have similar WTP distributions. More users prefer the verified certificate to the paper certificate for each course, which shows that the paying users care more about the quality of service when the course is popular and useful.

4.3 Offer the Same MOOC Repeatedly

Our model in Sect. 3 indicates that the WTP distribution remains the same in different settings. We observe the case of offering a MOOC multiple times. We select the most popular accounting course on the platform and compare the sales data in three semesters.

We can see that the proportional relations of the three values for each semester are almost the same, which shows that the WTP distribution for a course does not change when we offer it multiple times, and indicating that the WTP distribution may be affected more by the intrinsic properties of the course (e.g. quality, usefulness) instead of the external factors. On the other hand, we also see that the total number of paying users declines when we rerun a course multiple times, indicating the *law of diminishing returns* [15] in economics (Table 3).

Table 3. WTP distributions for the MOOCs with the highest payment rate.

Subject category	Completion rate	$WTP > 0$	$100 \leq WTP < 300$	$WTP \geq 300$
Financial engineering	0.24%	21	3	16
Computer science	0.45%	42	12	69
Mathematics	0.82%	9	3	5
Computer science	0.35%	29	8	17

4.4   MOOCs with the Highest Payment Rate

We then observe the sales data for the courses with highest *payment rate*, which is the proportion of paying users among the users completing the MOOC.

We observe that the MOOCs with highest payment rate are those science and engineering courses with high estimated efforts to complete. Also, the paying users for these courses have higher WTPs as they have already invested much time in the courses.

5   Future Work

In this section, we present our plans for improving our model for B2C markets by considering more factors affecting the MOOC profitability. We also propose the ideas of modeling the B2B course sub-licensing market.

5.1   Other Factors Affecting the B2C Markets

To make our model straightforward, we only adopt some basic variables to the models in Sect. 3. There are some other factors in consideration when we apply the models in industry and we will verify the relationship between these factors and the marketing performance in our future work.

Growing User Bases: The size of the user base of MOOC platform is dynamically increasing. In practice, we care more about the proportion of paying users among the active users than the actual number of paying users. Under this setting, the pricing strategy is a trade-off between the proportion of paying users and the net revenue gained from each user.

Competitions Among Platforms: The MOOC market is an oligopoly market with a limited number of competitive MOOC platforms, and each platform has millions of registered users, as small players lack the source of MOOC producers and can not afford the high production cost. There are always similar courses among different MOOC platforms (e.g. various kinds of data science MOOCs). The differentiation strategy is a fundamental way to attract the users. In addition, some platforms occasionally offer discounts for their value-added services, some platforms reorganize the courses into bundled courses, and some platforms offer more attractive value-added services.

Externalities: For MOOC services, We may also consider the case that users impose externalities on each other. For instance, there may be a positive externality in which a user's learning outcome improves as other users give her more help on the discussion forum. On the other hand, one could also observe negative externalities, for instance, one user's resource consumption may affect the other users' experience of watching MOOC videos.

Seasonality: MOOC services have a strong seasonality. In winter and summer vacations, there are almost no B2B sub-licensing services to offer since the institutional users are on vacation. On holidays, individual users have more time on the Internet and the motivated learners will spend more time on taking MOOCs. At the end of each semester, MOOCs may conflict with the in-classroom courses of the college users, and thus attract less active users.

5.2   Modeling the B2B Market

Although the B2B model arises less attention to both the industries and academics in the MOOC ecosystem, it also plays an important role for the MOOCs, and brings more revenue to some early-stage MOOC platforms than B2C services. In the B2B *course sub-licensing* market, the MOOC platform is the seller in the market, and users are organizations with the demand of using the MOOC content on the platform for education purpose. Since the copyright of the licensed MOOCs does not belong to the platform, the MOOC platform should first get sub-licensing approval from the MOOC producers and share revenue with them.

The sub-licensing service is an excellent way to help universities improving their teaching outcomes by importing high-quality MOOC content from MOOC platform. In practice, the B2B services always exist in the format of *SPOCs* (i.e. Small Private Online Courses) by using blended teaching and learning approaches. To guarantee the quality of service, the sub-licensing services are always dynamic with highly customized bundles including various education services, such as MOOC contents, teaching assistance services, SaaS services, technical supports and so forth. As the users' demands are dynamic and complicated, we can no longer use the flat-rate pricing for the B2B markets. The auction-based pricing scheme can better fit the market settings. We will formulate and analyze the auction-based pricing scheme for the B2B market in our future work.

Another cutting-edge B2B business model is *cross-platform MOOC exchange and internationalization*, which is a hybrid B2B2C (i.e. business-to-business-to-customer) business model. The two platforms both benefit from the content and revenue sharing collaboration, as it will help each platform make money from some hard-to-reach secondary markets. An experiment is that edX sub-license some of its courses to xuetangX, by allowing xuetangX to translate the materials into Chinese and provide localized teaching assistant services [16].

6   Conclusion Remarks

Working in the MOOC industry for the past three years, we realize the importance of designing appropriate marketing strategies to sustain our business. Even though there is few academic work on analyzing the business models for MOOCs in scientific and structured ways, we make our initial attempts to construct mathematical models to capture the insights of MOOC markets and use data-driven marketing approaches to verify our models. We hope that our current and future research will bring more ideas to both the industries and the academics for the sustainable development of MOOC ecosystems.

References

1. Morrison, D.: Need-to-Know MOOC News: MOOCs Find Their Niche and Business Model in 2016. https://onlinelearninginsights.wordpress.com/2016/02/03/need-to-know-mooc-news-the-mooc-business-model-gets-its-teeth-in-2016/. Accessed 2016
2. State of the MOOC 2016: A Year of Massive Landscape Change For Massive Open Online Courses. http://www.onlinecoursereport.com/state-of-the-mooc-2016-a-year-of-massive-landscape-chhange-for-massive-open-online-courses
3. Daniel, J., Cano, E.V., Cervera, M.G.: The future of MOOCs: adaptive learning or business model? Int. J. Educ. Technol. High. Educ. **12**, 64–73 (2015)
4. Bacsich, P.D.: Business models for opening up education (2016). http://www.dtransform.eu/wp-content/uploads/2016/04/O1-A2Business-models-edition-1-final.pdf
5. Sen, S., Joe-Wong, C., Ha, S., Chiang, M.: Smart data pricing: using economics to manage network congestion. Commun. ACM **58**, 86–93 (2015)
6. Zhang, L., Weijie, W., Wang, D.: Time dependent pricing in wireless data networks: flat-rate vs. usage-based schemes. In: Proceedings of the IEEE INFOCOM (2014)
7. Shakkottai, S., Srikant, R.: Economics of network pricing with multiple ISPs. IEEE/ACM Trans. Netw. (TON) **14**, 1233–1245 (2006)
8. Breidert, C., Hahsler, M., Reutterer, T.: A review of methods for measuring willingness-to-pay. Innov. Mark. **2**, 8–32 (2016)
9. Berry, S., Levinsohn, J., Pakes, A.: Automobile prices in market equilibrium. Econometrica **63**, 841–890 (1995)
10. Economies of scales. http://www.biu.ac.il/soc/ec/students/teach/953/data/Lec2_IRS.pdf
11. Stankova, K.: On Stackelberg and Inverse Stackelberg Games. http://stankova.net/game_theory/phd_stankova.pdf
12. Backward Induction and Subgame Perfection. http://www.econ.ohio-state.edu/jpeck/Econ601/Econ601L10.pdf
13. Shah, D.: How Does Coursera Make Money? https://www.edsurge.com/news/2014-10-15-how-does-coursera-make-money
14. The Lorenz Curve and Gini Coefficient. http://www.jslon.com/AP_Economics/Micro/ActAns/Micro5-7%20Answers.pdf
15. Shephard, R.W.: Proof of the law of diminishing returns (1969)
16. A deeper partnership with XuetangX to increase quality education for Chinese students. http://blog.edx.org/deeper-partnership-xuetangx-increase

Using Pull-Based Collaborative Development Model in Software Engineering Courses: A Case Study

Yao Lu$^{(\boxtimes)}$, Xinjun Mao, Gang Yin, Tao Wang, and Yu Bai

College of Computer, National University of Defense Technology, Changsha, China
{luyao08,xjmao,gangyin,taowang2005,baiyu}@nudt.edu.cn

Abstract. The pull-based development model is an emerging way of contributing to distributed software projects within the Open Source Software (OSS) communities. To train students' development skills with this modern paradigm and evaluate the effects in classroom settings, we designed a pull-based development model in classroom settings. In addition, we built the support environment for the process and integrated it in a popular teaching platform – *TRUSTIE*. With this platform, we further conducted a case study to investigate how the students benefit from this process and what challenges exist. In this experiment 22 students worked in 5 groups to independently complete an in-classroom programming project. Quantitative and qualitative results show some different characteristics of using pull-based work model from which in the OSS context, and also provide constructive advice for future practices.

1 Introduction

A critical goal of software engineering education is to train students to build the capability of collaboratively developing certain scale of software in the industrial context [1]. To achieve this goal, an important way is to train them necessary skills of using collaborative development tools, especially for modern industrial ones. However, previous research has shown that recent graduates struggle with using such configuration management systems [2,3]. To diminish the gap between classroom practice and industrial expectations, we seek for deploying a modern collaborative development paradigm—pull-based development model in students' practices for capstone projects.

The emergence of pull-based development has led to a new contributing paradigm for distributed software development, and a number of code hosting sites such as GitHub and Bitbucket provide support for pull-based development [4]. Based on the distributed version control system Git, the pull-based development paradigm allows developers to clone a copy version of the main repository

We gratefully acknowledge the financial support from Natural Science Foundation of China under Grant No. 61532004, and 61203064. We thank our students on their active participation in the course, and also want to thank *TRUSTIE* team for their kind help for the integration of the environment.

© Springer International Publishing AG 2017
Z. Bao et al. (Eds.): DASFAA 2017 Workshops, LNCS 10179, pp. 399–410, 2017.
DOI: 10.1007/978-3-319-55705-2_32

to his own working repository, so that the developers can modify the repository without being part of the development team [5]. After performing changes in his own repository, the developers can create a merging request to the core team. Through the reviewing process, the core team can decide whether to accept the merging request. Whereby decoupling the development effort from the decision to incorporate the results of the development of the code base [5], the pull-base development model improves the development efficiency, as well as controlling the quality of contributions [6]. Besides, the transparency of the contributing process and the clearness of authorship motivate external developers to contribute.

In this study, we built the execute environment and integrated it into a popular teaching platform—$TRUSTIE$[1]. Further, with this platform, we conducted an experiment to evaluate how the students benefit from the model, and what challenges exist. In this experiment, 22 students working in 5 teams were required to complete a software project. The group members used pull-requests to submit their code and documents to the group leaders. Quantitive and qualitative results show some different characteristics of using pull-based work model from which in the OSS context. We present several challenges of using pull-based work model in the classroom settings, and provide valuable advice for future replicated practices.

2   Related Work

In this section, we review related work on pull-based development in the Open Source Software (OSS) context, and the collaborative development models in classroom settings.

2.1   Pull-Based Development

Since the pull-based development model was first proposed and supported by GitHub, most of the existing work are based on this platform. Gousios et al. conducted a series studies to deeply understand this model [5,7,8]. Using their provided public GitHub data set GHTorrent [9], they explored how pull-based software development works, and found that the pull-request model offers fast turnaround, increased opportunities for community engagement and decreased time to incorporate contributions. They proposed a relatively small number of factors affecting both the decision to merge a Pull Request (PR) and the time to process it, and found that technical ones are only a small minority. In 2015 and 2016, they conducted surveys to GitHub developers to understand the work practices and challenges in pull-based development from the integrator's and contributor's perspectives, respectively. Main findings show that: integrators struggle to maintain the quality of their projects and have difficulties with prioritizing contributions that are to be merged; while contributors have a strong

---

[1] https://www.trustie.net/login.

interest in maintaining awareness of project status to get inspiration and avoid duplicating work, but they do not actively propagate information. In another work, they designed and initially implemented a prototype PR prioritization tool PRioritizer [10].

Yu et al. [11] proposed a reviewer recommender to predict highly relevant reviewers of incoming PRs. Using textual semantic of PRs and the social relations of developers, their approach can reach a precision of 74% for top-1 recommendation, and a recall of 71% for top-10 recommendation. Zhu et al. [12] analyzed the effectiveness of code contributions from patch-based and pull-based systems, and found that pull-based systems are associated with reduced review times and larger numbers of contributions. Moreira et al. [13] proposed the use of the extraction of association rules to find patterns that exert influence on the acceptance (merge) of a PR. Zhang et al. [14] conducted an exploratory study of @-mention in pull-request based software development, and found that @-mention is more likely to be used in those complex pull-requests which have more commits, more comments, more participants.

2.2 Models and Tools for Students' Collaborative Development

The ability to use version control systems is a highly desired skill in the software industry and is a basic skill for collaborative development. Teaching students such skills does benefits to teachers as well: it facilitates the teachers to monitor developing progress of students and evaluate their contributions. A number of literatures are around this subject.

Radermacher et al. [2] conducted an empirical study on industrial managers to investigate the skill gap between graduating students and industry expectations. They found that recent graduates struggle with using configuration management systems (and other software tools), effectively communicating with co-workers and customers, producing unit tests for their code, and other skills or abilities. Haaranen and Lehtinen [15] described how to incrementally present features of Git and incorporate them into the course workflow. They conducted a case study of running a 200-students course utilizing Git and evaluated the results both from instructors' and learners' point of view. Results showed that a distributed version control system can be used successfully to disseminate course materials and facilitate exercise submissions. Krusche et al. [16] described an informal review technique—branch based code reviews and applied this technique in a workflow during a project-based capstone course over the period of three semesters. They found that students do not longer see reviews as a bureaucratic burden and improve their skills through the comments of experienced team members. Feliciano et al. [17] conducted a case study where GitHub is used as a learning platform for two software engineering courses. Students' perceptions indicated that students do benefit from GitHub's transparent and open workflow. They also provide some recommendations for using GitHub in education context.

3 The Case Study

3.1 Course Background

Course Setup. In our college, the undergraduates majoring in Software Engineering had to take a series of three mandatory practice courses, which spread through the autumn, spring and summer semesters respectively. During the autumn semester, the students were required to read the code of an open source Android project, and then add some new features. During the spring and summer semesters, the courses were centered around a project: the students should develop an ingenious software project in teams of four or five using the *Iterative Development Model*. They were required to accomplish four iterations (two iterations in each semester). In each iteration, they should submit a executable version of software and related documents.

Participants. We had a total number of 22 students in the class, and two classroom teachers along with three teacher assistants. Each teacher or assistant supervised a project team on their progress of development, artifacts and tool use. The students were juniors majoring in Software Engineering. All of them were boy students. 50%, 22.73% and 27.27% of the students have respectively less one year, 1–2 years and 2–3 years of Java development experience.

Projects. At the beginning of the practice phase in the spring semester, each group was given one week to come up with innovative ideas and requirements of their projects. In the end, two groups decided to develop *Android* apps (one was for 3D navigation and the other was for campus information sharing) in Java; two groups focused on developing robot-based apps (one was for *Smart Home* and the other was for *Smart Library*) based on NAO[2] platform in Java, *Python* and *C++*; and the other group decided to develop a battle-simulation application in Java. In each of these projects, at least three open-sourced software or library were used, and the lines of code written by the students were more than 5000.

3.2 The Pull-Based Collaborative Development Environment

In the study, we leverage *TRUSTIE*[3] (*Trust*worthy *S*oftware *t*ools and *I*ntegration *E*nvironment) to support our course teaching and pull-based collaborative development. *TRUSTIE* is a popular platform among Chinese universities that enables collaborative learning (*e.g.*, resource sharing, homework assignments, discuss forum, mutual evaluation, program testing, etc.) and collaborative developing (*e.g.*, issue tracking, task assignment, version control, PRs & comments, quality analysis, etc.) [18].

The collaborative development environment builds on *Git* and *GitLab*. Having a similar feature of pull-request in GitHub, *TRUSTIE* supports students

---

[2] https://www.ald.softbankrobotics.com/en/cool-robots/nao.
[3] https://www.trustie.net/login.

ZhangYuanliang / SmartLibraryRobot   Stars 0   Members 7   Fork 3

News 12   Issue 6   Forum 3   Resources 2   Repository 203   Pull Requests 74   Milestones   Schedules   Graphs   Settings

Open 0   Merged 66   Closed 8   New pull request

🗍 **Merging branches to test automatic inspection**   💬 2
Created by Feng Chendong 4 months ago   updated 4 months ago   Merged to

🗍 **Submitting to master**   💬 2
Created by Feng Chendong 4 months ago   updated 4 months ago   Merged to

🗍 **Merging branches to test automatic inspection**   💬 1
Created by Feng Chendong 4 months ago   updated 4 months ago   Merged to

🗍 **Modifying the code of backgound**   💬 1
Created by Wang Teng 4 months ago   updated 4 months ago   Merged to

Fig. 1. The screenshot of dashboard page of PR in TRUSTIE

to look through modified details of commits and discuss pull-requests through comments. An example of dashboard for pull quests of the *Smart Library* group is shown in Fig. 1.

4   The Pull-Based Collaborative Development Process

Based on the pull-request workflow in GitHub, we designed a pull-based collaborative development process in classroom settings. Figure 2 depicts a typical process for a group. Initially, the group leader creates a *git* repository. Then he create two branches: a *develop* and a *master* branch: the *develop* branch is managed by the group leader and stores the up-to-date files of the group; while the *master* branch stores the stable releases of the project. After that, each group member forks his own branch from the *develop* branch. Each member clones the remote repository to his local *git* repository. After accomplishing a task, *e.g.*, a feature enhancement or a bug fixing, he should pull the up-to-date code on the

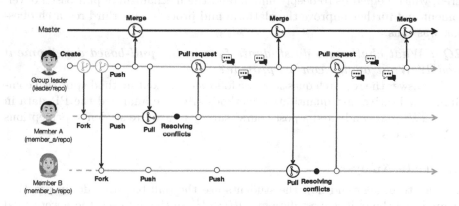

Fig. 2. The pull-based collaborative development process

develop branch, in order to resolve conflicts caused by concurrent code changes. Then he pushes the local code changes to the corresponding remote branch and creates a PR, requesting the group leader to merge his work. The group leader can review the PR or assign another member to review, to decide whether to accept the PRs. All group members can participate in review process and have a discussion through the pull quest comments in *TRUSTIE*.

5   Methodology

Applying pull-based development paradigm in students' in-classroom projects, we conducted a case study to investigate how this process influences their development. Firstly, we were concerned about the actual performance during the process. Hence, our first research question was:

RQ1: What are students' work practices in the pull-based collaborative development model?
More specifically, we wanted to answer *RQ1* from several aspects below:

RQ1.1: What do the students use PRs to do?
RQ1.3: How long do the students take to merge PRs?
RQ1.2: How do the students collaborate using PRs?

In the OSS context, pull-based contributing paradigm is an effective way to manage external contributions [5,12]. However, the classroom settings are totally different from the OSS context: the group members are co-located and their tasks are explicitly assigned. Intuitively, the transparency characteristic of pull-based development which plays a significant role in the OSS context, seems to be weakened in the co-located classroom settings. Therefore, our second research question was to validate whether and how pull-based development model benefited the students:

RQ2: How do students benefit from using pull-based development model in their courses?
Subsequently, we intended to understand what issues the students encountered, which helped us to deeply understand the mechanism of pull-based development and further improve the platform and process. Our third research question was thus:

RQ3: What challenges do students face in the pull-based development model? What advice can we provide?
To answer the research questions, we followed a mixed-method approach combining qualitative and quantitative methods [19]. We analyzed the PR data in *TRUSTIE*, and conducted a post-course survey to acquire students' perceptions on the process.

5.1   Data Analysis

In order to understand how the students use the pull-based model in practice, we analyzed the pull-request data in *TRUSTIE*. In the data set, there were total 98 PRs: 97 PRs had been merged while only one was still open.

5.2   Survey

At the final class of the course, we sent two online questionnaires to the group leaders and members respectively to get their perceptions of the process. The two questionnaires had both 14 questions, including single choice, Likert scale and open-ended questions. The questionnaires started with some simple questions such as their technical experiences. Then we asked them about their practical details through the process; their attitudes towards the pull-based model; and their encountered problems and advice on the process. The average answer time for a questionnaire was 176 s.

6   Results

This section presents the results of the qualitative and quantitative analyses. When quoting survey respondents, we refer to the group leader and member using the [Lx] and [Mx] notations respectively, where x is the respondent's ID, ordered by the submitted time of the questionnaire [8]. Codes resulting from coding open-ended answers are underlined [8].

6.1   RQ1: Work Practices in the Pull-Based Model

We sought to understand the students' work practices in the pull-based model. We analyzed the PR data in *TRUSTIE* to learn the PRs' intentions and their handling time for merging. Besides, we collected the students' response in the questionnaire to understand their collaboration practices in the process.

PR Intentions. Based on the titles and descriptions of the PRs, we categorized the PRs into 5 categories:

- *doc*, updating documents, *e.g.*, requirements specification, UML diagrams, etc.
- *feature*, adding feature enhancements.
- *fixing*, fixing bugs or code quality issues.
- *test*, adding testing files.
- *other*, PRs titled with unrecognized words, *e.g.*, "sadfas", which we suppose are to testing PR features in *TRUSTIE*.

The statistics are shown in Fig. 3. We can see that the students use PRs mainly to bring together feature enhancements and document updates. A small portion of them use PRs to submit defect fixes and testing files. We suppose this can be due to the students' programming tasks in educational context: students are more concerned about the functional implementation, compared with the code quality; and teachers tend to arrange tasks on documents to achieve corresponding teaching goals.

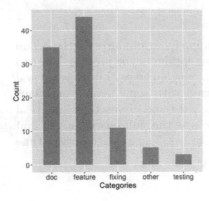

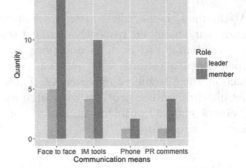

Fig. 3. The categories of PRs   **Fig. 4.** The communicating means for PRs

Handling Time for Accepting. We sought to understand how the students used PRs for collaboration. We firstly analyzed the handling time for accepting PRs by calculating the time span between the created time and closed time of a PR. Table 1 shows the distribution of handling time. We can see that the median time span for handling PRs is 11s, and over 75% of the PRs are merged less than one minute.

Table 1. Descriptive statistics of PR handling time[1]

Min	1st quarter	Median	Mean	3rd quarter	Max
2	6	11	15846	46	573361

[1] The unit of time is *second (s)*.

Collaboration Ways. We set a series of questions on their practical behavior of collaboration using pull-based model. A variety of survey questions led to this understanding and the answers are presented in Fig. 5. We can see that 76% of the group members tend to reminder their leaders after submitting PRs. However, because of the coupling of work, 70% of the respondents have encountered conflicts[4]. Besides, over 70% of the respondents would communicate with their group leaders or members, and would also care for the quality of PRs.

We also used a multi-choice subject to understand the students' communication ways on PRs. Some commonly-used channels were listed in the options, and as a complement, we provide an 'other' option to gather other alternatives. The results are summarized in Fig. 4. The students chose the communication means from the options we provided, without other options. We can conclude from the figure that in the in-classroom development settings, the students tend

---

[4] Conflicts occur when *git* try to merge two branches which have changed common code pieces.

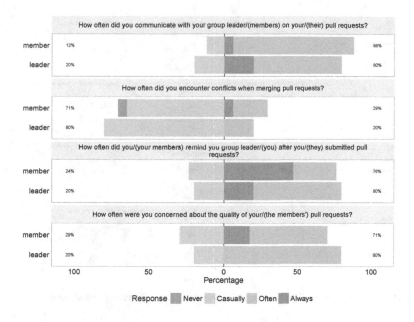

Fig. 5. Students' collaborating ways using pull-based model

to use immediate ways to communicate on PRs. For instance, they prefer to using face-to-face and IM tools for collaboration.

Synthesizing the results of *RQ1*, we find that the students perform differently when using pull-based collaboration model from the developers in the OSS communities: **In the classroom context, the students usually use face-to-face and IM tools for communication, and use PRs to merge contributions of code and doucuments. Besides, they tend to notify the group leader after submitting PRs, and the group leaders tend to accept PRs in a short time span, with few code reviews.**

6.2   RQ2: Benefits from the Pull-Based Model

To ask respondents about how they benefit from the pull-based work model, we provided a set of 5 questions, including four 5-level Likert scale questions. The results of Likert scale questions are shown in Fig. 6. Results show that, in general, 94% of the members and 80% of the leaders feel that the pull-based work model improves their development efficiency, and over 80% of the respondents believe that it is more convenient than merging branches using command lines. Besides, through visualizing contributions of the team, the pull-based model helps them to maintain awareness of the dynamic and progress of other members. This can be attributed to the pull-based development model, which is also a confirmed mechanism to attract external contributions in the OSS context [20]. Moreover, from the standpoint of the teachers, we also benefit from the 'transparency' characteristic of pull-based model. The PRs help us to quantify

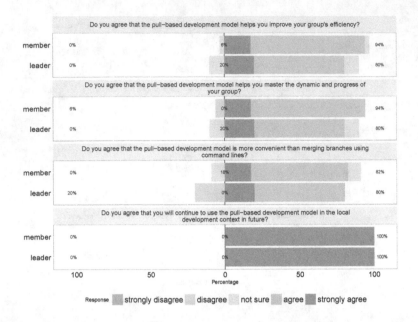

Fig. 6. The students' perceptions of pull-based model

students' work, and facilitates us to evaluate the quality and quantity of their work. Hence, we can reach a finding that: **(1) The students benefit from the transparency of PRs, which helps them maintain awareness of the developing progress of the team. (2) They also benefit from the convenience to gather members' contributions, compared with traditional ways using command lines. (3) Besides, the teachers benefit from the visualization and quantization of students' work, which facilitates their evaluation work.**

6.3   RQ3: Encountered Problems and Advice

To investigate what challenges the students have encountered in the process, we provided 3 open-ended questions for the students to understand: what problems they have encountered when using pull-based work model; and what advice they have for the process and the platform.

Problems. In the survey, 3 group leaders and 13 members described their encountered problems. Coincidentally, all 3 group leaders along with 2 members described the issue of conflicts when merging PRs, *e.g.*, *"resolving conflicts is a difficult work"* [L4], and *"conflicts would happen when multiple students submit PRs"* [M11]. 4 members expressed their difficulties in initially learning the tool use and the process, *e.g.*, *"I have difficulties in getting start to understand the process"* [M1], and *"when I started to practice, I was not familiar with the specific operations"* [M2]. Besides, all the remaining 7 members described the problems of the environment, e.g., *"sometimes, the tool TortoiseGit reports errors when executing pull or push operations"* [M10], *"I feel that the interface is not friendly enough, and the failures casually occur"* [M16].

Advice for the Pull-Based Model. There are one group leader and 10 members offering advice on the process. We present two representative and constructive ones: [M1] suggests *"integrating pull-based process and IM tools "reinforcing communications, (e.g. the Wechat app)"*. [M13], [M15] and [M17] suggest *among the members, in order to reducing conicts"*.

Advice for the Platform. Some respondents provide valuable advice for the platform improvement, e.g., *"[I hope] the platform can automatically recommend and assign PRs to different reviewers, including the group members"* [M2], *"[I hope that Trustie] can provide visualizations of handling progresses of the PRs"* [M13]. Besides, other two respondents also mentioned adding instant notifications in PRs, in order to improve the collaborative efficiency of the team. ([M11] and [M15])

7   Conclusion

We presented a case study of applying pull-based collaborative development model in students' practice courses. Through quantitive and qualitative analysis, we find some valuable findings, and also provide constructive advice for future practices. We make the following key contributions:

- We designed a pull-based collaborative development model in classroom settings, and built the execution environment.
- We conducted a case study in a software engineering course, and found some differences of pull-based development model between the OSS context and classroom settings.
- We validated the benefits of using such a model in educational context, and identified some challenges as well. Further, we provided valuable advice for future practices.

In terms of future work, we will improve our process and platform, and continue to conduct the experiment on the new group of students in the next semesters.

References

1. She, J., Tong, Y., Chen, L., Cao, C.C.: Conflict-aware event-participant arrangement and its variant for online setting. IEEE Trans. Knowl. Data Eng. **28**(9), 1–1 (2016)
2. Radermacher, A., Walia, G., Knudson, D.: Investigating the skill gap between graduating students and industry expectations. ACM (2014)

3. Radermacher, A., Walia, G.: Gaps between industry expectations and the abilities of graduates. In: Proceedings of the ACM Technical Symposium on Computer Science Education, pp. 525–530 (2013)
4. Barr, E.T., Bird, C., Rigby, P.C., Hindle, A., German, D.M., Devanbu, P.: Cohesive and isolated development with branches. In: Proceedings of International Conference on Fundamental Approaches to Software Engineering, pp. 316–331 (2012)
5. Gousios, G., Pinzger, M., Deursen, A.V.: An exploratory study of the pull-based software development model. In: Proceedings of International Conference on Software Engineering, pp. 345–355 (2014)
6. Vasilescu, B., Yu, Y., Wang, H., Devanbu, P., Filkov, V.: Quality and productivity outcomes relating to continuous integration in github. In: Proceedings of the 10th Joint Meeting on Foundations of Software Engineering, pp. 805–816 (2015)
7. Gousios, G., Zaidman, A., Storey, M.A., Deursen, A.V.: Work practices and challenges in pull-based development: the integrator's perspective. In: Proceedings of International Conference on Software Engineering, pp. 358–368 (2015)
8. Gousios, G., Storey, M.A., Bacchelli, A.: Work practices and challenges in pull-based development: the contributor's perspective. In: Proceedings of the 38th International Conference on Software Engineering, pp. 285–296. ACM (2016)
9. Gousios, G.: The GHTorent dataset and tool suite, pp. 233–236 (2013)
10. Veen, E.V.D., Gousios, G., Zaidman, A.: Automatically prioritizing pull requests. In: Proceedings of Mining Software Repositories, pp. 357–361 (2015)
11. Yu, Y., Wang, H., Yin, G., Ling, C.X.: Reviewer recommender of pull-requests in github. In: Proceedings of International Conference on Software Maintenance and Evolution, pp. 609–612 (2014)
12. Effectiveness of code contribution: from patch-based to pull-request-based tools. ACM (2016)
13. Daricélio, M., de Lima, M.L., Murta, L.: Acceptance factors of pull requests in open-source projects. In: Proceedings of the 30th ACM Symposium on Applied Computing, pp. 1541–1546 (2015)
14. Zhang, Y., Yin, G., Yu, Y., Wang, H.: A exploratory study of @-mention in github's pull-requests, vol. 1, pp. 343–350 (2014)
15. Haaranen, L., Lehtinen, T.: Teaching git on the side: version control system as a course platform. In: Proceedings of ACM Conference on Innovation and Technology in Computer Science Education, pp. 87–92 (2015)
16. Krusche, S., Berisha, M., Bruegge, B.: Teaching code review management using branch based workflows. In: Proceedings of the International Conference on Software Engineering, pp. 384–393 (2016)
17. Feliciano, J., Storey, M.A., Zagalsky, A.: Student experiences using github in software engineering courses: a case study. In: Proceedings of the International Conference on Software Engineering, pp. 422–431 (2016)
18. Wang, H., Yin, G., Li, X., Li, X.: TRUSTIE: A Software Development Platform for Crowdsourcing. Springer, Heidelberg (2015)
19. Jick, T.D.: Mixing qualitative and quantitative methods: triangulation in action. Adm. Sci. Q. 24(4), 602–611 (1979)
20. Dabbish, L., Stuart, C., Tsay, J., Herbsleb, J.: Social coding in github: transparency and collaboration in an open software repository. In: Proceedings of Computer Supported Cooperative Work, Seattle, WA, USA, February, pp. 1277–1286 (2012)

A Method of Constructing the Mapping Knowledge Domains in Chinese Based on the MOOCs

Zhengzhou Zhu[1(✉)], Yang Li[2], Youming Zhang[1], and Zhonghai Wu[1]

[1] School of Software and Microelectronics, Peking University, Beijing, China
zzzmad@163.com

[2] Department of Information Management, Peking University, Beijing 100871, China

Abstract. While the number of MOOCs users in China has been increasing dramatically, the users still face the risk to give up learning in the half way due to unfamiliarity with the course structure, prerequisite courses and so on. To resolve this problem, the MKD plays an important role by providing a clear structure map of the course, helping user to realize about the appropriate learning path as well as the knowledge relationship. In this article, a method of constructing a usable MKD in Chinese has been raised based on online courses. The online course data are obtained by web crawling from the MOOCs sites, then processed through the data clean and data fusion after which the MKD is extracted and evaluated. This method is applied to all the existing MOOCs courses thus has showed practical significance.

Keywords: MOOCs · Mapping Knowledge Domain · Data clean · Data fusion

1 Introduction

Comparing to the traditional open course and remote education, the MOOCs features not only the utility of the formal course [1], but also the openness, large scale, Self-Organizing and sociality. In recent years, with the maturity of the related Internet and cloud computing technologies, an extensive range of educational organizations and outstanding universities, both domestic and abroad, have recognized the concept of MOOCs and started to pay attention to the related practical activities [2].

Among the studies of MOOCs, [2] investigated the relationship between learner's activities, the effect on learning and the prediction on the learning achievements, for the first time from the perspective of the learners. Studying from a MOOCs platform course resource perspective, one particular course may be taught by different teachers from various MOOCs web sites. In this case, if we can integrate the resource of the same course from different MOOCs platforms and hence construct a Chinese Mapping Knowledge Domain based on the online courses, then it will produce directive significance to the MOOCs development, the related course design and students' learning activities.

In the conference of the "Mapping Knowledge Domains" held by the National Academy of Sciences at year 2003, the concept of Mapping Knowledge Domain (MPD) was first raised, and was also called as the Scientific Knowledge Map, the Knowledge

© Springer International Publishing AG 2017
Z. Bao et al. (Eds.): DASFAA 2017 Workshops, LNCS 10179, pp. 411–416, 2017.
DOI: 10.1007/978-3-319-55705-2_33

Field Visualization, the Knowledge Graph or the Knowledge Field Map. The concept consists of a series of graphs showing the development progress and the structural relationship of knowledge. By describing the resource of knowledge and its vehicle with visualization technology, it's capable of mining and analyzing the knowledge while constructing and displaying the relationship of knowledge units. The MPD obtains the property of 'graphic' as it contains visualized graphics, and it also has the property of 'pedigree' since it's a serialized pedigree of the knowledge [3]. There are two ways of constructing the MPDs: the Top-Down method and the Bottom-Up method. The Top-Down method extracts the body of the knowledge and its pattern from high quality resources in the structural data among encyclopedic web sites, and then stores the result in the knowledge repository. The Bottom-Up method on the other hand, obtains the pattern of the resources from public data and selects out the ones with relatively higher confidence level, and then the results are artificially verified before entering into the knowledge repository. Constructing the MPD with the Bottom-Up method can be considered as an iterative process. The key technologies to construct the MPD focus on solving the dimensionality reduction of the data. The conventional technologies to do this include the multivariable statistical analysis, which consists of factorial analysis, multidimensional analysis and cluster analysis. On the other hand, there have been more advanced technologies for doing this, like the Pathfinding network [6], Self-Organizing feature map (SOM) [7], Force directed placement algorithm [8], Latent semantic analysis [9], Minimum spanning tree algorithm [10], Triangulation measuring method, etc.

The paper's contributions are:

- We raised an approach for constructing the MPD in Chinese based on the MOOCs;
- We raised a new data fusion method based on the similarity, granularity and threshold value.

2   Data Crawling and Data Clean

Currently, most of the MOOC websites contain dynamic content, which means that all the course details are dynamically obtained from the backend and then displayed on the browser by using JavaScript. The advantage from using dynamic content is that, the data is displayed on the browser based on user operations without the need to reload the whole page, hence the user experience is improved. However, this brings difficulty to the data crawling process since the useful data cannot be directly obtained by parsing the static content from the web page.

In order to parse the dynamic content which is returned from the backend, a JavaScript executor is required. Therefore, in the web crawler we use the Google V8 JavaScript Engine to dynamically simulate the process of parsing the html data.

The curriculum data fetched from the web page contains the curriculum name, curriculum description, curriculum content, etc. Among them, the curriculum content is the HTML body where the HTML labels are excluded, and it contains the curriculum section information to build the knowledge map. Because the data fusion cannot deal with unstructured data in the HTML body, it is necessary to process the HTML body content information into structured JSON data and save it in a persistent way. In the data fusion process, if the course consists of the standard Four-Layer structure, then the course name is used as the top node while the chapter information and the section information can be obtained by traversing the corresponding JSON data. From the JSON data, the chapter information of the curriculum can be fetched with the key of this chapter. And then with the key of the sections under this chapter, we can access much more Fine-Grained information of this curriculum.

3 Data Fusion

3.1 Constructing the Curriculum Tree

The MOOCs generally consist of three layers: course, chapter, section. The process to construct the curriculum tree can be divided into two steps in the following.

Step 1. Convert the JSON data which are obtained from the data clean process into the tree structure, namely the structure of the curriculum tree. To do this, the JSON data is firstly transformed into a dictionary data structure.

Step 2. Convert the In-Memory curriculum tree into Two-Dimensional table. To achieve this work, we use one field of the node to record the IDs of all the children from this node.

3.2 Extract the Keywords of Titles from the Curriculum Tree Nodes

This phase determines the weight ratio of each keyword from the titles by using Baidu as tool, and paves the way for calculating the similarity of the titles as well as judging whether the data could be fused in the next phase. The steps in the following have been taken to achieve this work.

Step 1: Search each title of the courses in Baidu by traversing each course in the curriculum tree. The top 20 results in each search are stored as the baidu_list

Step 2: For each course, the Chinese text segmentation is performed. For each word from the segmentation, the frequency of occurrences in the corresponding baidu_list is calculated.

Step 3: For the words under each course, their frequencies of occurrences are normalized. In this way, the sum of squares of the words' frequencies under each course equals to one.

Step 4: Calculate the similarity of each course, which will be used as evidence to decide whether the data fusion should be carried out.

A lot of Chinese text segmentation platforms were used for the course titles, including the Jieba Chinese Text Segmentation. During the text segmentation, we unify the letter case and exclude some unnecessary words including "I", "and", "a", etc. When searching the titles on the Baidu web site, the useless punctuations are also removed by using regular expressions.

3.3 Define the Granularity of the Course Node

The granularity makes up the evidence by which the curriculum tree is layered. It's assumed that if the granularity of the node is greater, then the separability of the course presented by this node becomes greater, and vice versa. Prior to the data fusion, we investigated and classified the data in the nodes by its granularity. This had solved the problem when the nodes' data in the same level of the curriculum tree could not be fused, and when the nodes' data from different levels need to be fused.

3.4 The Integration of the Curriculum Tree

Data Fusion Procedure. The fusion procedure includes 4 steps which are described in the following.

Step 1: From the database, the curriculum data will be read into a tree structure.

Step 2: Select a certain node and find the node with the highest degree of similarity to this certain node.

Step 3: If the similarity between the two nodes is greater than the threshold, then proceed to the next step. Otherwise, return to Step 2 and select the next node for processing.

Step 4: After the similarity meets the criteria, the granularities of the nodes are compared. If the granularities of the two nodes are similar, the data in the two nodes can be merged.

Selection of the Fusion Thresholds. The threshold is chosen as the key point, and continuous experiments has been taken to detect and obtain the threshold values which meet the quality requirement.

Selection of the Post-Fusion Text Keyword. There are 3 strategies for selecting the node keywords after the fusion: (1) Randomly pick up a title from the fusion source; (2) All the title keywords from the fusion source are taken as a union; (3) Take the title (one or more) which contains the highest number of keywords from the fusion source.

Extraction of the Knowledge Level. This process could be divided into 3 steps as in the following.

Step 1: Count the granularity of the courses and chapters which have obvious differences. Then find the range of their granularities respectively to determine a threshold interval.

Step 2: Traverse the course nodes, if a course node is reached, then set the node as a course and set is_course_node = 1

Step 3: The sub-Node of the course node in Step 2. is lifted up to the level above, which means the corresponding sections become chapters. Meanwhile the following value is set: is_course_node = 0.

3.5   Evaluation of Fusion Results

The fusion results are evaluated according to the disciplinary criteria from the subject standards. This is done with two approaches.

The first approach traverses the list of the Post-Fusion curriculum trees. Each tree is compared with the tree structure established from the subject standard. The process could be divided into the following steps:

Step 1: Firstly, we compare the top layer from the subject standard tree, namely the knowledge field layer, with the top layer from the curriculum tree, which is the corresponding course name. The number of keywords exist in both the two trees (the firstSameWord) is counted. If the count result is greater than zero, then the top layers of the two trees are considered matching while the comparison continues for the layers underneath.

Step 2: The nodes of the second level are compared between the curriculum tree and the subject standard tree. The node from the curriculum tree is the name of the corresponding chapter, and the node from the subject standard tree is the knowledge unit layer. Pairwise comparison is conducted during the comparison, while the number of keywords matched between the chapters and the knowledge units is counted as the value secondSameWord.

Step 3: If a chapter matches with a knowledge unit, then the sections under the chapter and the knowledge points under the knowledge unit are taken for comparison from each pair. The number of the keywords matched between the sections and the knowledge points is counted as the value thirdSameWord.

In addition, regardless of whether the first layer node, the second layer node, or the third layer node is compared with the standard discipline, if the matching node is a fused node, then it means the original course source exists. In this case, the number of nodes from the original course source is defined as the TotalOriginalSize. Finally, after all the nodes have been traversed, the count value is generated in order to represent how the curriculum tree is similar to the subject standard tree.

The second way to evaluate the fusion result is first comparing the keyword occurrences between the course name and the corresponding knowledge field, so that their similarities are quantified and calculated. After that, the knowledge units and knowledge points are used to compare with the corresponding curriculum tree nodes respectively, by using the keyword occurrences in similar manner.

Based on the methods above, the final decision is made about whether the fusion result meets the quality requirement. In detail, this is done by combining both the result from the artificial recognition and the result from referencing the subject standard.

4   Conclusion

This approach for constructing Knowledge Mapping Domains, including the definition and calculation of the similarity, granularity and threshold value raised in this article, is derived from huge quantity of investigation and experimental data comparison, thus is of high utility value and is capable of creating a Chinese Knowledge Mapping Domain that can be put into practical usage. However, the evaluation of the similarity and granularity is a complicated process, which needs further investigation on the possible related factors in order to improve the accuracy. Second to that, the definition of the threshold value is also based on evaluating huge number of experimental data, therefore the setting about the threshold value may differ between different types of online courses on which the method is applied.

Acknowledgement. This work was supported by National Natural Science Foundation of China (Grant No. 61402020) and a grant from the Ph.D. Programs Foundation of Ministry of Education of China (No. 20130001120021). Thanks to Yipeng Liu, Na Li, Xiaoli He, Lu Liu, Yaoya Wang, Xiaozheng Zhang, Debin Zhang, Cong Sun, Yijun Wang, Tengyu Wang, Juan Gu, Meibing Sun and Zenghua Chen who have done a lot of work for the paper.

References

1. Li, X.: MOOC, showcase or shop? J. Commun. China Comput. Fedaration **9**(12), 24–28 (2013)
2. Jiang, Z., Zhang, Y., Li, X.: Learning behavior analysis and prediction based on MOOC data. J. Comput. Res. Dev. **3**, 614–628 (2015). doi:10.7544/issn1000-1239.2015.20140491
3. Jiang, L., Zhang, H.: Research on knowledge map of MOOC research hotspot and development. J. Distance Educ. **12**, 35–40 (2014). doi:10.3969/j.issn.1009-458X.2014.12.007
4. Roberts, E., Engel, G., Chang, C., et al.: Computing curricula 2001: computer science. J. IEEE Comput. Soc. **34**(1), 4–23 (2001) http://www.sigcse.org/cc2001
5. Wexler, M.N.: The who, what and why of knowledge mapping. J. Knowl. Manag. **5**(3), 249–264 (2001)
6. White, H.D.: Pathfinder networks and author cocitation analysis: a remapping of paradigmatic information scientists. J. Assoc. Inf. Sci. Technol. **54**(5), 423–434 (2003)
7. Chen, C.: Searching for intellectual turning points: progressive knowledge domain visualization. J. Proc. Natl. Acad. Sci. U.S.A. **101**(Suppl.), 5303–5310 (2004)
8. Xia, L., White, H.D., Buzydlowski, J.: Real-time author co-citation mapping for online searching. J. Inf. Proc. Manag. **39**(5), 689–706 (2003)
9. Lin, X.: Searching and Browsing on Map Displays. C Asis Meeting, pp. 13–18 (1995)
10. Moody, J., Bender-Demoll, S.: Dynamic network visualization 1. J. Am. J. Soc. **110**(4), 1206–1241 (2005)

Social Friendship-Aware Courses Arrangement on MOOCs

Yuan Liang[✉]

State Key Laboratory of Software Development Environment,
School of Computer Science, Beihang University, Beijing, China
liangyuan120@buca.edu.cn

Abstract. Massive open online courses (MOOCs) provide an opportunity for learners to access free courses offered by top universities in the world. However, with contrast to large scale enrollment, the completion rate of these courses is really low. One of the reasons for students to quit learning process is that they could not study the courses with their friends. In order to improve the completion rate, we address the importance of content interest and social friendship for courses arrangement for learners in MOOCs. We first develop a greedy algorithm to solve the arrangement according to the friendship of learners and the content of courses. Then we used the game theoretic framework to improve greedy algorithm performance. Finally, we verify the effectiveness and efficiency of the proposed solutions through extensive experiments on both real and synthetic datasets.

1 Introduction

In recent years, Massive open online courses (MOOCs) are a trend in online or distance education, such as Coursera[1], edX[2], and Udacity[3]. The MOOCs have attracted tremendous numbers of users and have played an increasingly important role in online learning. Millions of learners with different professional backgrounds and motivations came from different countries and gathered in the same "classroom" [1].

As introduced in [1], MOOCs has a large enrollment. However, Based on the data given by a MOOCs statistics project, the maximum completion rate in Coursera is less than twelve percent, and the average completion rate is only about six percent. Many reasons can cause the low completion rate [2], such as insufficient time for learners and language barrier for speaking learners, or lack of an available arrangement method for learners. Imagine the following scenario, Amy, Bob and Cathy usually more interactive in MOOCs, we can think of them as online friends. One day, Amy logs into the website and attend the *Machine Learning*, after a few lessons, she felt a little tedious in her own study, so she

---

[1] http://www.coursera.org/.
[2] http://www.edx.org/.
[3] http://www.udacity.com/.

© Springer International Publishing AG 2017
Z. Bao et al. (Eds.): DASFAA 2017 Workshops, LNCS 10179, pp. 417–422, 2017.
DOI: 10.1007/978-3-319-55705-2_34

want to attend *Machine Learning* with her friends. However, Bob and Cathy may be not know Amy has attended this course, so they may not attend the same course in most cases. Then it is raises a problem that most MOOSc platforms encounter: *how to arrange the courses to learners which they can complete the courses according to learners' interest and learners' social friendship?* Therefore, it is very necessary for us to develop an efficient arrangement method for learners to obtain maximum satisfaction (e.g. happiness).

To the best of our knowledge, as discussed later, this is the first work that studies the arrangement of the courses to the learners in MOOCs, and thus we should design efficient algorithms specifically for our problem. We make the following contributions.

(a) We identify a greedy algorithm of the arrangement problem that assign courses to learners.
(b) We develop a game theoretic approach that can improve the performance of greedy.
(c) we conduct extensive experiments on real and synthetic datasets which verify the efficiency and effectiveness of our proposed algorithms.

The rest of the paper is organized as follows. We present the problem definition in Sect. 2. The first algorithm of our problem is shown in Sect. 3 and the game theoretic approach is proposed in Sect. 4. We conduct an extensive performance study in Sect. 5 to evaluate the performance of our proposed solutions. Section 6 review the related literature. The conclusion of our work is shown in Sect. 7.

2 Problem Statement

Consider a setting where learners of a large gathering for a same courses, or a scenario that plan courses for its learners over the content information and social information in MOOCs platform (e.g. coursera). In both cases, there is a set of learners $L = (l_1, l_2, ..., l_n)$ who must be assigned to courses from a given set of possible courses $C = (c_1, c_2, ..., c_m)$.

We assume the presence of a social network, through which the friendships among those learners can be obtained. Let $G = (L, E, W)$ be the friendship graph induced on L, where L is the set of learners, E is the set of edges (i.e., social connections), and W is the set of edge weights (denoting the strength of social connections). The edges directed (e.g. two learners interact in MOOCs), and the weights may be binary (i.e., indicating simply be presence or absence of a friendship). Let C be a set of courses in MOOCs, our goal is to find an arrangement of each learner l to a course c that maximizes the following function:

$$Utility(L, C, \alpha) = \alpha \cdot \sum_{l \in L} s(l, c_l) + (1 - \alpha) \cdot \sum_{e=(l_i, l_j) \wedge c_{l_i} = c_{l_j}} w_e \quad (1)$$

The first sum in Eq. 1 represents the interest of learners between courses, while the second one corresponds to the social information. c_l presents learner l attend the course c. The *preference* parameter $\alpha \in (0, 1)$ adjusts the relative importance of the two factors. If $\alpha > 0.5$, the utility function should aim more at maximizing the interest between courses and learners. Our objective is to find a feasible arrangement that maximizes the utility between learners and courses.

3   Greedy

There are a set of learners and a set of courses in the MOOCs platform, and we want to find an arrangement method to assigned learners to courses so as to all learners will attain better satisfaction/utility. From Eq. 1, this is a linear programming problem, we can use GLPK[4] to solve this problem to obtain optimal solution, as the GLPK package is intended for solving large-scale linear programming. However, linear programming can not solve large-scale data problem and it has high complexity. Therefore, we propose a greedy algorithm to solve the arrangement of courses to learners.

Let H contains a tuple $\langle l, c, u \rangle$ representing the potential arrangements of pairs of learners and courses, and let $A(l)$ denotes the arrangement of learner l. If the potential arrangement utility only consider the learners' interest, and does not consider the friendships among learners. We will define the utility as $u(l, c|\emptyset) = s(l, c_l)$. If the utility includes the interest of learners and courses and the social friendship among all learners, the potential arrangement utility will defined as $u(l, c|c_l) = \frac{1}{2} \cdot s(l, c_l) + \frac{1}{2} \cdot \sum_{c_{l_i} = c_{l_j}} w(l_i, l_j)$.

First, we extract the pair with the largest $u(l, c|\emptyset)$ from the heap H, which stores a tuple containing learner l, course c and a utility value u, and we assigned leaner l to courses c. Then we will see the friend of learner l, and we will use l' denotes the friend of learner l. We extract the maximum $u(l, c|c_l)$ according to the heap H, and assign l' to a course. This process can be repeated as needed until either all learners are assigned or there are no more available courses.

Complexity Analysis. The Greedy algorithm takes at most $O(|m||n|)$ time to initialize the heap H and insert the utilities of all learner-course pairs into the heap, where m is the number of courses in MOOCs, n is the number of learners in online MOOCs. In the following iterations, at most $|m||n|$ learner-course pairs are extracted from H, but only $|m|$ pairs are inserted into the arrangement. Along with each insertion operation, at most d elements in H are updated, leading to $O(d \log(|m||n|))$ swapping-element operations in H, where d is the degree of $l \in L$. The above analysis indicates that the final time complexity is $O(|m||n| + |m|d_{max} \log(|m||n|))$.

4   A Game Theoretic Approach

When learners choose a courses in MOOCs, they will have a game on the pros and cons of the process according to all aspect. Therefore, the game theoretic approach can be used to solve the arrangement of courses to learners. In strategic games, we assume each learner is a player, and all players will compete with each other over the same courses in order to optimize their individual objective function that join in a course with high utility with their close friends.

Initially, learners are randomly assigned to courses and then they start changing courses according to their best response until they reach a Nash equilibrium.

---

[4] https://www.gnu.org/software/glpk/.

In the best response of learners, each learner l will consider his/her own decision that maximizes his/her own utility as expressed by Eq. 1. Then, the learner will repeat the best response according to $U_l = \alpha \cdot s(l, c_l) + (1 - \alpha) \cdot w_e$. Next, we will consider change a learner from c_i to c_j with the maximum utility. The algorithm terminates when there is no courses change for any learner during the repeat.

Complexity Analysis. We analyze the running time of GTA. Let $m = |C|$ be the number of courses in MOOCs, $|L|$ be the number of learners in MOOCs and $|E|$ be the number of edges among of the social graph. The best response of a learner l requires initializing the utility for m decisions, which takes constant time. Then we will compute the number friends of learner l, it take d_l time, where d_l is the degree of l. Finally, for all courses, compute the utility of l. Summarizing the total utility for a learner takes $2 \cdot m + d_l$. Repeating the same process for all learners, and we have $\sum_{l \in L}(2 \cdot m + d_l) = 2 \cdot m \cdot |L| + \sum_{l \in L} d_l = 2 \cdot m \cdot |L| + 2 \cdot |E| = \ominus(m \cdot |L| + |E|)$. Finally, the final time complexity is $\ominus(m \cdot |L| + |E|)$.

5   Experimental Evaluation

5.1   Experiment Setup

In this subsection, we evaluate our proposed algorithms. We use both real and synthetic datasets for experiments.

We used the Meetup dataset from [3] as the real dataset. Then we evaluated our algorithms in terms of arrangement utility, running time and memory cost and studied the effects of varying the parameters on the performance of the algorithms. The synthetic datasets were created in Python, and all algorithms were implemented in C++ and executed under the Linux Ubuntu operating system. The experiments were performed on a computer with an Intel Xeon E5620 with a 2.40 GHz 16-core CPU and 12 GB of memory.

5.2   Experiment Results

In this section, we evaluated the proposed algorithms in terms of arrangement utility, running time and memory cost. Then, we tested the performances of the proposed algorithms by varying the parameters as follows: the size of L, the social degree of d, the size of C and the balance parameter α.

Effect of $|L|$. Next, we set the number of learners as {1000, 2000, 3000, 4000, 5000}, and the the number of courses as 200. And We show the effect of varying $|L|$. Figures 1a to c show the arrangement utility, running time and memory cost, respectively. We can make the following observations. First, the arrangement utility increases when $|L|$ increases, because we must compute more learners' arrangement utility when $|L|$ increases. Second, the running time increases as $|L|$ increases. This is because when $|L|$ is larger and $|C|$ is fixed, more learners must be calculated. Third, the memory cost increases as $|L|$ increases, because as $|L|$ increases, it requires more memory.

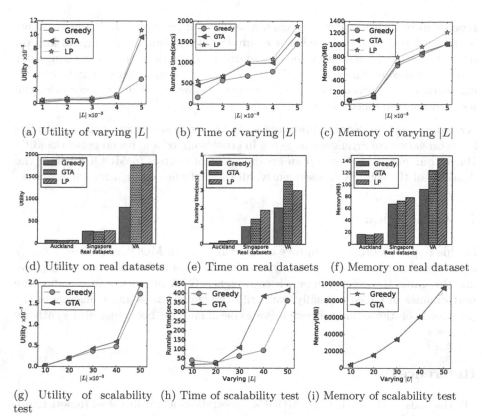

(a) Utility of varying $|L|$ (b) Time of varying $|L|$ (c) Memory of varying $|L|$

(d) Utility on real datasets   (e) Time on real datasets   (f) Memory on real dataset

(g) Utility of scalability test   (h) Time of scalability test   (i) Memory of scalability test

Fig. 1. Results of varying $|C|$, $|L|$, and $|d|$

Real Dataset. Figures 1d to f show the results of the arrangement utility, running time and memory cost, respectively, on the real dataset (VA, Auckland and Singapore) [3]. The results on this real dataset present patterns similar to the results with synthetic data.

Scalability. Finally, we studied the scalability of all the proposed algorithms. Specifically, we set $|C|$ to 100 and $|L|$ to $10k$, $20k$, $30k$, $40k$, and $50k$. The results are shown in Figs. 1g to i in terms of arrangement utility, running time and memory cost, respectively. We can observe that the arrangement utility, running time, and memory cost of all algorithms grow linearly with the size of the data. In addition, the result shows that all the algorithms are scalable in terms of both time and memory cost.

6   Related Work

In this section, we will review the related works in two categories, recommendation and arrangement in social networks.

Recommendation. In recent years, a number of recommendation works have been developed and evaluated. For example, [4] have analysed learners logs on several MOOCs promoted by the course designer and developed a recommender system for encouraging learners. [5] propose a recommendation system to recording learners experiences in different courses. However, [5] did not consider the social friendship among learners.

Arrangement in Social Networks. In recent years, many existing studies focus on various of arrangement issues in crowdsourced and social networks [3,6–10]. In our paper, we develop an arrangement of courses in MOOCs before the deadline of the courses. Consequently, all these studies differ from our research.

7   Conclusion

In this paper, in order to improve completion rate in MOOCs, we first identify a greedy algorithm to solve the arrangement of courses to learners. Then, we use the game theoretic framework to design a GTA algorithm to enhance the performance of Greedy. Finally, we verify the effectiveness and efficiency of the proposed solutions through extensive experiments on both real and synthetic datasets.

References

1. Hollands, F.M., Tirthali, D.: MOOCs: expectations and reality. Full Report (2014, online submission)
2. Vaquero, L., Cebrian, M.: The rich club phenomenon in the classroom. Sci. Rep. **3** (2013). Article ID 1174
3. Liu, X., He, Q., Tian, Y., Lee, W., McPherson, J., Han, J.: Event-based social networks: linking the online and offline social worlds. In: SIGKDD, pp. 1032–1040 (2012)
4. Zhuhadar, L., Butterfield, J.: Analyzing students logs in open online courses using SNA techniques. In: Proceedings of AMCIS (2014)
5. Yang, D., Wen, M., Rose, C.: Peer influence on attrition in massively open online courses. In: Educational Data Mining 2014 (2014)
6. She, J., Tong, Y., Chen, L.: Utility-aware event-participant planning. In: SIGMOD, pp. 1617–1628 (2015)
7. She, J., Tong, Y., Chen, L., Cao, C.: Conflict-aware event-participant arrangement and its variant for online setting. IEEE Trans. Knowl. Data Eng. **28**(9), 2281–2295 (2016)
8. Tong, Y., She, J., Ding, B., Wang, L., Chen, L.: Online mobile micro-task allocation in spatial crowdsourcing. In: ICDE, pp. 49–60 (2016)
9. Tong, Y., She, J., Ding, B., Wang, L., Chen, L., Wo, T., Xu, K.: Online minimum matching in real-time spatial data: experiments and analysis. Proc. VLDB Endow. **9**(12), 1053–1064 (2016)
10. Tong, Y., She, J., Meng, R.: Bottleneck-aware arrangement over event-based social networks: the max-min approach. World Wide Web J. **19**(6), 1151–1177 (2016)

Quality-Aware Crowdsourcing Curriculum Recommendation in MOOCs

Yunpeng Gao[✉]

State Key Laboratory of Software Development Environment, Beihang University,
Beijing, China
atomgyp@gmail.com

Abstract. With larger and larger numbers of students participating in Massive Open Online Courses (MOOCs), finding top-k suitable courses increasingly becomes a challenging issue for students in terms of course quality, which is hard for computer to compare. Thanks to emerging crowdsourcing platforms, the crowd are assigned to compare the objects and infer the $top - k$ objects based on the crowdsourced comparison results. In this paper, we focus on one such function, $top - k$, that finds the former k ranked objects. We then provide heuristic functions to recommend the $top-k$ elements given evidence. We experimentally evaluate our functions to highlight their strengths and weaknesses.

1 Introduction

Massive Open Online Courses (MOOCs) have rapidly moved into a place of prominence in the media in recent years. MOOC platforms, such as Coursera[1] and EdX[2], are faced with course registration and participation in the hundreds of thousands, and potentially have even larger student populations [1]. As class sizes grow, the number course choice increases rapidly. Consequently, it becomes more difficult for students to find what suitable for them in terms of course quality. However, it is a big challenge for computer to compare quality.

A crowdsourcing database system uses people to perform data cleansing [6], collection or filtering tasks that are difficult for computers to perform. There are many hard tasks for analysis software to identify, but could be identified relatively easily by people who take the courses recently.

In particular, in this paper we focus on the $top-k$ function: The database has a set of objects, where conceptually each object has an intrinsic quality measure (e.g., how clear is the course video, how suitable the course for a beginner takes, how about is the lecturers ability). Of the set of objects, we want to find the one with the $top - k$ quality measure.

This problems is quite challenging because there may be many objects in the database, and because there are many possible votes to invoke. The Judgment Problem draws its roots from the historical paired comparisons problem,

---

[1] https://www.coursera.org/.
[2] https://www.edx.org/.

© Springer International Publishing AG 2017
Z. Bao et al. (Eds.): DASFAA 2017 Workshops, LNCS 10179, pp. 423–428, 2017.
DOI: 10.1007/978-3-319-55705-2_35

wherein the goal is to find the best ranking of objects when noisy evidence is provided. The problem is also related to the Winner Determination problem in the economic and social choice literature.

2   Judgement Problem

2.1   Problem Setup

Objects and Rankings: We are given a set O of n objects $\{o_1, o_2, \ldots, o_n\}$, where each object o_i is associated with a latent quality c_i, with no two c's being the same. If $c_i > c_j$, we say that o_i is greater than o_j. Let π denote a ranking, e.g., a bijection from N to N, where $N = 1, 2, \ldots, n$. We use $\pi(i)$ to denote the rank, or index, of object o_i in permutation π, and $\pi^{-1}(i)$ to denote the object index of the ith position in ranking π.

Problem (Judgement): Given W, predict the $top - k$ objects in O, $\pi^{-1}(1)$, $\pi^{-1}(2), \ldots, \pi^{-1}(k)$.

Representation: We represent the evidence obtained as an $n \times n$ vote matrix W, with w_{ij} being the number of votes for o_j being greater than o_i. Note that $w_{ii} = 0$ for all i. The evidence can also be viewed as a directed weighted graph $G_v = (V, A)$, with the vertices being the objects and the arcs representing the vote outcomes. Figure 1 displays a sample vote matrix and equivalent graph representation.

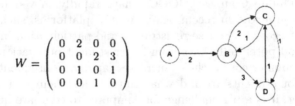

$$W = \begin{pmatrix} 0 & 2 & 0 & 0 \\ 0 & 0 & 2 & 3 \\ 0 & 1 & 0 & 1 \\ 0 & 0 & 1 & 0 \end{pmatrix}$$

Fig. 1. Vote matrix (left) and equivalent graph representation (right). Arc weights indicate number of votes.

2.2   Maximum Likelihood Formulation

We denote the probability of a given ranking π_d given the vote matrix W as $P(\pi = \pi_d | W)$. To derive the formula for $P(\pi_d | W)$, we first apply Bayes theorem,

$$P(\pi_d | W) = \frac{P(W | \pi_d) P(\pi_d)}{P(W)} = \frac{P(W | \pi_d) P(\pi_d)}{\sum_j P(W | \pi_j) P(\pi_j)} \tag{1}$$

The permutations optimizing Eq. 4 are also known as Kemeny rankings [2]. Let $\pi^{-1}(i)$ denote the position of object i in the ranking associated with random variable π. we have: $P(\pi^{-1}(k) = j | W) = \frac{\sum_{d:\pi_d^{-1}(k)=j} P(W | \pi_d)}{\sum_l P(W | \pi_l)}$.

Algorithm 1. Maximum Likelihood

Input: n objects, probability p, vote matrix W
Output: ans = maximum likelihood maximum objects

1: $s[\cdot] \longleftarrow 0$ {$s[i]$ is used to accumulate the probability that i is the maximum object}
2: **for** each permutation π of n objects **do**
3:      $prob \longleftarrow 1$ {prob is the probability of permutation π given vote matrix W}
4:      **for** each tuple $(i,j) : i < j$ **do**
5:          **if** $\pi(i) < \pi(j)$ **then**
6:              $prob \longleftarrow prob \times \binom{w_{ij}+w_{ji}}{w_{ij}} p^{w_{ji}}(1-p)^{w_{ij}}$
7:          **else**
8:              $prob \longleftarrow prob \times \binom{w_{ij}+w_{ji}}{w_{ji}} p^{w_{ij}}(1-p)^{w_{ji}}$
9:          **end if**
10:     **end for**
11:     $s[\pi^{-1}(1)] \longleftarrow s[\pi^{-1}(1)] + prob$
12: **end for**
13: $ans \longleftarrow \arg\max_i s[i]$

---

3 Solutions

3.1 Maximum Likelihood (ML)

The ML scoring function is computationally inefficient and also requires prior knowledge of p, the average worker accuracy, which is not available to us in real-world scenarios. We take it as a baseline. We next investigate the performance and efficiency of four heuristic strategies, each of which runs in polynomial time.

3.2 Degree Score (DEG)

The first heuristic we consider is an Degree scoring function proposed by Coppersmith et al. [3] to approximate the optimal feedback arc set in a directed weighted graph where arc weights l_{ij}, l_{ji} satisfy $l_{ij} + l_{ji} = 1$ for each pair of vertices i and j.

Algorithm 2. Degree Score(DEG)

Input: n objects, probability p, vote matrix W
Output: ans = maximum likelihood maximum objects

1: $s[\cdot] \longleftarrow 0$
2: **for** $i : i, 2, \ldots, n$ **do**
3:      **for** $j : 1, 2, \ldots, n, j \neq i$ **do**
4:          $s[i] \longleftarrow s[i] + l_{ji}\{l_{ji} = P(\pi(i) < \pi(j)|w_{ij}, w_{ji})\}$
5:      **end for**
6:      $s[\pi^{-1}(1)] \longleftarrow s[\pi^{-1}(1)] + prob$
7: **end for**

Algorithm 3. Two Steps Away(TSA)

    Input: n objects, probability p, vote matrix W
    Output: ans = maximum likelihood maximum objects
1: $wins[\cdot], losses[\cdot], s[\cdot] \longleftarrow 0$
2: **for** each tuple (i,j) **do**
3:   $wins[j] \longleftarrow wins[j] + w_{ij}$
4:   $losses[i] \longleftarrow losses[i] + w_{ij}wins[j] + w_{ij}$
5: **end for**
6: **for** $i : 1, 2, \ldots, n$ **do**
7:   $s[i] \longleftarrow wins[i] - losses[i]$
8:   **for** $j : 1, 2, \ldots, n, j \neq i$ **do**
9:     **if** $w_{ij} < w_{ji}$ **then**
10:       $s[i] \longleftarrow s[i] + wins[j]$
11:     **else**
12:       $s[i] \longleftarrow s[i] - losses[j]$
13:     **end if**
14:   **end for**
15: **end for**

3.3 Two-Step Away (TSA)

The Degree Score method is simple to compute, but only takes into account local evidence. That is, the score of object o_i only depends on the votes that include o_i directly. We now consider a Two Steps Away method, which considers evidence two steps away from o_i. This heuristic is based on the notion of wins and losses, defined as follows: $wins(i) = \sum_j w_{ji}$, and $losses(i) = \sum_i w_{ij}$. More formally, score $s(i)$ is defined as follows: $s(i) = wins(i) - losses(i) + \sum_j [1(w_{ji} > w_{ij}wins(j))] - \sum_j [1(w_{ij} > w_{ji}losses(j))]$.

4 Experiments

In this section, we evaluate the performance of our algorithms. All experiments are implemented in C++, and are performed on an Intel i5-4590 processor equipped with 8 GB RAM. We experimentally compare our strategies: Maximum Likelihood, Degree Score, Two Steps Away. However, since ML is computationally expensive, we only do this comparison on a small scenario. For our experiments, we synthetically generate problem instances, varying: n (the number of objects in O), v (the number of votes we sample for W), and p (average worker accuracy).

In our base experiments, we vary the number of sampled votes v, from 0 to $5n(n-1)$ and vary worker accuracy p from 0.55 to 0.95. As a point of reference, we refer to $\frac{n(n-1)}{2}$ votes as $v = 1x$ Edge Coverage, e.g. each pair of objects is sampled approximately once. So $5n(n-1)$ votes is equivalent to $v = 10x$ Edge Coverage in our experiments. Comparing the predicted ranking with π^* we record both (a) a "yes" if the predicted maximum agrees with the true maximum, and (b) reciprocal rank, the inverse rank of the true maximum object in the predicted ranking.

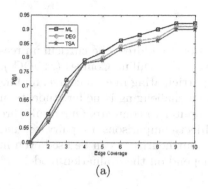 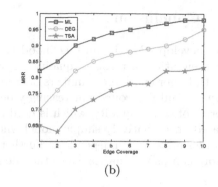

(a)   (b)

Fig. 2. Prediction vs. Edge coverage. 5 objects, $p = 0.75$. P@1 (left), MRR (right).

As a first experiment, we consider the prediction performance of Maximum Likelihood and the four heuristics for a set of 5 objects with $p = 0.75$, displayed in Fig. 2. We choose a small set of objects, so that ML can be computed. Looking at Fig. 2 (left), we find that as the number of votes sampled increases, the $P@1$ of all heuristics increase in a concave manner, approaching a value of 0.9 for 10x Edge Coverage.

As expected, ML performs the best in Fig. 2, but recall that ML requires explicit knowledge of p, and it is computationally very expensive. For a larger experiment, we consider the problem of prediction for $n = 100$ objects in Fig. 3. The strength of the TSA comes from its ability to leverage the large number of redundant votes, in order to iteratively prune out lower-ranked objects until there is a predicted maximum.

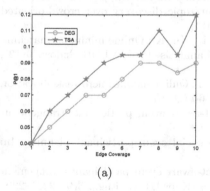 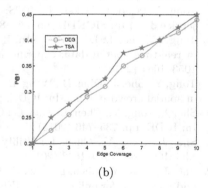

(a)   (b)

Fig. 3. Precision at 1 (P@1) vs. Edge coverage. $p = 0.55$ (left), $p = 0.75$ (right).

5   Related Work

In this section, we review related works from two categories, theory community regarding ranking in the presence of errors [2,4,5], data-driven crowdsourcing [7–9] and event recommendation on social networks [10–13].

6   Conclusion

We develop a Maximum Likelihood Formulation to address the curriculum recommendation in terms of quality for Massive Open Online Courses (MOOCs). With larger and larger numbers of students participating in each course, finding $top - k$ suitable courses increasingly becomes a challenging issue for students in terms of course quality, which is hard for computer to compare. Our results are based on a relatively simple model where object comparisons are pairwise, and worker errors are independent. Furthermore, our heuristics can be used even in more complex scenarios, since they do not depend on the evaluation model.

References

1. Alario-Hoyos, C., Pérez-Sanagustín, M., Mar, N., Kloos, C., Oz-Merino, P.: Recommendations for the design and deployment of moocs: insights about the mooc digital education of the future deployed in miradax. In: ICTEEM, pp. 403–408 (2014)
2. Conitzer, V., Davenport, A., Kalagnanam, J.: Improved bounds for computing kemeny rankings. In: AAAI, pp. 620–627 (2006)
3. Coppersmith, D., Fleischer, L., Rudra, A.: Ordering by weighted number of wins gives a good ranking for weighted tournaments. In: SODA, pp. 145–148 (2006)
4. Conitzer, V., Davenport, A., Kalagnanam, J.: How to rank with few errors. In: STOC, pp. 95–103 (2007)
5. Ailon, N., Charikar, M., Newman, A.: Aggregating inconsistent information: ranking and clustering. In: STOC, pp. 123–128 (2005)
6. Tong, Y., Cao, C., Zhang, C., Li, Y., Chen, L.: CrowdCleaner: data cleaning for multi-version data on the web via crowdsourcing. In: ICDE, pp. 1182–1185 (2014)
7. Tong, Y., Cao, C., Chen, L.: TCS: efficient topic discovery over crowd-oriented service data. In: SIGKDD, pp. 861–870 (2014)
8. Tong, Y., She, J., Ding, B., Chen, L., Wo, T., Xu, K.: Online minimum matching in real-time spatial data: experiments and analysis. Proc. VLDB Endow. **9**(12), 1053–1064 (2016)
9. Tong, Y., She, J., Ding, B., Wang, L., Chen, L.: Online mobile micro-task allocation in spatial crowdsourcing. In: ICDE, pp. 49–60 (2016)
10. She, J., Tong, Y., Chen, L., Cao, C.: Conflict-aware event-participant arrangement. In: ICDE, pp. 735–746 (2015)
11. She, J., Tong, Y., Chen, L.: Utility-aware social event-participant planning. In: SIGMOD, pp. 1629–1643 (2015)
12. She, J., Tong, Y., Chen, L., Cao, C.: Conflict-aware event-participant arrangement and its variant for online setting. IEEE Trans. Knowl. Data Eng. **28**(9), 2281–2295 (2016)
13. Tong, Y., She, J., Meng, R.: Bottleneck-aware arrangement over event-based social networks: the max-min approach. World Wide Web J. **19**(6), 1151–1177 (2016)

Crowdsourcing Based Teaching Assistant Arrangement for MOOC

Dezhi Sun[1(✉)] and Bo Liu[2]

[1] State Key Laboratory of Software Development Environment, Beihang University,
Beijing 100191, People's Republic of China
csusun12@gmail.com
[2] Shandong Computer Science Center (National Supercomputer Center in Jinan),
Shandong Provincial Key Laboratory of Computer Networks, Jinan, China
liubo@sdas.org

Abstract. With the development of new web technologies, the Massive Open Online Course (MOOC) which aims at unlimited participation and access is emerging. In contrast to traditional education, learners could get access to filmed lectures and tests online anytime and anywhere. However, there still exists some problems with MOOCs. Currently, one major problem is the imbalance between teachers and learners online. In many courses, thousands of learners enroll in a class with a single instructor which could lead to bad learning effect and very low completion rates, so we propose crowdsourcing based teaching assistant assignment for MOOC in order to optimize the reasonable disposal of manpower. We present effective algorithms for the selection and assignment of teaching assistants. With experiments on various datasets, we verify the effectiveness of our proposed methods.

1 Introduction

Information technology has played a key role in the modern education. It becomes convenient for people to get access to learning materials online, such as instructional video, teaching Power Point, and so on. At the same time, people could comment on the forum of the online study platform and interact with each other. However, there are many issues with MOOCs that we shoud concern. As [1] pointed out, MOOCs require rigorous attention and should pay long hours to prepare for the lecture. It is a big challenge for instructors. Meanwhile, the success rate of many MOOCs is very low and many online learners would lose interest after a few weeks of course study [2]. The MOOC format itself suffers from weaknesses around access, content, quality of learning, accreditation, pedagogy, poor engagement of weaker learners, exclusion of learners without specific networking skills [3]. There have been some suggestions on how to improve the quality of online learning [1,4,5], such as careful preparation or design for online courses. However, enhancement of the quality of teaching should not only consider the factor of instructor. In this paper we propose there is a need to set up online teaching assistants which selected from learners proficient in the course

Z. Bao et al. (Eds.): DASFAA 2017 Workshops, LNCS 10179, pp. 429–435, 2017.
DOI: 10.1007/978-3-319-55705-2_36

based on crowdsourcing. There are many advantages for the setting of the teaching assistants, first it would reduce the stress of instructors directly and it would promote the communication with learners and enhance learning efficiency. In the long run, it could be useful to increase the completion rate of learners.

2   Related Work

The first MOOCs emerged from the open educational resources [6] and there have been much research about the development and issues of MOOCs [1,2,5]. All course content was open to online student. Teachers of MOOCs should consider more on the organization of online course and formulation of discussions and prepare for the lectures of the filmed video. However, it would be time consuming for teachers and there would not be enough time for them to answer questions about online courses. At the same time there are other problems like very low success rate. Reasons include the high drop-out rate in many types of learning, and the evidence that with no penalty for exit or entry [3].

Crowdsourcing is an effective way to solve problems which cannot be effectively addressed by computers. An important problem in crowdsourcing is task assignment, which assigns tasks to appropriate human workers [7–12].

As mentioned above, there is a need to set up teaching assistants for online courses. At the same time, a flexible strategy of assigning online teaching assistants is needed. First, the teaching quality of assistants should keep high level. Second, teaching assistants should keep online as the course is open. It is reasonable for us to consider the crowdsourcing mode to achieve that goal. Because the number of learners is large and it would be reasonable to find many volunteers who are willing to be teaching assistants.

3   Problem Statement

Definition 1 (Teaching Assistant Assignment). A feasible assignment is a two dimension matrix A, which $A_{ij} = 1(0)$ represents learner i is (not) assigned to be the teaching assistant of course j, and the assignment of teaching assistants is based on the learners' teaching level which will be discussed in Sect. 4.

Definition 2 (CTAA Problem). Given a set of learners, every learner could participate in the open online courses any time any where and many of which are willing to be crowdsourcing workers. We calculate each workers' (learners') teaching level of every course first, which is a two dimensional matrix denoted by T. Based on T we make an assignment for each worker available. The problem of crowdsourcing based teaching assistant assignment is to find an optimal feasible assignment A that $A = \text{argmax}(T^*A^T)$.

4   Characterizing the Teaching Level

In this section, we first present solutions for ranking learners' teaching level. First we compute the expertise level part, which could be computed by the interaction between learners. [13] has proved that relying on content such as word and document frequencies is limited. In other words, solutions to leverage the expertise levels is more needed so as to distinguish their expertise levels. Inspired by [14], we use a pageRank [15] similar algorithm to compute the expertise level of learners. Specifically, we define expertise level of learner l on course c is el(l,c), and e(l) is the corresponding column vector of learner l's expertise level. We denote the number of learners who helps learner l_i as H(l_i), it can be computed by the answer he accepts on the study platform or other formulation.

$$el(l, c) = \beta * (el(l1, c)/H(l_1) + ... + el(l_N, c)/H_N) + (1 - \beta)/N$$

The expertise level rank can be computed iteratively through each time step which is similar to [14]. With the computation of expertise level rank, we can get one learner's teaching level by combining his test score. More specifically, for every course c, we compute the average test score for learner l, detonated by LL(l,c). And we use a balance factor α to compute the teaching level of each learner. All the learners' teaching level of every courses formulate a two dimension matrix T. The default value of α is 0.5.

$$T(l, c) = \alpha * LL(l, c) + (1 - \alpha) * EL(l, c)$$

5   Teaching Assistant Assignment

5.1   Offline Analysis

Because the characteristic of MOOCs, the learners could participate in courses at anytime they want. The online study platform do not have knowledge about learners. At every instance of time, there are new learners and courses updated. Since a global optimal assignment for learners is not feasible, it is more realistic to optimize every instance locally at every time. We can solve the teaching assistant assignment problem in a greedy way, which find the feasible assignment for every instance that maximum the total teaching level of assistants. Based on the teaching level computation, the teaching assistant assignment problem can be reduced to the maximum flow problem [16]. We can run maximum flow algorithm at every time step and get the feasible arrangement set and we defined this method as teaching assistant assignment algorithm. However, learners could participate in or exit the study platform at any time they want. We have to consider the offline rate of learners. We define it as the rate of log out times divided by his online time. Then minimum cost maximum flow algorithm (MCMF) is used to compute the result which cost equals the offline rate and the step is similar to the teaching assistant assignment algorithm.

5.2   Online Analysis

In the offline version, the behavior of learners is assumed to be stable and we can run the maximum flow algorithm at every time step. However, the time of learners participate in is random and there is much knowledge about future. Consequently, we design two approximate algorithms for the online version. The first algorithm is Greedy algorithm. At every time step or someone log out, the system dispatch the course with the learner with the highest teaching level interactivity until exceed the budget of learners or courses.

Greedy algorithm computes the maximum tl(c,l) every time in a greedy way, but it can not achieve a global optimal result. Because if learner L1 arrives before L2, he would be selected as the teaching assistant even if his teaching level is lower than L2, when L2 comes, there would no course need a teaching assistant at that time. It is a waste of human resources. Inspired by [7], we divided learner into two equal groups, the first group takes greedy strategy while the second group learners find a feasible arrangement through maximum flow algorithm. The second half of learners who are not assigned to course c is put into a vector set LM. We run the teaching assistant assignment when size of LM is bigger than the size of course set C. Our proposal is based on the following observation, if one learner L1 keep waiting until the next time step, it may be a waste of time because there would be no learners have higher teaching level than L1. In the meanwhile, we should to try to optimize teaching arrangement problem in global, so when the size of LM is larger than the threshold we execute the teaching assistant assignment algorithm for learners in vector LM. We defined this method as partition based algorithm.

6   Experimental Study

Experimental Configuration. Although MOOCs is becoming popular nowadays, it not easy to get real data in practical. For the synthetic, we use some data generating algorithms of [17] to produce the synthetic dataset we want to test. We generated synthetic data with uniform (UNI) and mixture of uniform and gaussian (MIX) distributions.

Algorithms Evaluating. In this paper we adopt four methods, which is teaching assistant assignment algorithm (TAA), minimum cost maximum flow algorithm based Assignment (MMA), and the greedy algorithm (GD). The partition based algorithm (PB). Our major concern is the average maximum of total teaching level and running time of each algorithm.

Running Time of Varying $|T|$. We first evaluate the proposed algorithms by varying the number of leaders under different distribution. That is, with different distribution of uniform (UNI) and mixture of uniform and gaussian (MIX), we vary the number of learners from 2 K to 10 K. From Fig. 1 we can see that GD algorithm runs faster than PB, in the meanwhile both TAA and MMA is slower. This is because the process of matching algorithm is time consuming than others. And we can see that there is little difference in running time under different distributions. The effect of changing the number of courses is similar to change the number of learners, so we omit it here (Fig. 1).

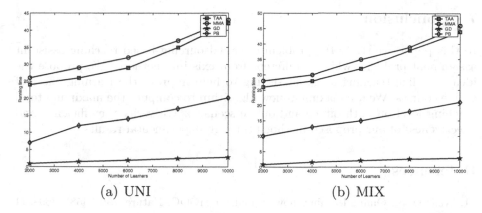

(a) UNI   (b) MIX

Fig. 1. Running time of varying number of learners

Total Teaching Level. We check the maximum total teaching level by varying the number of learners under different distribution and different offline rate. First we keep the offline rate stable, which is 0.2, and vary the number of learners from 2 K to 10 K, and we can find the total teaching level is increasing with the raising of number of leaders. We can see that MMA which considers offline rate plays better than the other. And greedy which selects the highest teaching level first is not stable. Then, we keep the number oflearnerss stable and vary the offline rate, we can see that with the increase of offline rate, total teaching level is decreasing. PB algorithm is relatively stable on both conditions which is reflected in Fig. 2.

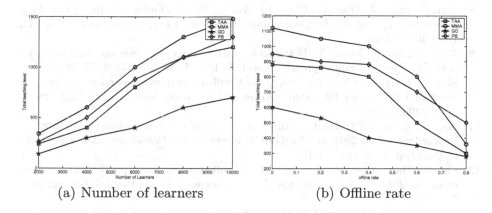

(a) Number of learners   (b) Offline rate

Fig. 2. Results of total teaching level

Experimental Environment: We implement our algorithms in C++, and the experiments were performed on a machine with Intel i5-5200 CPU @ 2.2 GHZ and 8 GB memory.

7 Conclusion

In this paper, we study the problem of crowdsourcing based teaching assistant assignment problem, which is different from existing methods. We propose the idea of online teaching assistant setting to help improve the learning effect of online courses. We use maximum flow algorithm to compute the maximum total teaching level on both offline and online scenario. We verify the efficiency and effectiveness of our proposed methods through experimental results.

References

1. Kellogg, S.: Online learning: how to make a MOOC. Nature **499**(7458), 369–371 (2013)
2. Rai, L., Chunrao, D.: Influencing factors of success and failure in MOOC and general analysis of learner behavior. Int. Inf. Educ. Technol. **6**(4), 262 (2016)
3. https://www.gov.uk/government/uploads/system/uploads/attachment_data/file/240193/13-1173-maturing-of-the-mooc.pdf
4. Bartoletti, R.: Learning through design: MOOC development as a method for exploring teaching methods. Curr. Issues Emerg. eLearn. **3**(1), 2 (2016)
5. Daniel, J., Cano, E.V., Cervera, M.G.: The future of MOOCs: adaptive learning or business model? Revista de Universidad y Sociedad del Conocimiento **12**(1), 64–73 (2015)
6. https://en.wikipedia.org/wiki/Massive_open_online_course
7. Tong, Y., She, J., Ding, B., Wang, L., Chen, L.: Online mobile micro-task allocation in spatial crowdsourcing. In: IEEE 32nd International Conference on Data Engineering (2016)
8. Tong, Y., She, J., Ding, B., Chen, L., Wo, T., Xu, K.: Online minimum matching in real-time spatial data: experiments and analysis. Proc. VLDB Endowm. **9**(12), 1053–1064 (2016)
9. Tong, Y., She, J., Meng, R.: Bottleneck-aware arrangement over event-based social networks: the max-min approach. World Wide Web **19**(6), 1151–1177 (2016)
10. She, J., Tong, Y., Chen, L., Cao, C.C.: Conflict-aware event-participant arrangement and its variant for online setting. IEEE Trans. Knowl. Data Eng. **28**(9), 2281–2295 (2016)
11. She, J., Tong, Y., Chen, L.: Utility-aware social event-participant planning. In: Proceedings of the 2015 ACM SIGMOD International Conference on Management of Data, pp. 1629–1643 (2015)
12. She, J., Tong, Y., Chen, L., Cao, C.C.: Conflict-aware event-participant arrangement. In: 2015 IEEE 31st International Conference on Data Engineering, pp. 735–746 (2015)
13. Littlepage, G.E., Mueller, A.L.: Recognition and utilization of expertise in problem-solving groups: expert characteristics and behavior. Group Dyn. Theory Res. Pract. **1**(4), 324 (1997)
14. Zhang, J., Ackerman, M.S., Adamic, L.: Expertise networks in online communities: structure and algorithms. In: Proceedings of the 16th International Conference on World Wide Web, pp. 221–230. ACM (2007)
15. Page, L., Brin, S., Motwani, R., Winograd, T.: The pagerank citation ranking: bringing order to the web (1999)

16. Kazemi, L., Shahabi, C.: GeoCrowd: enabling query answering with spatial crowd-sourcing. In: Proceedings of the 20th International Conference on Advances in Geographic Information Systems, pp. 189–198. ACM (2012)
17. To, H., Asghari, M., Deng, D., Shahabi, C.: SCAWG: a toolbox for generating synthetic workload for spatial crowdsourcing (2016)

Quantitative Analysis of Learning Data in a Programming Course

Yu Bai, Liqian Chen, Gang Yin$^{(\boxtimes)}$, Xinjun Mao, Ye Deng, Tao Wang, Yao Lu,
and Huaimin Wang

College of Computer, National University of Defense Technology,
Changsha, China
gangyin@nudt.edu.cn

Abstract. Online learning platform, which has taken higher education by storm, provides an opportunity to track students' learning behaviors. The vast majority of educational data mining research has been carried out based on the online learning platform in Europe and America but few of them use the data from programming courses with large scale. In this paper, we track students' code submissions for assignments in a programming course and collect totally 17,854 submissions with the help of TRUSTIE, a famous online education platform in China. We perform a preliminary exploratory inspect for code quality by SonarQube from the code submissions. The analysis results reveal several interesting observations over the programming courses. For example, results show that logical training is more important than grammar training. Moreover, the analysis itself also provides useful feedback of students' learning effect to instructors for them to improve their teaching in time.

Keywords: Quantitative analysis · Learning data · Trustie · Programming course

1 Introduction

Many students struggle with some computer science courses that require basic practical skills for interacting with the learning environments, such as programing courses and experimental courses. Web-based online learning platforms, such as CodingBat [1], CloudCoder [2,3] and Trustie [4], provide useful ways for instructors to give the students opportunities that allow them to experience the basic concepts and techniques directly and interactively, and to help students achieve better outcomes.

Notably, in Trustie, instructors can assign different kind of exercises to students, especially the programing exercises that require each student to write a

We gratefully acknowledge the financial support from Natural Science Foundation of China under Grant Nos. 61303064, 61432020, 61472430, 61502512, 61432020 and 61532004. We thank our students on their active participation in the course, and the cooperation of TRUSTIE.

Z. Bao et al. (Eds.): DASFAA 2017 Workshops, LNCS 10179, pp. 436–441, 2017.
DOI: 10.1007/978-3-319-55705-2_37

small piece of code in C, C++, Java or Python. Instructors need to construct a test set (contains pairs of inputs and correct outputs) for each exercise. The correctness of the student's code is judged automatically after it is submitted to Trustie. In addition to the pedagogical benefits of using programming exercises in Trustie, the data collected by online learning platform offers detailed data set that reflects the students' habits and behaviors. In this paper, we examine the data collected in the Trustie programming exercise websites in CS1 (introductory computer science course) taught at National University of Defense Technology. In this work, we collected 17,854 code submissions in CS1 from Trustie. Then, we perform a preliminary exploration of code quality by using SonarQube to check these code pieces. Numerical experiments suggest that, logical training is more important than grammar training and "code smell" will affect the quality of the program.

The rest of paper is organized as follows. In Sect. 2, the related work is given. In Sect. 3, the online learning platform Trustie and programming analysis tool SonarQube are introduced, then the method that used to analyze the data collected from the Trustie programming exercise. Some discussions and lessons are given after we conduct the experiments in Sect. 4, and we conclude in Sect. 5.

2   Related Work

With the development of technologies and intelligent devices, the forms of educational resources have become diverse, the corresponding tools or platforms also have become rich. These tools and platforms can be used to collect data and assist researchers to understand the students encounter. The most often mentioned data collection instruments were WebCAT [5], BlueJ [6] or an extension of BlueJ [7], and CodeWrite [8] or its plugin. For example, article [9] introduces the work that blackbox collects information about submissions made by students in the BlueJ environment into a central repository. But BLUEJ is an integrated development environment (IDE) just for the Java programming language. And language limitations and restrictions on data collection will narrow the scope of observation.

The data reported in this paper was collected using an online learning platform Trustie and quality management platform SonarQube. The data reported in this paper was collected using an online learning platform Trustie and quality management platform SonarQube. Students can make multiple attempts for submission and typically take advantage of the feedback from Trustie.

3   The Foundation of Work

3.1   Platform

– **Trustie**
  TRUSTIE[1] (**Trust**worthy Software tools and **I**ntegration **E**nvironment) is an online learning platform. One of the goals of the Trustie project is to

---

[1] https://www.trustie.net.

provide a platform for collecting data on how students learn to program. What makes it special is that it contains a web-based programming exercise system and a version control system. It supports exercises in several programming languages, including (at the time of writing) C/C++, Python, Java, and Ruby.

Assignments are created or selected from exercise repository by instructors and distributed to their students using the assignment post feature. In the platform instructors can identify a due date and associate attachments from their attachment repository or computer. And the most important thing is that they must also provide several sets of test cases (input and output), which are used to verify solutions to the problem before publishing the assignment. Trustie collects data from the process that students undertake when they doing their online homework.

- **Sonarqube**
SonarQube is an open source platform for continuous inspection of code quality which contained minute quantities of rules on java and c#, therefore majority of detection rules are provided by plugins. Through the plugin mechanism it can be integrated with different test tools, code analysis tools and continuous integration tools. More than 20 programming languages are covered through plugins including Java, C/C++, Objective-C, C#, PHP, Flex, Groovy, JavaScript, Python, PL/SQL, COBOL, Swift, etc. There are three SonarQube quality models, bugs, vulnerabilities and code smells, respectively.

3.2 Profile of a Programming Course

In this paper, we track the learning data of a programming course namely CS1 in National University of Defense Technology in China, which is deployed on Trustie. CS1 is opened for the first-year undergraduate students, and around 120 undergraduate students follow this course each year in recent two years. For this course, instructors build an on-line class and deploy all programming assignments on Trustie. Trustie provides a web-based automatic grading system of source code submissions to automatically evaluate the correctness of students' code submissions. For each programming assignment, instructors design a set of test cases to assess the correctness of code submissions. Those code submissions are tested automatically by a test suite deployed inside Trustie. The automatic grading system will give different grades for submitted programs that pass different portions of test cases deployed by instructors. When a submitted program passes all test cases, the automatic evaluation system will give full mark. The main benefit of using the automatic grading system lies in that it improves significantly the speed, quality and objectivity of the evaluation for assignments and provides real-time feedback to the students. To get full mark, students will complete each assignment in a trial-and-error manner interactively, until they pass all the test cases for that assignment. Trustie records all the versions of students' code submissions using a database. Instructors could see all the versions as well as the corresponding testing results of each student's submissions. On the other hand, the automatic grading system offers a great relief for the instructors, since manually evaluating code submissions from around 120 students indeed takes time. Thus, instructors can concentrate more on giving comments and suggestions in time for those students that need help to complete assignments.

4   Analysis of The Results

In this section, we will analyze the code quality by SonarQube from the original data based on the solution introduced above.

To display the reason for delayed submission and no submission, we first show the relationship among times of submissions (x axis), numbers of delayed submissions (y axis) and numbers of no submissions (z axis) in Fig. 1.

Significantly, more times of submissions indicate that the work is more difficult. We find that, in most cases, there are few students who submit their works late and unsuccessfully. Unexpectedly, the maximum number of delay and no submission assignments are not the most difficult. As shown in Fig. 1, red and yellow lines represent the 26th and 27th assignment (last two assignments), respectively. And blue lines represent the 15th and 20th assignments.

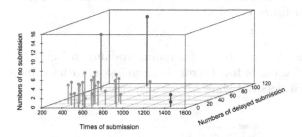

Fig. 1. The relationship among times of submissions, numbers of delayed submissions and numbers of no submissions (Color figure online)

We found that the last two assignments were published and closed on the same day. And in the meanwhile, other exams may be affect the submission since the last two assignments are published at the end of the term.

We can thus draw conclusion that the frequency of issuing online assignments has the strong effect on the quality of students' submissions. We suggest that instructors should keep a proper frequency of issuing assignments, and it's better not to issue twice within five days. Especially at the end of the semester, students will have less time to finish their assignments because they have to prepare for exams of other courses.

In Fig. 2, we show the number of logic errors and compile fails in order to investigate the impact of logical training and grammar training on the code correction. Red, blue and black lines represent the times of submissions, the number of errors and compiling fails, respectively. From Fig. 2, we find that compiling fails are less than logic errors as the students did more programming work. It indicates that logical training is more important than grammar training.

It can be seen that logical training is more important than grammar training for a programming course. When students pay more attention to logic training, the learning of programming will be more successful from a long-term view. Hence, instructors should concentrate more on logic training in order to improve students' ability of logical thinking, which is also very helpful for students to building a solid foundation of programming and even for their future career.

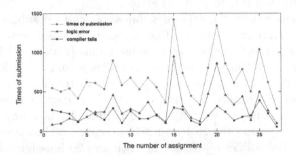

Fig. 2. The impact of logical training and grammar training on the correction of programming (Color figure online)

We visualize the error in programming training in Fig. 3. We observed that "variableScope" is one of top 4 errors in submission, with the increasing difficulty of assignment, it does not show an upward trend.

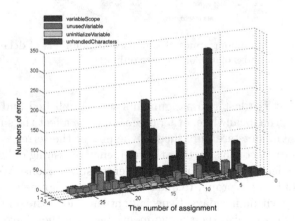

Fig. 3. The type with the largest number of error

The analysis results of the learning data provide an objective feedback for instructors, which are very helpful for them to evaluate the teaching effectiveness during the programming course. With the help of analysis results of the intermediate learning data, the instructors can adjust their teaching plan in time.

5 Conclusions

In this paper, we track the learning data of students in a programming course and collect totally 17,854 code submissions, based on the online education platform TRUSTIE. Then, we analyze the learning data from different aspects. E.g., we investigate the code quality of students' code submissions by using SonarQube. And we make interesting observations according to the analysis results. E.g., logical training is more important than grammar training. The analysis results are also very useful for instructors to improve their teaching, since instructors can see more intermediate learning data of students thank to the online education platform.

References

1. Phatak, D.B.: Tools for Programming in MOOCs (Assess Student's Knowledge)
2. Hovemeyer, D., Hertz, M., Denny, P., et al.: CloudCoder: building a community for creating, assigning, evaluating and sharing programming exercises. In: Proceeding of the 44th ACM Technical Symposium on Computer Science Education, p. 742. ACM (2013)
3. Hovemeyer, D., Spacco, J.: CloudCoder: a web-based programming exercise system. J. Comput. Sci. Coll. **28**(3), 30 (2013)
4. Zhang, X., Zheng, L., Sun, C.: The research of the component-based software engineering. In: Sixth International Conference on Information Technology: New Generations, ITNG 2009, pp. 1590–1591. IEEE (2009)
5. Edwards, S.H., Perez-Quinones, M.A.: Web-CAT: automatically grading programming assignments. ACM SIGCSE Bull. **40**(3), 328 (2008). ACM
6. Van Haaster, K., Hagan, D.: Teaching and learning with BlueJ: an evaluation of a pedagogical tool. In: Information Science + Information Technology Education Joint Conference, Rockhampton, QLD, Australia, pp. 455–470 (2004)
7. Patterson, A., Kölling, M., Rosenberg, J.: Introducing unit testing with BlueJ. ACM SIGCSE Bull. **35**(3), 11–15 (2003)
8. Denny, P., Luxton-Reilly, A., Tempero, E., et al.: CodeWrite: supporting student-driven practice of java. In: Proceedings of the 42nd ACM Technical Symposium on Computer Science Education, pp. 471–476. ACM (2011)
9. Brown, N.C.C., Kölling, M., McCall, D., et al.: Blackbox: a large scale repository of novice programmers' activity. In: Proceedings of the 45th ACM Technical Symposium on Computer Science Education, pp. 223–228. ACM (2014)

Author Index

Printed in the United States
by Bookmasters

Printed in the United States
By Bookmasters